W0042461

**Reviews of Plasma Physics**

VOPROSY TEORII PLAZMY

ВОПРОСЫ ТЕОРИИ ПЛАЗМЫ

Translated from Russian by J. G. Adashko

# Reviews of Plasma Physics

Edited by Acad. B. B. Kadomtsev

Volume 13

CONSULTANTS BUREAU · NEW YORK-LONDON

The Library of Congress cataloged the first volume of this title as follows:

Reviews of plasma physics. v. 1–
New York, Consultants Bureau, 1965–
/v. illus. 24 cm.
Translation of Voprosy teorii plazmy.
Editor: v. 1– M. A. Leontovich.

1. Plasma (Ionized gases)—Collected works. I. Leontovich, M. A., ed. II. Consultants Bureau Enterprises, Inc., New York. III. Title: Voprosy teorii plazmy. Eng.
QC718.V63 64-23244

The original text, published by Energoatomizdat in Moscow in 1984,
has been corrected and updated by the authors.

ISBN-13: 978-1-4615-7780-5 e-ISBN-13: 978-1-4615-7778-2
DOI: 10.1007/978-1-4615-7778-2


Softcover reprint of the hardcover 1st edition 1987
A Division of Plenum Publishing Corporation
233 Spring Street, New York, N.Y. 10013

## CONTENTS

### PARTICLE DYNAMICS IN MAGNETIC TRAPS

B. V. Chirikov



### TRANSPORT PROCESSES IN AXISYMMETRIC OPEN TRAPS

D. D. Ryutov and G. V. Stupakov

CLASSICAL LONGITUDINAL PLASMA LOSSES FROM OPEN ADIABATIC TRAPS

V. P. Pastukhov



SPECTRAL-LINE BROADENING IN A PLASMA

V. I. Kogan, V. S. Lisitsa, and G. V. Sholin



ELECTRON CYCLOTRON PLASMA HEATING IN TOKAMAKS

A. D. Piliya and V. I. Fedorov

# PARTICLE DYNAMICS IN MAGNETIC TRAPS

B. V. Chirikov

## 1. Introduction. Budker's Problem [1]

The investigation of the dynamics of individual (noninteracting) charged particles in a magnetic trap is probably the simplest of the problems of prolonged plasma containment for controlled thermonuclear fusion. Nonetheless, even this "simple" problem is quite rich in content and is far from completely solved, notwithstanding many years' efforts in this direction (see, e.g., [2-4]). In addition, the dynamics of an individual particle is an integral part of the more complicated problem of collective processes in a plasma. Finally, the problem of containing a single particle in a magnetic trap must be faced every time when a new scheme or a substantial modification of an old method of magnetic confinement of a plasma appears.* An example is Dimov's ambipolar (tandem mirror) trap [7]. It is one of the so-called open systems of plasma confinement, or traps with "magnetic mirrors," which will be discussed below. We shall refer to them for brevity simply as traps.

The dynamics of a particle in a trap can be regarded as a problem of nonlinear oscillations induced possibly by some external perturbation in a system having, generally

*Certain very simple particle-interaction effects can be accounted for also within the scope of the single-particle problem, for example the change of the magnetic-field configuration on account of plasma diamagnetism or particle scattering in a plasma.

speaking, three degrees of freedom. The term "oscillations" is used here in a broad sense as a synonym for finite motion of a particle in a bounded region of space (particle containment). In other words, we assume here that the construction of the trap (the magnetic-field configuration) ensures "locking" of the particle for a time equal at least to several passes of the particle through the trap. Under these conditions, the main factor that determines the character of a prolonged evolution of the oscillations are the resonances, or the commensurability of the periods of various degrees of freedom of the oscillations, as well as commensurability with the external perturbation. The mechanism whereby the resonances act is particularly clear (and well known) in the simplest case of linear oscillations. Nonlinear dynamics is much more complicated, but in this case resonant processes are unique for the very same reason: small resonant perturbations repeat in time and accumulate, leading to a greater long-time effect than nonresonant perturbations. This important property can be used to define resonant processes in a broad sense.

Such a "resonant" approach means, figuratively speaking, that before some dynamic problem is solved it must be checked for the presence of resonances, even if it seems obvious at first glance that none exist here. A splendid example of such a situation is the action exerted on a system by an adiabatic (i.e., a very-low-frequency) perturbation, and is in fact the subject of this article.

We shall accordingly pay principal attention below to a fullest possible analysis of various resonances and to the interaction between them. The latter means the effect of the joint action of several resonances, which does not add up simply to the sum of the effects due to individual resonances, since the equations of motion are nonlinear and the superposition rule is thus invalid.

It turns out that, under certain conditions, the interaction of the resonances alters radically the character of the motion and transforms it from the well-known regular or quasiperiodic oscillations into the relatively little-known "random (stochastic) walk" of the particle in its phase space (see, e.g., [8, 9]). The latter is very similar to (and in some cases indistinguishable from) random oscillations, i.e., oscillations caused by some external (relative to the dynamic system) random perturbation or "noise," although such a perturbation need not necessarily be present. Strange as it may seem, this unusual motion regime is quite widespread for nonlinear oscillations in general, and for a particle in a magnetic field in particular. From the standpoint of prolonged containment of particles in a trap, the onset of random oscillations is harmful, since it leads as a rule to particle loss through diffusion in phase space.

We confine ourselves in this review to a discussion of the simplest case of nonlinear oscillations with two degrees of freedom, when a random regime of motion in a conservative system is still possible. For particle motion in a trap this presupposes some symmetry of the magnetic field, meaning an additional integral of motion (besides the energy). In the case of axial symmetry, for example, the additional integral is the component of the particle's canonical momentum along the symmetry axis. The motion is then over the intersection, in phase space, of the constant-energy surface and the constant-value surface of the additional integral.

We actually consider below a certain quite special case, when the connection between the two degrees of freedom is adiabatic, i.e., the ratio of the fundamental unperturbed frequencies is very large (or small). This special problem, which stems from an analysis of the operation of a trap with magnetic mirrors (the Budker problem), is nonetheless connected with one classical problem of mechanics, viz., adiabatic invariance of the action variables. It is worthwhile noticing that it is precisely research into particle containment in traps which made possible considerable progress toward the solution of this general problem, too. It became clear, in particular, that, under certain conditions, an adiabatic invariant becomes an exact integral of the motion [10].

These and other results are discussed below using as an example particle dynamics in several quite simple yet typical modifications of a magnetic trap. The problem can be solved in some cases completely; i.e., it is possible to obtain in explicit form the particle containment conditions, on the one hand, and the statistical properties of the motion in the unstable region, on the other [2].

The author is indebted to G. I. Dimov, D. D. Ryutov, and D. L. Shepelyanskii for interesting discussions of the questions touched upon in the article, and to G. B. Minchenkov for help with the numerical simulation.

## 2. Choice of Unperturbed System

As a rule, an analytic investigation is possible only by using some approximate methods or perturbation theory. The first step is therefore to divide the investigated system into an unperturbed part and a perturbed one. This division is not unique and is dictated by the type of physical problem.

In the present case we might, for example, choose as the unperturbed system a particle in a uniform magnetic field, and include the spatial inhomogeneity of the field as a whole in the perturbation. Although the unperturbed motion is in this case extremely simple, this is not the best choice. The point is that such an unperturbed motion is infinite and, therefore, differs qualitatively from the actual (perturbed) motion. The unperturbed motion is said in this case to be degenerate (one of the fundamental frequencies of the system is zero), and an arbitrarily small perturbation (field inhomogeneity of suitable configuration) leads to bifurcation.

To choose an unperturbed system more suitable for investigation we recall that particle containment in an open trap is the result of approximate conservation of the particles' magnetic moment $\mu = mv_\perp^2/2B$, which is proportional to an adiabatic invariant (the action variable $J_\perp$ of the Larmor rotation). Taking as the basic units the speed of light, the charge, and the particle mass ($c = e = m = 1$), we get

$$J_\perp = \frac{1}{2\pi} \oint (\mathbf{p}_\perp + \mathbf{A})\, d\mathbf{r} = v_\perp^2/2\omega = \mu. \tag{2.1}$$

The subscript $\perp$ labels here quantities that characterize motion in a plane perpendicular to the magnetic field $\mathbf{B} = \operatorname{rot} \mathbf{A}$; $\omega = B$ is the Larmor frequency. We shall be interested below in the nonrelativistic case ($v \ll 1$), but all the relations remain in force at arbitrary velocities if the substitution $m \to \gamma m$, $\gamma = (1 - v^2)^{-1/2}$ is made. Expression (2.1) for the action $J_\perp$ is exact only in a uniform field. We can, however, define the unperturbed system by stipulating just this quantity to be an exact integral:

$$\mu \equiv \text{const}. \tag{2.2}$$

The identity sign indicates that the last condition is an arbitrary choice of the unperturbed system rather than the property of the quantity $\mu$. We refer all the changes of $\mu$ to the perturbation, which we assume to be small enough. The last condition is known to be met for a sufficiently strong magnetic field or, equivalently, for a sufficiently small Larmor radius of the particle (see Section 4 below).

For an axisymmetric magnetic trap, neglecting the particle drift velocity, the unperturbed Hamiltonian is

$$H^0(p, s) = v^2/2 = p^2/2 + \mu\omega(s), \tag{2.3}$$

where s is the coordinate along the magnetic line, and $p = v_{\parallel} = \dot{s}$ is the conjugate momentum (and the longitudinal velocity) of the particle. It can be seen that an unperturbed system defined with the aid of (2.2) is convenient also because the effective potential energy of the longitudinal motion turns out, in this case, to be simply proportional to the specified magnetic field strength. Knowing the Hamiltonian (2.3), we can, in principle, obtain the unperturbed longitudinal motion of the particle.

## 3. A Few Examples

1. Auxiliary System. To demonstrate most simply the main features of particle dynamics in a magnetic trap, we consider, besides different trap variants, also an auxiliary model with two degrees of freedom, specified by the Hamiltonian

$$H(x, y, p_x, p_y) = \frac{p_x^2 + p_y^2 + (1 + x^2)^2 y^2}{2}, \tag{3.1}$$

where $p_x = \dot{x}$; $p_y = \dot{y}$. The equipotentials of this system $[(1 + x^2)y = \text{const}]$ are not closed, so that the energy conservation law does not by itself ensure as yet that the motion is finite. Nonetheless, as will be shown below (Section 10), the oscillations of this system turn out under certain conditions to be rigorously bounded. If the frequency of the oscillations along x (the x oscillations) is much lower than that of the y oscillations, the system (3.1) simulates approximately the motion of a particle in an axisymmetric magnetic trap. Indeed, y oscillations with variable frequency $\omega_y = 1 + x^2$ correspond to Larmor rotation in an inhomogeneous magnetic field $B = 1 + s^2$. The action variable for these oscillations is

$$J_y = \omega_y a_y^2/2 = (1 + x^2)\, a_y^2/2 \equiv \text{const}, \tag{3.2}$$

where $a_y$ is the amplitude of the y oscillations. The last condition, just as (2.2), defines an unperturbed system whose Hamiltonian we obtain with the aid of the condition

$$\frac{p_y^2}{2} + \frac{\omega_y^2 y^2}{2} \equiv \omega_y J_y, \tag{3.3}$$

which is equivalent to (3.2). From (3.1) we get

$$H^0 = p_x^2/2 + J_y(1+x^2) = J_y + J_x \sqrt{2J_y}. \tag{3.4}$$

The unperturbed Hamiltonian depends only on the action variables, which are therefore integrals of the motion, and describes independent oscillations in two degrees of freedom, with frequencies

$$\omega_x = \partial H^0/\partial J_x = \sqrt{2J_y}, \qquad \langle\omega_y\rangle = \partial H^0/\partial J_y = 1 + J_x/\sqrt{2J_y}. \tag{3.5}$$

The quantity $\langle\omega_y\rangle$ has the meaning of the time-averaged frequency of the y oscillations and can, of course, be obtained also by directly averaging the quantity $(1 + x^2)$.

An important property of the considered unperturbed oscillations is that they are not isochronous – the oscillation frequencies depend on the actions (or amplitudes), although both the x and the y oscillations are almost harmonic (the frequency of the y harmonics varies slowly with time).

2. Short Magnetic Trap. Assume that the trap field has not only a symmetry axis ($r = 0$ in the cylindrical coordinates $z$, $r$, $\varphi$) but also a symmetry plane ($z = 0$). Let the field on the trap axis be given as

$$B_z(z, 0) = B_{00} f(z), \quad f(0) = 1, \qquad f(-z) = f(z), \quad B_r(z, 0) = 0, \tag{3.6}$$

where $B_{00}$ is the field at the trap ($z = r = 0$). The field configuration $f(z)$ on the axis depends generally speaking both on the external currents in the trap windings and on the currents in the plasma. In a sufficiently close vicinity of the axis, the vector potential of the field (3.6) can be written in the form

$$A_\varphi(z, r) \approx B_{00}\left[\frac{rf(z)}{2} + r^3 g(z)\right]. \tag{3.7}$$

There is no term proportional to the $r^2$ because of the axial symmetry of the field. Given $f(z)$, the function $g(z)$ can be arbitrary, depending on the currents in the plasma. In the absence of the latter ("vacuum" field) we have $g(z) = -f''/16$ from $\Delta \mathbf{A} = 0$ or rot $\mathbf{B} = 0$. At the accuracy assumed, the field next to the axis is

$$\left.\begin{aligned} B_z(z, r) &\approx B_{00}[f(z) + 4r^2 g(z)], \\ B_r(z, r) &\approx -B_{00}\left[\frac{rf'}{2} + r^3 g'\right]. \end{aligned}\right\} \tag{3.8}$$

It is important that, given the field on the axis, the currents in the plasma add only small corrections of order $r^2$. Neglecting these, we can write for the dependence of the field strength along a magnetic line

$$B(s) \approx B_0 f(s), \quad B_0 \approx B_{00}[f(0) + 4r_0^2 g(0)] \approx B_{00}. \tag{3.9}$$

The zero subscripts denote here and elsewhere the values of the corresponding quantities in the symmetry plane $z = 0$.

An example of a short trap is a field configuration corresponding to

$$f(s) = 1 + (s/L)^2, \tag{3.10}$$

where L is the longitudinal scale of the trap. This configuration is a good approximation of the central part of a classical trap with magnetic mirrors. The meaning of the term "short trap" will be explained below (see Subsection 3).

Using the results of Subsection 1 of the present section, we can write right away the unperturbed Hamiltonian for the field (3.10):

$$H^0 = p^2/2 + \mu\omega_0\left(1 + \frac{s^2}{L^2}\right) = \mu\omega_0 + J\sqrt{2\mu\omega_0}/L \tag{3.11}$$

and the frequencies

$$\left.\begin{aligned} \Omega(\mu) = \partial H^0/\partial J &= \sqrt{2\mu\omega_0}/L, \quad \langle\omega(\mu, J)\rangle = \omega_0 + \frac{J\sqrt{\omega_0}}{L\sqrt{2\mu}} \\ &= \omega_0\left(1 + \frac{a^2}{2L^2}\right) = \frac{\omega_0}{2}\left(1 + \frac{1}{\sin^2\beta_0}\right). \end{aligned}\right\} \tag{3.12}$$

Here $a$ is the amplitude of the longitudinal oscillations of the particle; $J = \Omega a^2/2$ is the longitudinal action; $\beta_0$ is the angle between the particle-velocity vector and the magnetic line (at $s = 0$).

The foregoing example (3.10) could be called also a harmonic trap (in view of the waveform of the longitudinal oscillations).

3. Long Magnetic Trap. This configuration is different in that the field is almost constant over most of the trap, whose mirrors are relatively short and steep. Such a field is a feature of Dimov's ambipolar trap [7]. We describe it by the relation [4]

$$f(s) = 1 + (s/L)^n, \tag{3.13}$$

where n is some even number. As before, L is the length of the entire trap, whereas the length of the mirror is of the order of

$$l = L/n \tag{3.14}$$

(a short trap corresponds to n = 2 and $\ell \sim L$, while for a long one we have $n \gg 1$ and $\ell \ll L$); i.e., the mirrors are much shorter than the entire trap. This is the distinction between "short" and "long" traps.

At $n \gg 1$, the effective potential energy of the longitudinal motion can be approximately represented by a rectangular well of length 2L. The unperturbed Hamiltonian (2.3) takes then the form

$$H^0 = p^2/2 + \mu\omega_0 f(s) \approx J^2/2M + \mu\omega_0. \tag{3.15}$$

Recall that $\omega_0$ does not depend on s on the given magnetic line. We have $M = (2L/\pi)^2$, and the longitudinal action is

$$J = \frac{1}{2\pi} \oint p ds \approx \frac{2L}{\pi} p. \tag{3.16}$$

The fundamental unperturbed frequencies are

$$\Omega = \frac{J}{M} = \frac{\pi}{2L} p, \quad \langle\omega\rangle \approx \omega_0. \tag{3.17}$$

4. Multimirror Trap [11]. Such a trap is a chain of ordinary traps; in other words, a long trap with a corrugated magnetic field whose strength varies along the trap periodically (in the simplest case). We specify the field configuration in such a trap in the form

$$f(s)=\frac{1}{2}\left[(\lambda+1)-(\lambda-1)\cos\left(\frac{\pi s}{L}\right)\right], \tag{3.18}$$

where $\lambda = f_{max}/f_{min}$ is the mirror ratio. The unperturbed Hamiltonian (2.3) now becomes

$$H^0=\frac{p^2}{2}+\frac{\mu\omega_0}{2}\left[(\lambda+1)-(\lambda-1)\cos\left(\frac{\pi s}{L}\right)\right]. \tag{3.19}$$

The equations of motion of such a system are known to be fully integrable in terms of elliptic functions (see, e.g., [8, 12-14]). The modulus k of the elliptic integrals is given in terms of the system parameters in the form

$$k^2=\begin{cases}\dfrac{W}{\mu\omega_0(\lambda-1)}, & H^0<\lambda\mu\omega_0,\\[2ex] \dfrac{\mu\omega_0(\lambda-1)}{W}, & H^0>\lambda\mu\omega_0,\end{cases} \tag{3.20}$$

where $W = H^0 - \mu\omega_0$ is the unperturbed energy reckoned from the minimum of the potential ($\mu\omega_0$). The first regime of the motion corresponds to trapped particles that execute oscillations limited in s, and the second to untrapped particles. The limiting trajectory that separates the two regimes ($H^0 = \lambda\mu\omega_0$) is called the separatrix. A diagram of the phase trajectories of the system (3.19) is shown in Fig. 1.

For trapped particles, the longitudinal-oscillation frequency is

$$(H^0)=\frac{\pi\Omega_0}{2\mathcal{K}(k)}, \tag{3.21}$$

where $\mathcal{K}$ is a complete elliptic integral of the first kind, and

$$\Omega_0=\frac{\pi}{L}\sqrt{\frac{\mu\omega_0(\lambda-1)}{2}} \tag{3.22}$$

is the frequency of the small oscillations. For untrapped (denoted by subscript "un") particles the fundamental frequency of the longitudinal motion can be represented in a form similar to (3.21):

$$\Omega_{un}\equiv 2\pi/T_{un}=\pi\Omega_r/2\mathcal{K}(k), \tag{3.23}$$

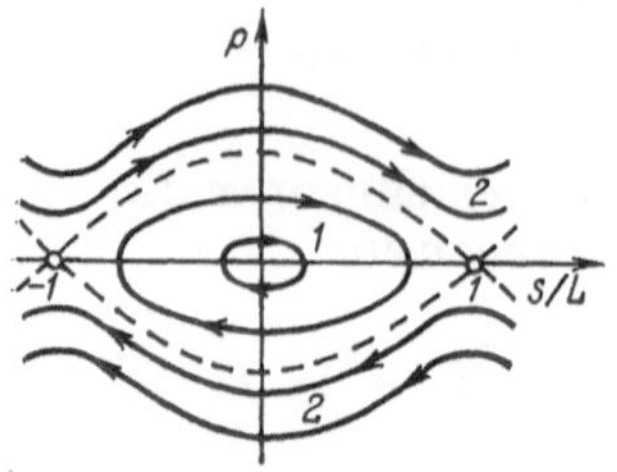


Fig. 1. Diagram of unperturbed phase trajectories in a multimirror trap: 1) trapped particles; 2) untrapped particles; the separatrix is shown dashed.

where $T_{un}$ is the time to travel over one period of the field (3.18), and

$$\Omega_r = \frac{\pi}{L}\sqrt{2W} \tag{3.24}$$

is the free motion frequency (at $\lambda = 1$). Note that, in the case considered, we have $k = 2\Omega_0/\Omega_r$ (3.20).

We shall hereafter be particularly interested in motion near the separatrix, the energy distance to which will be characterized by the dimensionless quantity

$$w = (H^0 - \lambda\mu\omega_0)\frac{2}{(\lambda - 1)\mu\omega_0}. \tag{3.25}$$

A close proximity to the separatrix corresponds to $|w| \ll 1$, with $w < 0$ and $w > 0$ for trapped and untrapped particles, respectively. In both regimes we have $k^2 \to 1 - |w|/2$, $\Omega_r \to 2\Omega_0$, and

$$\Omega(w) \approx \frac{\Omega_{un}(w)}{2} \to \frac{\pi\Omega_0}{\ln\left(\frac{32}{|w|}\right)}. \tag{3.26}$$

Although the frequencies in the two regimes differ by a factor of 2 [the period of the untrapped particle is exactly half that of the trapped (see Fig. 1)], the motions in both cases are close to each other and to the motion along the separatrix, the latter expressed in terms of elementary functions:

$$\left.\begin{aligned} \dot{s}_c &= p_c(s) = \pm \frac{2}{\pi} L\Omega_0 \cos\left(\frac{\pi s}{2L}\right), \\ \frac{s_c(t)}{L} &= \frac{4}{\pi} \arctan\left(e^{\Omega_0 t}\right) - 1. \end{aligned}\right\} \tag{3.27}$$

The phase trajectory on the separatrix (the first expression) is shown in Fig. 1. In the second expression, the time origin ($t = 0$) corresponds to the minimum of the field ($s = 0$). The significant difference between the motion along the separatrix from that along neighboring trajectories reduces only to the frequency, which is exactly zero on the separatrix [cf. (3.21) and (3.23)].

The action is easiest to determine by integrating expression (3.21) for the frequency. In the case of trapped particles

$$J(H^0, \mu) = \int_0^W \frac{dW}{\Omega(W, \mu)} = \frac{8}{\pi^3} L^2\Omega_0 [E(k) - (1 - k^2)\mathcal{K}(k)], \tag{3.28}$$

where we have transformed to integration with respect to dk with the aid of (3.20); E(k) is a complete elliptic integral of the second kind. This expression defines implicitly the function $H^0(J, \mu)$. As $W \to 0$ (small oscillations), we have $J \to W/\Omega_0$ [cf. (3.11)]. The derivative $\partial H^0(J, \mu)/\partial J = [\partial J(H^0, \mu)/\partial H^0]^{-1}$ yields, of course, again the frequency (3.21). To calculate the average Larmor-rotation frequency $\langle\omega\rangle = \partial H^0(J, \mu)/\partial\mu$, we write the differential

$$dJ(H^0, \mu) = \frac{\partial J}{\partial H^0} dH^0 + \frac{\partial J}{\partial \mu} d\mu = 0.$$

Hence

$$\langle\omega\rangle = -\frac{\partial J/\partial\mu}{\partial J/\partial H^0} = \omega_0\left[\lambda - (\lambda - 1)\frac{E(k)}{\mathcal{K}(k)}\right] \to \lambda\omega_0\left[1 - \frac{2(\lambda - 1)/\lambda}{\ln(32/|w|)}\right]. \tag{3.29}$$

The later expression gives the asymptote of $\langle\omega\rangle$ near the separatrix ($|w| \ll 1$). For small oscillations we have $\langle\omega\rangle \to \omega_0 + W/2\mu$ [cf. (3.12)]. When calculating (3.29) it must be recognized that $\partial W(H^0, \mu)/\partial\mu = -\omega_0$.

For the untrapped particles we obtain analogously the action in the form

$$J_{\text{un}}(H^0, \mu) = J^c_{\text{un}} + \int\limits_{W_c}^{W} \frac{dW}{\Omega_{\text{un}}(W, \mu)} = \frac{4L^2\Omega_0}{\pi^3} \frac{E(k)}{k}. \tag{3.30}$$

Here $W_c = \mu\omega_0(\lambda - 1)$ and $J_{\text{un}} = 4L^2\Omega_0/\pi^3$ are the energy and action on the separatrix. We note that the second of these quantities is half the value (3.28) for trapped particles. The action as a function of the energy has thus a discontinuity on the separatrix, for the same reason as the frequency [see (3.26)]. The mean value of the Larmor frequency is

$$\langle\omega\rangle = \omega_0\left[1 + \frac{\lambda - 1}{k^2}\left(1 - \frac{E(k)}{\mathcal{K}(k)}\right)\right] \to \lambda\omega_0\left[1 - \frac{2(1 - 1/\lambda)}{\ln(32/|w|)}\right]. \tag{3.31}$$

We point out that the asymptote of $\langle\omega\rangle$ as $w \to 0$ is the same on both sides of the separatrix [cf. (3.29)]. At $W \gg W_c$ we have $\langle\omega\rangle \approx \omega_0(1 + \lambda)/2$; i.e., it is simply equal to the average over s.

5. Field-Reversed Mirrors, Planar Geometry. A diagram of the magnetic lines of such a field is shown in Fig. 2a. The vector **B** is in the (x, y) plane ($B_z = 0$) and does not depend on z. We note that owing to this symmetry we have here, as in an axisymmetric trap, an additional exact integral of motion (z component of the canonical momentum of the particle), and the problem reduces to two degrees of freedom. The configuration of the considered field is the same in all four quadrants of the (x, y) plane, so that it suffices to consider one of them, say the first (x, y > 0). In addition, the field is symmetric about the bisector of the angles between the coordinate axes ($y = \pm x$). We specify the magnetic line in terms of the minimum distance $r_0$ to the origin (Fig. 2a). The vector potential of this field can be chosen in the form (see, e.g., [15])

$$A_z(x, y) = Cxy, \tag{3.32}$$

where C is a certain constant to be determined below. Then

$$B_x = Cx, \quad B_y = -Cy, \quad B = Cr, \quad r^2 = x^2 + y^2. \tag{3.33}$$

We denote by $B_0$ the minimum field along the magnetic line at the point $\mathbf{r}_0$; then $C = B_0/r_0$; i.e., the minimum field is proportional to $r_0$. In the case considered, the paraxial approximation used in the preceding magnetic-trap examples is quite unsuitable, since the field on the trap axis (x, y =

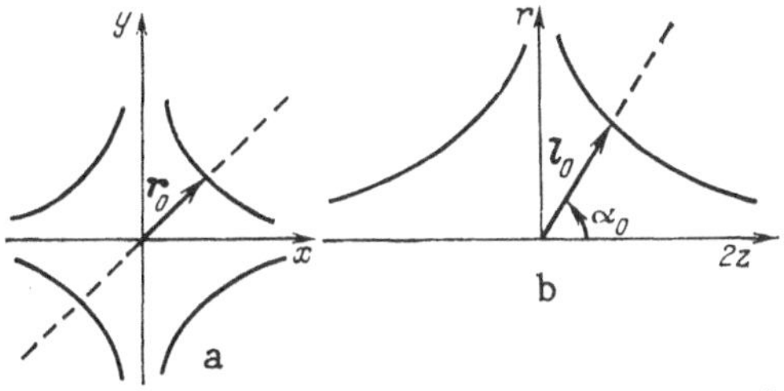


Fig. 2. Diagram of magnetic lines of field-reversed mirrors: a) planar geometry (the field is independent of z); b) cylindrical (axisymmetric) geometry $r^2 = x^2 + y^2$, $\tan\alpha_0 = \sqrt{2}$ ($\alpha_0 \approx 55°$). The field-maximum line is shown dashed. The vectors $\mathbf{r}_0$ and $\mathbf{l}_0$ specify the magnetic line.

0) is zero. To find the function B(s) along the magnetic line we determine first the length of the line s, measured from the point $\mathbf{r}_0$. The equation of the line is $A_z$ = const or $2xy = r_0^2$, or else $r_0^2 = r^2 \sin 2\varphi$ (in the polar coordinates r, $\varphi$). The coordinate along the line is

$$s(r) = \int_{r_0}^{r} \frac{r^2 dr}{\sqrt{r^4 - r_0^4}} \to \begin{cases} \sqrt{r_0(r-r_0)}, & r - r_0 \ll r_0, \\ r + \frac{r_0}{6} \approx r, & r \gg r_0. \end{cases} \tag{3.34}$$

This integral can be expressed in terms of elliptic functions, but we confine ourselves to the two indicated extreme situations. The first corresponds to the central section of the magnetic line near the field minimum. Taking (3.33) into account, we get in this region

$$f(s) \approx 1 + \frac{s^2}{r_0^2}, \qquad \omega(s) = \omega_0 f(s), \qquad \omega_0 = B_0, \tag{3.35}$$

i.e., the field of the short trap (3.10) with characteristic length $L = r_0$. The second expression in (3.34) describes the edges of the magnetic lines. Here

$$f(s) \approx |s|/r_0. \tag{3.36}$$

This configuration is new to us, and we shall examine it in greater detail. The unperturbed Hamiltonian is of the form

$$H^0 = \frac{p^2}{2} + \frac{\mu\omega_0 |s|}{r_0} = \left(\frac{3\pi}{4\sqrt{2}} \frac{\omega_0}{r_0} \mu J\right)^{2/3}, \tag{3.37}$$

where J is the longitudinal action. The unperturbed frequencies are

$$\left.\begin{aligned} \Omega &= \frac{2}{3}\left(\frac{3\pi}{4\sqrt{2}} \frac{\omega_0}{r_0}\right)^{2/3} \frac{\mu^{2/3}}{J^{1/3}} = \frac{\pi}{2}\sqrt{\frac{\mu\omega_0}{2r_0 a}}, \\ \langle\omega\rangle &= \frac{2}{3}\left(\frac{3\pi}{4\sqrt{2}} \frac{\omega_0}{r_0}\right)^{2/3} \frac{J^{2/3}}{\mu^{1/3}} = \frac{2}{3} \frac{\omega_0 a}{r_0}, \end{aligned}\right\} \tag{3.38}$$

where $a$ is the amplitude of the longitudinal oscillations.

6. Field-Reversed Mirrors, Cylindrical Geometry. We consider now the magnetic-field configuration produced by axisymmetric field-reversed mirrors, say by two identical cylindrical coils with oppositely directed currents and magnetic fields (Fig. 2b). The vector potential of such a field can be chosen in the form (see, e.g., [15])

$$A_\varphi = Czr,$$

whence

$$B_z = C2z, \quad B_r = -Cr, \quad B = Cl, \tag{3.39}$$

where the vector $\mathbf{l}$ with components (2z, r) characterizes the position of the point on the (2z, r) plane (see Fig. 2b).

The main difference between this trap and the preceding one is that the magnetic lines are not symmetric about the field on the line. The minimum-field line is, as before, a straight line, but it makes now an angle $\alpha_0 \approx 55°$ [$\tan \alpha_0 = \sqrt{2}$ with the z axis on the (2z, r) plane]. The equation of the magnetic line is obtained from $rA_\varphi$ = const, or

$$zr^2 = (l_0/\sqrt{3})^3, \tag{3.40}$$

where the vector $\mathbf{l}_0$ defines the position of the field minimum on this line. Accordingly, the constant is $C = B_0/\ell_0$.

The coordinate s along the magnetic line is reckoned as before from the point $\mathbf{l}_0$:

$$s = \int_{z_0}^{z} dz \sqrt{1 + \frac{b^3}{4z^3}} \to \begin{cases} -\frac{b^{3/2}}{\sqrt{z}}, & z \ll z_0, \\ \sqrt{3}\,(z - z_0), & z \approx z_0, \\ z, & z \gg z_0. \end{cases} \tag{3.41}$$

Here $z_0 = b/2$; $b = \ell_0\sqrt{3}$. Combining this with the expression $B = \omega = C\ell$ (3.39), we get

$$f(s) \approx \begin{cases} |s|/l_0, & z \to 0, \\ 1 + 2s^2/l_0^2, & z \approx z_0, \\ 2\,|s|/l_0, & z \to \infty. \end{cases} \tag{3.42}$$

As in the preceding case, near the minimum of the field ($z \approx z_0$) we can use the short-trap approximation with scale $L = \ell_0/\sqrt{2}$, while far from the minimum the effective potential is linear in s. It is significant, however, that in contrast to the preceding example the slope of the potential curve (i.e., the effective longitudinal force) is different on the two sides of the minimum of the potential. This is the result of the aforementioned asymmetry of the trap's cylindrical geometry. The cause of the symmetry can be illustratively explained in the following manner: as $z \to \infty$ the bundle of magnetic lines is compressed in two directions (x, y), whereas as $r \to \infty$ the compression is only along one direction (z).

To calculate the frequencies $\langle\omega\rangle$ and $\Omega$, we can use expressions (3.38), taking the arithmetic means of $\langle\omega\rangle$ and $1/\Omega$ for both slopes of the potential at one and the same energy $H^0$.

## 4. Adiabatic Perturbation

Our principal task is to investigate the effect of the perturbation, i.e., of the change of the particle's magnetic moment $\mu$, which was assumed in the unperturbed system to be constant [Eq. (2.2)]. In this section we obtain an equation for $\dot{\mu}$.

We begin with the simple model system (3.1). We obtain an expression for $\dot{J}_y$ by differentiating (3.3) and using the exact equations of motion from the Hamiltonian (3.1):

$$\dot{J}_y = \dot{x}x\left(y^2 - \frac{\dot{y}^2}{\omega_y^2}\right) = 2J_y \frac{\dot{x}x}{1 + x^2} \cos(2\theta). \tag{4.1}$$

Here $\omega_y = 1 + x^2$, and we have put

$$y \equiv a_y \cos\theta, \quad \dot{y} \equiv \omega_y a_y \sin\theta, \qquad (4.2)$$

defining by the same token new variables $a_y$ and $\theta$ (the amplitude and phase of the y oscillations), in terms of which $J_y = \omega_y a_y^2/2$ [see (3.2) and (3.3)]. Expression (4.1) is indeed the sought-for equation for $J_y(t)$. It will be integrated in Sec. 7. It is useful, however, to note right away some of its properties. The right-hand side of the equation is the product of functions that vary with time at different frequencies: low [the factor $S(t) = \dot{x}x/(1 + x^2)$ has the same frequency $\Omega$ as the longitudinal oscillations] and high [the factor $F(t) = \cos(2\theta)$ has double the frequency of the Larmor rotation, viz., $2\langle\omega\rangle$]. In first-order approximation $J_y$ will therefore execute bounded and small (if the ratio $\Omega/2\langle\omega\rangle$ is small) oscillations. The cumulative changes of $J_y$, all that interests us here, are made possible only by the resonance between the two motions, i.e., by the sufficiently high harmonics of the low frequency $\Omega$. The smaller the frequency ratio $\Omega/2\langle\omega\rangle$, the higher the harmonics needed for the resonance and the smaller their amplitudes and, consequently, a certain average rate of change of $J_y$. It is natural, therefore, to choose in this problem as the parameter characterizing the smallness of perturbation simply the frequency ratio:

$$\varepsilon = \Omega/\langle\omega\rangle. \qquad (4.3)$$

This quantity is usually called the adiabaticity parameter, since its smallness is in fact the main condition for adiabatic invariance of the action variables, i.e., the condition for their approximate conservation with time. This question will be discussed in greater detail below (see Section 10).

Proceeding to the discussion of the magnetic traps themselves, we note first that the small adiabaticity parameter (4.3) is determined by the second derivative of the magnetic field with respect to the coordinates; this derivative determines the frequency $\Omega$ of the longitudinal oscillations. The first derivative (field gradient) used sometimes for estimates has by itself no physical meaning as an adiabaticity parameter. Thus, for example, in the case of converging or diverging straight magnetic lines, the magnetic moment of the particle is exactly conserved, although the field gradient differs from zero. What is essential is the bending of the magnetic lines, their curvature. It turns out that in a mag-

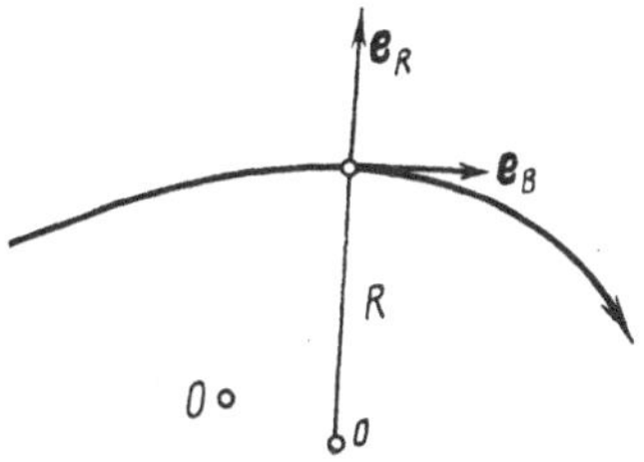


Fig. 3. Geometry of planar magnetic line: $\mathbf{e}_B$, $\mathbf{e}_R$ are the tangential and normal unit vectors; O is the center of the trap; 0 is the curvature center corresponding to the given point of the magnetic line; R is the curvature radius at this point (negative in this case; see the text).

netic trap it is precisely the curvature of the magnetic line that makes the main contribution to the change of μ, as was first elucidated in [16].

Let us examine this mechanism in greater detail. Bending (turning) of the magnetic line alters μ even if the velocity vector **v** remains constant, since μ depends only on the projection of v on a plane perpendicular to the vector **B**, and this plane rotates independently of **v**. Let us find the derivative $\dot{v}_\perp^2$ due to this effect. Assume that the magnetic line is a planar curve (has no torsion), and let $\mathbf{e}_B = \mathbf{B}/B$ be a unit tangent vector and $\mathbf{e}_R$ a unit outward normal (relative to the trap center) to the magnetic line (Fig. 3). We then have, from the expression $v_\perp^2 = v^2 - (\mathbf{v}, \mathbf{e}_B)^2$,

$$\dot{v}_\perp^2 \to -2v_\parallel (\mathbf{v}, \dot{\mathbf{e}}_B) = -2v_\parallel^2 v_n/R,$$

where we have assumed $\mathbf{v}$ = const; $v_\parallel = \mathbf{v}, \mathbf{e}_B$; $v_n = \mathbf{v}, \mathbf{e}_R = \mathbf{v}_\perp, \mathbf{e}_R$, and R is the curvature radius of the magnetic line and is assumed positive if the line is convex relative to the trap center. The instantaneous angular velocity $\mathbf{e}_B$ of the rotation as the particle moves along the magnetic line is

$$\Omega = \mathbf{e}_B \times \mathbf{e}_R \frac{v_\parallel}{R}.$$

At a sufficiently small Larmor radius, the projection of the particle velocity is $v_n \approx -v_\perp \sin\theta$, where $\theta$ is the Larmor phase reckoned from the direction of the vector $\mathbf{e}_R$ (a more accurate expression for $v_n$ is given in [8]). We ultimately get

$$\dot{\mu} \to \sqrt{2\mu}\,\frac{v_\parallel^2}{R\sqrt{\omega}}\sin\theta. \tag{4.4}$$

The arrow indicates here that the equation given here for $\dot{\mu}$ contains only the principal term that determines the cumulative resonance changes of $\mu$. The complete equation for $\dot{\mu}$ is given in [5]. It includes, in particular, also terms of form (4.1), which are proportional to $\cos(2\theta)$. It will be shown below, however (see Section 7), that they make only exponentially small additions, since the adiabaticity parameter (4.3) is only half as large. A detailed discussion of the separation of the principal term of the perturbation (4.4) is contained in [5]. We shall return to this question in Section 7.

## 5. Insignificant Effect of Perturbation

Our principal task is to integrate a perturbed equation such as (4.4) or (4.1), i.e., to determine the effect of the perturbation, meaning the change of the unperturbed integral of motion $\mu$ (or $J_y$). Such equations are quite frequently integrated by some asymptotic method [12], i.e., by constructing the solution in the form of an asymptotic series in powers of a small parameter of the problem, which in our case is the adiabaticity parameter (4.3). The term "asymptotic" means that the remainder term of such a series does not decrease in general with increasing number of its terms, and decreases only together with the smallness parameter. This approach was used in many studies also to investigate the dynamics of a particle in a magnetic trap (see, e.g., [17, 18]). As applied to equations such as (4.4) or (4.1), it means, roughly speaking, integration by parts. During each step the high-frequency factor F(t) of the right-hand side is integrated and the low-frequency side S(t) is differentiated. As a result, each step increases the degree of the small parameter (4.3) by unity. For Eq. (4.1), for example, the first step yields

$$\delta J_y = J_y\,\frac{\dot{\omega}_y}{2\omega_y^2}\sin(2\theta). \tag{5.1}$$

As soon as the unperturbed solution, which is substituted in the right-hand side of (4.1), becomes quasiperiodic, the variations of $J_y(t)$ given by the asymptotic series also become quasiperiodic, with fundamental frequencies $\Omega$ and $\langle\omega\rangle$ that are close the unperturbed ones. These variations are consequently bounded and small at a sufficiently small adiabaticity parameter. In this sense, such quasiperiodic variations of the unperturbed action are an insignificant effect of the perturbation. What is meant by insignificant? In the language of asymptotic expansions what is significant is the remainder term of such an expansion, and we shall proceed to determine it.

To conclude this section, we note that quasiperiodic oscillations of the unperturbed action (5.1) can be used to introduce a "more precise" action $J_y^{(n)}$, whose nonresonant changes will be smaller than for $J_y$ (n is the order of the precision). In first order, for example, we obtain from (5.1)

$$J_y^{(1)} = J_y - \delta J_y = J_y\left(1 - \frac{\dot{\omega}_y}{2\omega_y^2}\sin(2\theta)\right)$$

$$= J_y\left(1 \mp \sqrt{2}\,\frac{x\sqrt{H^0 - J_y(1+x^2)}}{(1+x^2)^2}\sin(2\theta)\right). \qquad (5.2)$$

In the last expression we used the unperturbed relation $\dot{x} = \pm\sqrt{2(H^0 - J_y\omega_y)}$ [see (3.4)]. The derivative is $\dot{J}_y^{(1)} \sim \varepsilon^2$, i.e., it is already of second order in the adiabaticity parameter (4.3). Introduction of a more precise action is equivalent to the canonical transformation of variables, which is widely used in nonlinear mechanics and eliminates the nonresonant terms of a perturbation. This procedure was first used for magnetic traps in [19] for a model of type (3.1) and in [17] for an arbitrary magnetic field (see also [18, 20]). A more precise $\mu$ is useful for a comparison of the theory with results of numerical simulation (see [20] and Section 7 below).

## 6. Nonlinear Resonances

As already noted, nonperiodic changes of $\mu$, which can be cumulative, are due to resonances between the Larmor precession of the particle and the higher harmonics of the longitudinal oscillations. To analyze these resonances we transform in the right-hand side of Eq. (4.4) for $\dot{\mu}$ to the unperturbed action variables $\mu$ and J and to their canonically conjugate phases $\varphi$, $\psi$. Taking the Fourier transforms with respect to the phases, we get

$$\dot{\mu} = \frac{1}{2} \sum_n f_n(\mu, J)\, e^{i(\varphi - n\psi)} + \text{c.c.}, \tag{6.1}$$

where $\dot{\varphi} = \langle\omega\rangle$; $\dot{\psi} = \Omega$, and n is an arbitrary integer. Generally speaking, this is a double Fourier series; i.e., it includes also harmonics of the phase $\varphi$. Since, however, $\langle\omega\rangle / \Omega \gg 1$, the amplitudes of the resonance harmonics $m\varphi$ become quite small compared with $f_n$, and we neglect them. The resonance conditions are (n is a positive integer)

$$\langle\omega\rangle = n\Omega. \tag{6.2}$$

Generally speaking, this condition is not met. Since, however, the unperturbed oscillations are not isochronous, i.e., their frequencies ($\langle\omega\rangle$, $\Omega$) depend on the actions $\mu$, J (see Section 3), there are always special (resonant) values $\mu = \mu_r$, $J = J_r$, for which the resonance condition (6.2) will be satisfied with some n = r.

If the amplitudes $f_n$ are small enough (a smallness condition will be obtained below; see Section 10), we need retain in (6.1) only the resonant term

$$\dot{\mu} \approx |f_r(\mu_r, J_r)| \cos\psi_r, \tag{6.3}$$

where we have introduced the resonant phase

$$\psi_r = \varphi - r\psi + \psi_r^0, \tag{6.4}$$

and $\psi_r^0(\mu_r, J_r)$ is some constant: $f_r = |f_r| \exp(i\psi_r^0)$. At exact resonance (6.2) we have $\psi_r$ = const. However, in view of the change of $\mu$ and of the dependence of the frequencies $\langle\omega\rangle$, $\Omega$ on $\mu$ and J, the resonant phase also varies with time. The equation for $\psi_r$ can be written as

$$\dot{\psi}_r \approx \langle\omega\rangle - r\Omega. \tag{6.5}$$

This equation is approximate, since $\psi_r$ is altered not only by the frequency change but also directly by the perturbation. The latter effect, however, is small if the perturbation is small, and we shall neglect it (this question is discussed in greater detail in [8]).

The perturbation alters not only $\mu$ but also J. In a purely magnetic field, however, the particle energy is conserved and the change of J can be expressed in terms of the change of $\mu$. We assume, therefore, that both the frequencies $\langle\omega\rangle$ and $\Omega$, and the Fourier amplitudes $f_n$ depend only on $\mu$. Equations (6.3) and (6.5) constitute then a complete system that describes the dynamics of one resonance. Equation (6.5) can be simplified by expanding the right-hand side near the resonance:

$$\dot{\psi}_r \approx \left.\frac{d\omega_r}{d\mu}\right|_{\mu=\mu_r} (\mu - \mu_r), \quad \omega_r(\mu) = \langle\omega\rangle - r\Omega. \tag{6.6}$$

Introducing the quantity $\nu = \mu - \mu_r$, and noting that $\dot{\nu} = \dot{\mu}$, we obtain a pair of canonical equations

$$\left.\begin{aligned} &\dot{\nu} = |f_r(\mu_r)| \cos\psi_r, \\ &\dot{\psi}_r = \omega_r' \nu, \quad \omega_z' \equiv \left.\frac{d\omega_r}{d\mu}\right|_{\mu=\mu_r} \end{aligned}\right\} \tag{6.7}$$

with a Hamiltonian

$$H_r(\psi_r, \nu) = \frac{\omega_\nu' \nu^2}{2} - |f_r| \sin\psi_r. \tag{6.8}$$

It can be easily seen that this is the Hamiltonian of a pendulum, say of unit length with mass $1/\omega_r'$ in a gravitational field $mg = |f_r|$, where $\nu$ is the angular momentum of the pendulum and $\psi_r$ is the angle of its inclination from the horizontal. The Hamiltonian (6.8) is therefore said to describe a nonlinear resonance in the pendulum approximation [8]. The conditions of this simple approximation are discussed in detail in [8]. What matters is just that the perturbed oscillations are not isochronous: $\omega_r' \neq 0$. The oscillations of $\nu$ are therefore bounded near resonance ($|\Delta\nu| \leq 2\sqrt{|f_r|/|\omega_r'|}$), and are small at small $|f_r|$, so that one can use for $\dot{\psi}_r$ the expansion (6.6) in terms of $\nu$.

The limited size of the oscillations of $\nu$ (hence of $\mu$), and the ensuing limited (and insignificant) energy exchange between the degrees of freedom of the system in question, constitute the essential difference between a nonlinear and a linear resonance. In the latter case, the exchange would be complete. The nonlinearity (nonisochronism) of the oscillations is said to stabilize the resonant perturbation. A transition to linear resonance in the Hamiltonian (6.8) corresponds to $\omega_r' \to 0$ and $\Delta\nu \to \infty$ (the pendulum approximation, naturally, then becomes invalid). We note also that the dynamics of a nonlinear resonance are similar to unperturbed motion of a particle in the periodic field of a multimirror trap [cf. (6.8) and (3.19)]. We shall need this analogy later (see Section 10).

Returning to our problem, we see that one nonlinear resonance cannot inhibit particle containment in a magnetic trap, since the oscillations of $\mu$ on it are bounded. In fact, however, there are many resonances (6.1). If the remaining terms in (6.1) are not neglected, the Hamiltonian (6.8) takes the more complicated form

$$H_r(\psi_r, \nu) \approx \frac{\omega_r' \nu^2}{2} - \sum_n |f_n| \sin(\psi_r - (n-r)\Omega t + \psi_n^0 - \psi_r^0), \tag{6.9}$$

where we have put approximately $\psi \approx \Omega t$. Although different terms of the last sum cause resonances under different initial conditions (different $\mu$), is it always possible to neglect the nonresonant perturbation, as was done in (6.8)? We shall consider this important and, in essence, central question somewhat later (see Section 10), and proceed now to calculate the resonant amplitudes $f_n$. We can use for this purpose the relation

$$\Delta\mu = T\,|f_r|\cos\psi_r, \tag{6.10}$$

which is obtained by integrating (6.1) over the period $T = 2\pi/\Omega$ of the longitudinal oscillations. On the other hand, we can obtain $\Delta\mu$ by directly integrating Eq. (4.4) for $\dot{\mu}$. We shall name the quantity (6.10) the resonance $\Delta\mu$.

## 7. Resonant $\Delta\mu$

Thus, our next task is to calculate the total change of the unperturbed action ($\Delta\mu$) within the period of the low-frequency (longitudinal) oscillations. We shall do this below for the set of typical examples described in Section 3. We begin, as usual, with the simple auxiliary system (3.1). In this case we must calculate the integral [see (4.1)]

$$\frac{\Delta J_y}{J_y} \approx \Delta \ln J_n = 2\int dt\, \frac{x\dot{x}}{1+x^2}\cos 2\theta \approx 2\,\mathrm{Re}\int d\theta e^{2i\theta}\frac{x\dot{x}}{(1+x^2)^2}, \tag{7.1}$$

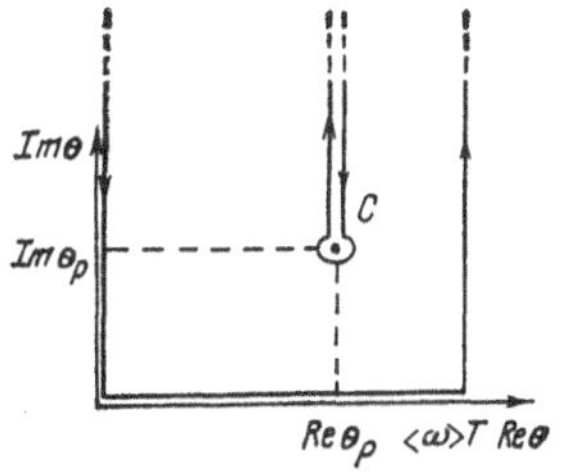


Fig. 4. Integration contour in the complex θ plane when calculating $\Delta J_y$ and $\Delta\mu$; C – bypass of the singularity along the cut; $\theta_p$ – location of singularity.

furthermore not asymptotically in terms of the small adiabaticity parameter $\varepsilon = \Omega/\langle\omega\rangle$ (4.3), but in a certain sense "exactly." Let us formulate the problem more accurately. Since we do know the integrand as a function of time, the use of some successive-approximation method is unavoidable. We therefore replace the integrand by an unperturbed solution, in particular $J_y$ = const, something already taken into account in the relation $d\theta = (1 + x^2)dt$ for the last representation of the integral (7.1). In the integration by parts in Section 5 we have integrated in essence only the high-frequency factor, "lumping" all the effects of the low-frequency part (including its high harmonics) in the remainder term which was left undetermined. Now, however, we integrate the unperturbed integrand exactly (with all its harmonics). This can be done in a number of cases by analytically continuing the integrand into the complex plane of the integration variable. If the latter is chosen to be the phase θ, as was done, e.g., in [5], we can close the integration contour in the upper half-plane of θ and neglect the contribution made to the integral by the infinitely remote part of this contour ($\mathrm{Im}\,\theta \to +\infty$). The value of the integral is then determined by the singularities of the integrand, which are bypassed by branch cuts (Fig. 4).

If the resonance condition (6.2) is satisfied, the sum of the integrals over the two vertical lines is also zero, since the quasiperiodic time dependence of the unperturbed solution (the two incommensurate frequencies $\langle\omega\rangle$ and $\Omega$) becomes at resonance periodic with a period T ($\langle\omega\rangle T = 2\pi n$). Off resonance, this sum is generally speaking not zero because of the different values of the function exp $(2i\theta)$ on the two straight lines. This leads to a quasiperiodic increment to $\Delta\mu$ (see Section 5) which, however, is of no interest to us.

For the integral (7.1), the only singularity in the upper $\theta$ half-plane is a pole at $x_p = i$. We note right away that the frequency at this singularity is $\omega_y = 0$, and the singularity is also a saddle point (col) of the function exp $(2i\theta)$.

Since the singularity is bypassed twice during one period of the x oscillations, it suffices to calculate the integral over the half-period T/2. Near the singularity we have $\omega_y(x_p) = 1 + x_p^2 = 0$ and $\theta = \theta_p$:

$$\theta - \theta_p \approx \left(\frac{d\theta}{dx}\right)_p (x - x_p) + \frac{1}{2}\left(\frac{d^2\theta}{dx^2}\right)_p (x - x_p)^2 = \frac{x_p}{v_0}(x - x_p)^2, \quad (7.2)$$

since $(d\theta/dx)_p = \omega_y(x_p)/\dot{x}_p = 0$, while $(d^2\theta/dx^2)_p = (d\omega_y/dx)_p/\dot{x}_p$ and $\dot{x}_p = \sqrt{2H0} = v_0$ [see (3.4)], where $v_0^2 = \dot{x}_0^2 + \dot{y}_0^2$ is the total velocity at the minimum of the potential energy (x = 0). Next,

$$(1 + x^2)^2 \approx (2x_p)^2 (x - x_p)^2 \approx 4x_p v_0 (\theta - \theta_p).$$

Substituting this expression in (7.1) and calculating the residue at the pole, we obtain

$$\Delta \ln J_y = \pi \operatorname{Re}\left(ie^{2i\theta_p}\right). \quad (7.3)$$

Note that were we to calculate the integral in the complex t or x plane, we would obtain the different (incorrect) result:

$$\Delta \ln J_y = 2 \operatorname{Re} \int \frac{x dx}{1 + x^2} e^{2i\theta} = 2\pi \operatorname{Re}\left(ie^{2i\theta_p}\right).$$

The reason for the difference is that the function exp $(2i\theta \cdot (x))$, generally speaking, does not vanish as $\operatorname{Im} x \to \infty$, so that the integral does not reduce to a residue at a pole. Finally, we can proceed as follows. We assume that the last integral is taken along a contour in the $\theta$ plane, but we simply replace the integration variable by x. As $x - x_p \propto \sqrt{\theta - \theta_p}$, when the singularity is bypassed in the $\theta$ plane [the complex vector $(\theta - \theta_p)$ is rotated through an angle

$2\pi$], the vector $(x - x_p)$ is rotated only through an angle $\pi$. Therefore, the value of the integral over dx is equal to half the residue at the pole, and this coincides exactly with (7.3).

We now find $\theta$, which we can represent in the form

$$\theta_p = \theta_0 + \int_0^{t_p} \omega_y dt = \theta_0 + \int_0^{x_p} \frac{\omega_y dx}{\dot{x}}, \qquad (7.4)$$

where $\theta_0 = \mathrm{Re}\,\theta_p$. To calculate this integral we use an unperturbed solution, for example in the form [see (3.4)]

$$\dot{x} = \sqrt{2J_y(a_x^2 - x^2)}, \qquad (7.5)$$

where $a_x$ is the amplitude of the longitudinal oscillations. Recognizing that $\omega_y = 1 + x^2$ and $x_p = i$, we obtain from (7.4)

$$\theta_p = \theta_0 + \frac{i\chi(\sin\beta_0)}{2v_0}, \qquad \chi(u) = \frac{1}{u^2}\left(\frac{1+u^2}{2u}\ln\frac{1+u}{1-u} - 1\right), \qquad (7.6)$$

where $\sin\beta_0 = \dot{y}_0/v_0$. In this form, the expression for the exponent that determines the order of the resonant perturbation was obtained (for a magnetic trap) in [21] and later in [5]. It is valid, of course, only for harmonic longitudinal oscillations, a fact used explicitly in the integration of (7.4). On the other hand, relation (7.6) is too complicated and inconvenient for further use. A simple approximation for $\theta_p$ is obtained by putting $\dot{x} \approx \dot{x}_0 = \mathrm{const}$ in (7.4). Then

$$\theta_p \approx \theta_0 + \frac{2i}{3v_0\cos\beta_0}. \qquad (7.7)$$

The condition $\dot{x} \approx \dot{x}_0$ will be called the small $\beta_0$ approximation since it is valid only if $\dot{x}_0 = v_0\cos\beta_0 \approx \dot{x}_p = v_0$. Hence $\cos\beta_0 \approx 1$ or $\beta_0 \ll 1$. Expression (7.7) can, of course, be obtained also directly from (7.6) as $\sin\beta_0 \to 0$. In a more general case, however, the expression for $\theta_p$ in the small $\beta_0$ approximation depends on the form of the function $\omega_y(x)$ (see below). For the harmonic oscillations considered here, a comparison of (7.6) and (7.7) shows that the difference between them does not exceed 10% at $\beta_0 \leq 50°$. We confine ourselves hereafter, for simplicity, to the small-$\beta_0$ approximation.

In this approximation we obtain from (7.3) for the considered model system

$$\Delta \ln J_y = -\pi \exp\left[-\frac{4}{3v_0 \cos\beta_0}\right]\sin(2\theta_0). \qquad (7.8)$$

If the singularity of the integrand is more complicated than a pole, for example a branch point, the integration method described above cannot be used. A more general method, used in particular in [15], is to reduce the integral along the cut (see Fig. 4) to a Γ function with the aid of the relation (see, e.g., [22]):

$$\int_C \frac{e^{-u}du}{(-u)^p} = \frac{2\pi i}{\Gamma(p)}, \qquad (7.9)$$

where, in our case, $u = -i(\theta - \theta_p)$, $p > 0$, and the integration contour C is shown in Fig. 4 (it is drawn in the negative direction). Using (7.2), we reduce the integral (7.1) (with respect to $d\theta$) to the form (7.9) with $p = 1$, obtaining again (7.3).

In the problem of particle motion in a magnetic trap we must calculate the integral [see (4.4)]

$$\frac{\Delta\mu}{\sqrt{2\mu}} \approx \Delta\sqrt{2\mu} = \int dt\, \frac{v_\parallel^2}{R\sqrt{\omega}} \sin\theta \approx \operatorname{Im}\int d\theta e^{i\theta}\, \frac{v_\parallel^2}{R\omega^{3/2}}. \qquad (7.10)$$

The singularities of the integrand are determined now not only by the function $\omega(s)$ but also by the function $R(s)$. It is therefore necessary first to find a sufficiently simple expression for the curvature radius of the magnetic line.

In the paraxial approximation, the vector potential of the field is given by relation (3.7), and the equation of the magnetic lines is $rA_\varphi = \text{const}$, or

$$r^2 f(s) = r_0^2, \qquad (7.11)$$

where $r_0$ is the distance from the symmetry axis to the magnetic line at the minimum of the field [$s = 0$, $f(0) = 1$]. Hence

$$\frac{1}{R} \approx \frac{d^2r}{ds^2} = \frac{3r_0}{4}\frac{(f')^2}{f^{5/2}} - \frac{r_0}{2}\frac{f''}{f^{3/2}} \to \frac{3r_0}{4}\frac{(f')^2}{f^{5/2}}, \qquad (7.12)$$

assuming that $(dr/ds)^2 \ll 1$ (paraxial approximation). We note first that in this approximation the singularity of R coincides with the singularity of $\omega$, since both quantities are proportional to $f_p$. It is furthermore clear that the main contribution is made by the term with the strongest singularity (maximum p, see below). We have therefore retained in (7.12) only the first term. Relation (7.2) near the singularity now takes the form

$$\theta-\theta_p \approx \left(\frac{d\theta}{ds}\right)(s-s_p)+\frac{1}{2}\left(\frac{d^2\theta}{ds^2}\right)_p (s-s_p)^2$$
$$=\frac{\omega_0 f_p'}{2v}(s-s_p)^2 \approx \frac{\omega_0}{2vf_p'}f^2, \qquad (7.13)$$

since $d\theta/ds = \omega/\dot{s} = 0$ at $s = s_p$; $(d^2\theta/ds^2)_p = (d\omega/ds)_p/\dot{s}_p = \omega_0 f'_p/v$ [see (2.3)]. In (7.13) we used the relation $f(s) \approx f(s_p) + f_p'(s - s_p) = f_p'(s - s_p)$. Note that the factor $\omega_0/vf_p' \sim \omega_0 L/v \sim 1/\varepsilon$ in (7.13) is $\gg 1$ [see (4.3)]. Therefore, each extra power of f in the denominator of the integral (7.10) increases its value by $\sim\varepsilon^{-1/2}$. This is why allowance for only the principal singularity of the integrand is justified.

This explains also why the remainder term of the asymptotic series in $\varepsilon$ for $\delta\mu$ (see Section 5) does not decrease with increasing number of terms of such a series. On the one hand, the time derivative of each succeeding more precise $\mu^{(n)}$ decreases by a factor $\varepsilon$, but, on the other hand, the denominator of the expression for $\dot{\mu}^{(n)}$ acquires thereby an extra $\omega^2$ [one $\omega$ on account of integration of $\exp(i\theta)$, and the other on account of the differential denominator]. As a result, the integral of $\dot{\mu}^{(n)}$ does not depend on the degree of precision and it is simplest to calculate it with unperturbed $\mu$.

Gathering together all the relations, we reduce the integral (7.10) to the form (7.9) with p = 2 and obtain

$$\left.\begin{aligned} \Delta\sqrt{\mu} &= -\frac{3\pi}{8\sqrt{2}} r_0 \sqrt{\omega_0}\, \mathrm{Im}\,(e^{i\theta_p}), \\ \frac{\Delta\mu}{\mu} &\approx -\frac{3\pi}{4}\frac{r_0\omega_0}{v_{\perp 0}} \mathrm{Im}\,(e^{i\theta_p}). \end{aligned}\right\} \qquad (7.14)$$

Since p is an integer in this case, the integral can, of course, be expressed also in terms of a residue at a pole.

Note that in the considered paraxial approximation the magnetic-field configuration influences only the value of $\theta_p$ (if $f'_p$ is finite; see below). A more general and correspondingly more complex expression for $\Delta\mu$, not restricted to the paraxial region, was obtained in [5] [see Eq. (32)]. Its analysis shows that the main condition of the paraxial approximation is the inequality

$$\frac{r_0^4}{L\rho_m^3} \ll 1, \qquad (7.15)$$

where $\rho_m = v/\omega_0$ is the maximum Larmor radius of the particle (at $\beta_0 = \pi/2$) and L is the longitudinal scale of the trap (see Section 3). Expression (4.4), on which the derivation of (7.14) is based, was obtained under the additional simplifying assumption that $\rho_m \ll r_0$ (see Section 4). A more detailed analysis [8] shows, however, that in the paraxial approximation all the relations remain unchanged also for $r_0 < \rho_m$. In particular, at $r_0 = 0$, when the center of the Larmor circle moves along the symmetry axis of the field, we have $\Delta\mu = 0$, for in this case $\mu$ coincides with an exact integral, viz., the canonical angular momentum conjugate to the azimuthal angle $\varphi$ [see (2.1)]:

$$(\mathbf{r} \times (\mathbf{p} + \mathbf{A}))_z = r(\mathbf{p}_\perp + \mathbf{A})_\varphi = \frac{v_\perp^2}{2\omega} = \mu.$$

Note also that expressions (7.14) are equally valid during both half-cycles of the longitudinal oscillations.

The calculation of $\theta_p$ in (7.14) is in accordance with an equation similar to (7.4):

$$\theta_p = \theta_1 + \omega_0 \int_{s_1}^{s_p} \frac{f(s)\,ds}{\dot{s}(s)}, \qquad (7.16)$$

where $\theta_1 = \theta(s_1)$ and $s_1 = \mathrm{Re}(s_p)$. In the general case we have $s_1 \neq 0$ (see below).

For a short trap (see Subsection 2 of Section 3), $f(s) = 1 + s^2/L^2$ and $s_p = iL$; $s_1 = 0$; $\theta_1 = \theta_0$. In the small-$\beta_0$ approximation $[\dot{s}(s) \approx \dot{s}_0]$ we obtain, in analogy with (7.7),

$$\theta_p \approx \theta_0 + \frac{2i}{3} \frac{L}{\rho_m \cos \beta_0}. \tag{7.17}$$

The same expression is valid (with a certain effect $L_{ef}$) also in the case of a multimirror trap (see Subsection 4 of Section 3). The value of $L_{ef}$ is obtained by expanding f(s) in the vicinity of s = 0. From (3.18) we have

$$f(s) \approx 1 + (\pi s/2L)^2 (\lambda - 1), \tag{7.18}$$

whence

$$L_{ef} = \frac{2L}{\pi \sqrt{\lambda - 1}}. \tag{7.19}$$

Since the singularity is located at the point $s_p = iL_{ef}$, at a larger mirror ratio $\lambda \gg 1$ we have $|s_p| \ll L$ and the expansion (7.18) can actually be used. The same condition ($\lambda \gg 1$) ensures smallness of $\beta_0$ for particles near the separatrix of a multimirror trap, for which $\sin \beta_0 = 1/\sqrt{\lambda} \ll 1$ (see Section 3). It is important that, in this case, there are no singularities other than $s_p \approx iL_{ef}$. This can be seen from the exact expression for $s_p$:

$$\cos(\pi s_p/L) = (\lambda + 1)/(\lambda - 1) \text{ [see (3.18)]}.$$

The situation is different for a long trap in the model (3.13). The upper s plane has in this case n/2 singularities (n is even) at the points

$$s_k = L \exp\left[i \frac{\pi}{n}(1 + 2k)\right], \tag{7.20}$$

where k takes on integer values from 0 to (n/2 − 1). The largest contribution to $\Delta\mu$ is made by the two singularities closest to the real axis and corresponding to k = n/2 − 1 and k = 0:

$$s_p = \mp L \cos(\pi/n) + iL \sin(\pi/n) \approx \mp L + i\pi L/n. \tag{7.21}$$

The last approximate expression is valid at $n \gg \pi$, i.e., precisely for a long trap.

From the connection (7.16) between $\theta$ and s in the small-$\beta_0$ approximation ($\dot{s} \approx v$),

$$\theta = \frac{\omega_0 s}{v}\left[1 + \frac{1}{n+1}\left(\frac{s}{L}\right)^n\right] + \text{const} \tag{7.22}$$

it can be seen that the singularities $\theta_k$ in the $\theta$ plane also lie on the circle $(s_k^n = -L^n)$, and the singularities closest to the real axis are well separated from one another. Thus, for the singularity (7.21) which neighbors on $s_p$, the value of Im $\theta_k$ is three times larger than for $s_p$. This justifies the neglect of all singularities except $s_p$ (7.21). The constant in (7.22) is determined by the values of $\theta_1$ and $s_1 \approx \mp L$. We ultimately get

$$\theta_p \approx \theta_1 \pm \frac{2n}{n+1}\frac{l}{\rho_m} + i\pi\frac{n}{n+1}\frac{l}{\rho_m} \approx \theta_1 \pm \frac{2l}{\rho_m} + i\pi\frac{l}{\rho_m}. \quad (7.23)$$

We see that the order of $\Delta\mu$ is determined now by the size $\ell = L/n$ of the mirror, and not by that of the whole trap. In addition, the phase on which $\Delta\mu$ depends is no longer equal to the Larmor phase $\theta_1$ at $s = s_1$, but is shifted by the large value $2\ell/\rho_m \gg 1$, and furthermore to opposite sides for the two singularities. Expression (7.14) for $\Delta\mu$ remains valid also in this case for each of the two main singularities. In contrast to the preceding examples, however, the changes of $\mu$ on two half-cycles of longitudinal oscillations (i.e., after a round trip of the particle over the trap) are not fully symmetric. Namely, the sign of the additional phase shift $2\ell/\rho_m$ is reversed.

In place of a phase shift by $\pm 2\ell/\rho_m$ at the point $s = s_1$ one can shift by a distance s the point at which the phase $\theta_1$ is chosen:

$$\left.\begin{aligned} &\theta_1(s_1) \pm 2l/\rho_m \rightarrow \theta_1(s_1 + \Delta s_1), \\ &\Delta s_1 \approx \pm (2l/\rho_m)/(d\theta/ds) \approx \pm 2l, \quad \Delta\,|\,s_1\,| \approx -2l. \end{aligned}\right\} \quad (7.24)$$

The point of each mirror shifts toward the center of the trap. The field at the shifted points

$$\omega_1/\omega_0 \approx 1 + (1 - 2/n)^n \approx 1 + e^{-2} \approx 1 \quad (7.25)$$

differs little from the field at the center of the trap. It can be stated that the shifted points are located at the inner edges of the mirrors.

For traps with field-reversed mirrors (see Subsections 5 and 6 of Section 3), there is no paraxial approximation, and $\Delta\mu$ must be calculated anew. This problem was solved in [15].

We consider initially a planar geometry of opposing mirrors (see Subsection 5 of Section 3). In contrast to a multimirror trap (see Subsection 4), the quadratic expansion of the field (3.35) near the minimum is not suitable in this case, since its singularity is too far away, at $|s_p| = r_0$, where the expansion is no longer valid [see (3.34)]. To get around this difficulty, we note that although the integration contour (7.10) should be taken in the $\theta$ plane, the integration variable and (or) the argument of the integrand can be arbitrary. To choose the most suitable variable, we express the Larmor frequency in the form [see (3.33)]

$$\omega = \frac{\omega_0}{r_0}\sqrt{x^2+y^2} = \frac{\omega_0 r_0}{2x}\sqrt{1+\frac{4x^4}{r_0^4}}, \tag{7.26}$$

where we use the magnetic-line equation $2xy = r_0^2$. If we now choose the argument of the function $\omega(\xi)$ to be, for example,

$$\xi = 2x^2/r_0^2, \quad \omega(\xi) = \omega_0\sqrt{(1+\xi^2)/2\xi}, \tag{7.27}$$

the position of the singularity in the complex $\xi$ plane will be $\xi_p = i$, and $\xi_0 = 1$ [at the minimum of $\omega(\xi)$ on the given magnetic line].

We find first the connection between the new variable $\xi$ and the phase $\theta$. We have

$$\theta = \int \omega dt = \int \frac{\omega(\xi)}{\dot{s}(\xi)}\frac{ds}{d\xi}d\xi \approx \frac{q_2}{2}\left(\xi - \frac{1}{\xi}\right). \tag{7.28}$$

The last expression is valid in the approximation of small $\beta_0$, when $\dot{s}(\xi) \approx \dot{s}_0 \approx v$, and we have used the relation

$$\frac{ds}{d\xi} = \frac{ds}{dx}\frac{dx}{d\xi} = \frac{r_0}{2\sqrt{2}}\frac{\sqrt{1+\xi^2}}{\xi^{3/2}}, \tag{7.29}$$

while $q_2 = \omega_0 r_0/2v = r_0/2\rho_m$ is the large parameter of the problem. In the same approximation we have

$$\theta_p = \theta_0 + iq_2. \tag{7.30}$$

The curvature of the magnetic line is obtained from the equation $2xy = r_0^2$ of the line:

$$\frac{1}{R(\xi)} = \frac{d^2y/dx^2}{\left[1 + \left(\frac{dy}{dx}\right)^2\right]^{3/2}} = \frac{r_0^2}{(x^2+y^2)^{3/2}} = \frac{\omega_0^3}{r_0 \omega^3(\xi)}. \tag{7.31}$$

Just as in the preceding examples, the curvature is expressed in terms of the field, and has a singularity at the very same point $\xi = \xi_p = i$. At the minimum of the field the curvature radius is $R_0 = r_0$; i.e., it is equal to the minimum distance from the line to the center $x = y = 0$ of the trap (see Fig. 2a).

We choose the same integration variable as before:

$$u = -i(\theta - \theta_p) \approx \frac{iq_2}{2\xi_p^3}(\xi - \xi_p)^2. \tag{7.32}$$

The last expression is obtained by expanding the function $\theta(\xi)$ (7.28) near the singularity. In the same region we have

$$\omega \approx \omega_0 (\xi - \xi_p)^{1/2} \approx \left(-\frac{2u}{q_2}\right)^{1/4} \tag{7.33}$$

[see (7.27)]. Substituting all these expressions in the integral (7.10) for $\Delta\mu$, we reduce it to the form (7.9) with $p = 9/8$. We obtain ultimately

$$\Delta\sqrt{\mu} = \frac{\pi}{2^{13/8}\Gamma(9/8)} \frac{v}{\sqrt{\omega_0}} q_2^{1/8} e^{-q_2} \sin\theta_0 \approx 1.08 \frac{v}{\sqrt{\omega_0}} q_2^{1/8} e^{-q_2} \sin\theta_0. \tag{7.34}$$

This is exactly the result of [15] (at $\beta_0 \ll 1$) corrected for the misprints there. We note that Howard [15] used a somewhat different method in which the integration variable was the scalar potential of the magnetic field.

In the case of cylindrical geometry of the field-reversed mirrors (Subsection 6), the procedure for calculating $\Delta\mu$ remains the same. We choose as the argument of the integrand the quantity $\xi = r^3/2b^3$, where b is a parameter of the magnetic line $zr^2 = b^3 = (\ell_0/\sqrt{3})^3$ (see Fig. 2b). Then

$$\omega(\xi) = \omega_0 \frac{2^{1/3}}{\sqrt{3}} \frac{\sqrt{1+\xi^2}}{\xi^{2/3}} \tag{7.35}$$

and

$$\theta(\xi) = 2^{-1/3} q_3 \xi^{2/3} \left(1 - \frac{1}{2\xi^2}\right), \qquad (7.36)$$

and $q_3 = \omega_0 \ell_0 / 3v = \ell_0 / 3\rho_m \gg 1$. Using the same integration variable $u = -i(\theta - \theta_p)$, we arrive at a singularity of the same order $p = 9/8$ and get

$$\Delta\sqrt{\mu} = \frac{\pi \cdot 3^{1/8}}{2^{11/8}\Gamma\left(\frac{9}{8}\right)} \frac{v}{\sqrt{\omega_0}} q_3^{1/8} e^{-\alpha q_3} \sin(\theta_0 - \Delta)$$

$$= 1.07 \frac{v}{\sqrt{\omega_0}} q_3^{1/8} e^{-\alpha q_3} \sin(\theta_0 - \Delta), \qquad (7.37)$$

$$\Delta = \frac{11}{48}\pi + 0.155 q_3, \qquad \alpha = 1.031 \approx 1.$$

We note first of all that, at a suitable choice of the parameters q, the expressions for $\Delta\mu$ are very close to each other for field-reversed mirrors of both types. The strongest difference is due to the phase shift $\Delta$ in the last case. Its cause is the same as for a long trap, viz., the asymmetric configuration of the field relative to the singularity. Here, too, one can introduce an equivalent shift of the point at which one takes the Larmor phase that determines $\Delta\mu$: $\theta_0(s = 0) \to \theta_1(s = s_1)$, where

$$s_1 \approx -v\Delta/\omega_0 \approx -l_0/19. \qquad (7.38)$$

Note that in both cases the shift is toward a smaller field gradient.

Expression (7.37) agrees with the results of [15], except for a constant phase shift $11\pi/48$, which differs in [15] by $+\pi/4$. This difference is possibly due to the fact that in [15] the integration is over a contour in the complex plane of z and not $\theta$.

In all the cases considered, the change of $\Delta\mu$ turns out to be exponentially small relative to the adiabaticity parameter $\varepsilon = \Omega/\langle\omega\rangle \sim \Omega/\omega_0 \sim \rho_m/L$. This raises the serious question of how reliable are estimates of so small a quantity, especially in view of the neglect of the much larger ($\sim\varepsilon$) quasiperiodic variations of $\mu$. Similar misgivings were raised many times in the literature concerning this problem and similar ones. This is seemingly confirmed by the fact that different preexponential factors in the expression for resonant $\Delta\mu$ were obtained by different workers. Some differ greatly enough to doubt the validity of the power-law (in $\varepsilon$) correction to $\Delta\mu$. The latter, however, is excluded completely for the following reasons. It is shown in [17] that asymptotic corrections to $\mu$, in any order in $\varepsilon$, are quasiperiodic. Consequently, the nonperiodic (resonant) change of $\mu$ decreases with $\varepsilon$ more rapidly than any power of $\varepsilon$, i.e., exponentially. The same conclusion can also be reached in much simpler fashion, by starting from the exact expression (4.4) for $\Delta\mu$. Indeed, in the case of an analytic time dependence of the right-hand side on the time or on the phase $\theta$ (and this is always the case for real field), we can shift the integration contour (with respect to $\theta$) along the imaginary axis ($\mathrm{Im}\,\theta = 0 \to \mathrm{Im}\,\theta = \theta_s = \mathrm{const}$), and this leads to the appearance of a constant factor $\exp(-\theta_s)$ [after making the substitution $\sin\theta \to \mathrm{Im}\exp(i\theta)$]. But since $\theta_s \propto \omega \propto 1/\varepsilon$, we have $\Delta\mu \propto \exp(-A/\xi)$; i.e., it decreases exponentially as $\varepsilon \to 0$.

A more complicated problem is that of the accuracy of the expressions obtained above for resonant $\Delta\mu$. The difficulty lies here in the fact that in the higher approximations the harmonics of the low frequency $\Omega$ are, generally speaking, enhanced, and can in principle alter somehow the argument of the exponential and (or) the preexponential factor. Nonetheless, it can apparently be stated that the relative correction to $\Delta\mu$ is small as $\varepsilon \to 0$. This is due to the specific structure of the integral for $\Delta\mu$ in (7.10). Although we do not know the integrand as a function of t, we can express it in terms of the exactly known functions $\omega(s)$ and $R(s)$ (the specified configuration of the magnetic field) and, generally speaking, the (exactly) unknown function $\mathbf{v}(s)$. But the latter is needed only in the singularities, where $\omega = \omega(s_p) = 0$ and $v_{\parallel}(s_p) = v$; $v_{\perp}(s_p) = 0$ in view of $v_{\perp}^2 = 2\mu\omega$. It is important that this result [$v_{\perp}(s_p) = 0$] does not depend on the corrections to $\mu$, provided they are small, i.e., at sufficiently low $\varepsilon$. In this way all the complexities of the $v(s)$ dependence in the higher approximations do not affect the integral value of $\Delta\mu$. This means physically that the contributions of the higher approximation to the $\mu(t)$ dependence cancel out over the half-period $T/2$. We note incidentally that, owing to $v_{\perp}(s_p) = 0$, we can discard in the equation for $\dot{\mu}$ not only the terms with $\sin(2\theta)$, but also the terms of form $v_{\perp} \sin\theta$, as was done in (4.4) (cf. the total expression in [5]).

It remains to discuss the calculation of $\theta_p$ by means of (7.16). In the small-$\beta_0$ approximation, the corrections to $\dot{s}$ are immaterial. In the general case [see, e.g., (7.6)], the correction to $\theta_p$ is proportional to the integral of $\delta\mu$. The latter is described by an expression such as (5.1), so that $|\delta\mu| \sim \varepsilon$ and is proportional to the rapidly oscillating function $\cos\theta$ (for a particle in a trap). The integral of $\delta\mu$ is therefore of the order of $\varepsilon^2$, and the correction to $\theta_p$, notwithstanding the large factor $1/\varepsilon$ in (7.16), turns out to be small ($\lesssim\varepsilon$).

Results of numerical simulation show that the typical first-approximation error of $\Delta\mu$ is of order 10% (see, e.g., [5]). In some special cases the accuracy can be even higher [15]. These figures include also the errors in the separation of the resonant $\Delta\mu$ against the background of much larger, generally speaking, quasiperiodic, oscillations $\delta\mu$ (see Section 5). As already noted above, the use of explicit expressions for a more precise $\mu$ greatly facilitates this task (see [15, 20]). An even more effective method can be the use of resonant trajectories for numerical simulation. In this case, the quasiperiodic variations cancel out at both reflection points.

Note that an exponentially small resonant $\Delta\mu$ was first obtained in [16] from the solution of the corresponding quantum-mechanical problem. The relatively simple classical-mechanics technique described above was proposed in [20] and then developed in a number of papers (see, e.g., [5, 15, 21]). The most extensive calculations were carried out in [5].

## 8. Mapping

Taken by itself, the quantity $\Delta\mu$ obtained in the preceding section not only fails to solve the problem of prolonged particle confinement in a trap, but is meaningless, for in typical cases it is much smaller than the quasiperiodic oscillations $\delta\mu$ of $\mu$. The central question in the problem of prolonged containment of particles in a trap is the accumulation of successive changes of $\mu$. This problem can be answered in two ways. The first is to analyze the Hamiltonian (6.9) that describes a system of interacting resonances with amplitudes determined by the quantity $\Delta\mu$(6.10). The second is to describe the particle motion by mapping or transforms, i.e., not continuously but at certain finite time intervals. We use below the second way, which turned out to be simpler and more convenient (see [2]). It is convenient to choose as the characteristic time interval the longitudinal-oscillation half-period T/2 to which the quantity $\Delta\mu$ pertains. This very quantity characterizes the variation of one of the dynamic variables of the system, that of the action $\mu$ in this interval. This, however, is not enough for a complete description of the motion since $\Delta\mu$ depends on another dynamic variable, the Larmor phase $\theta$ at a certain point of the magnetic line (in the simplest case – at the minimum of the field). We must therefore first find the change of $\theta$ between successive passes through this point. This problem is generally solved in various ways, depending on the actual magnetic-field configuration.

We begin with the model (3.1). According to the results of Section 7, the quantity $\Delta J_y$ is determined in this case by the value of the phase $\theta$ at the minimum of the potential ($x = 0$). The change of $\theta$ between successive passes through this minimum, during the half-period of the x oscillations, can therefore be written in the form

$$\Delta\theta_0 \approx \frac{\pi}{\omega_x}\langle\omega_y\rangle = \pi\left(\frac{1}{\sqrt{2J_y}} + \frac{J_x}{2J_y}\right). \qquad (8.1)$$

In the last expression we used the connection between the frequencies $\omega_x$, $\langle\omega_y\rangle$ and the actions $J_x$, $J_y$ (3.5).

In this form, however, relation (8.1) still does not solve our problem, since a new dynamic variable appears ($J_x$), and calls for one more equation. Instead of finding an equation for $\Delta J_x$, however, we can express $J_x$ in terms of $J_y$ by using the energy conservation law $H^0(J_x, J_y) =$ const. From (3.4) we get

$$J_x = (H^0 - J_y)/\sqrt{2J_y} = (v_0^2 - 2J_y)/2\sqrt{2J_y}, \qquad (8.2)$$

whence

$$\Delta\theta_0 = \frac{\pi}{2}\left(\frac{1}{\sqrt{2J_y}} + \frac{v_0^2}{(2J_y)^{3/2}}\right). \qquad (8.3)$$

Now let $J_y$ and $\theta_0$ be the values of the dynamic variables at some pass through the minimum of the potential, and $\overline{J}_y$, $\overline{\theta}_0$ the values of the same variables during the next pass. Mapping is defined as the connection between these two pairs: $J_y, \theta_0 \to \overline{J}_y, \overline{\theta}_0$. Introducing for convenience a new phase $\vartheta = 2\theta$, we can write this connection, in our case, in the form

$$\left.\begin{aligned}\bar{J}_y &= J_y + (\Delta J_y)_m \sin\vartheta_0,\\ \bar{\vartheta}_0 &= \vartheta_0 + G(\bar{J}_y),\\ (\Delta J_y)_m &\approx -\pi J_y \exp\left[-4/3v_0 \cos\beta_0\right],\end{aligned}\right\} \tag{8.4}$$

[see (7.8)], and take the function $G(J_y) = 2\Delta\theta_0(J_y)$ at $J_y = \bar{J}_y$, i.e., for the value of $J_y$ after the first pass, the very thing that determines the difference between two successive values of the phase $\vartheta_0$.

Mapping should describe completely the evolution of the considered model in its phase space ($J_y$, $\vartheta_0$). This space is, in this case, a semicylinder, since $J_y \geq 0$, while the phase $\vartheta_0$ is defined accurate to an integer multiple of $2\pi$. Given the energy $H^0 = v_0^2/2$, the motion is confined to the region $J_y \leq H^0$ (8.2). The term mapping stems from the fact that the difference equations (8.4) transform (or map) the phase semicylinder on itself; i.e., each point of this semicylinder goes over into one of the points of the very same surface.

The mapping (8.4) is not canonical; i.e., it does not conserve the measure (area) of the phase surface. Indeed, if we put for simplicity $\cos\beta_0 = 1$ ($\beta_0 \ll 1$), the Jacobian of the mapping is

$$\frac{\partial(\bar{J}_y, \bar{\vartheta}_0)}{\partial(J_y, \vartheta_0)} = 1 + \frac{\partial(\Delta J_y)_m}{\partial J_y}\sin\vartheta_0 \neq 1. \tag{8.5}$$

Since the initial system (3.1) is canonical, or Hamiltonian, the result (8.5) means simply that the mapping (8.4) is not exact. In fact, relation (7.8) defines the change not of $J_y$ but of $\ln J_y$. If we assume as before that $\cos\beta_0 = 1$ and introduce a new variable $P = \ln J_y$, the mapping

$$\bar{P} = P + (\Delta P)_m \sin\vartheta_0, \quad \bar{\vartheta}_0 = \vartheta_0 + G(\bar{P}) \tag{8.6}$$

now becomes canonical, since $(\Delta P)_m = \pi\exp(-4.3v_0)$ is independent of P.

Note that were we to retain the dependence of the argument of the exponential in $(\Delta P)_m$ on $\beta_0$ [$\cos\beta_0$ or, in the general case, the dependence (7.6)] and, consequently, also on $J_y$, the last mapping would again become noncanonical, since $\partial(\Delta P)_m/\partial J_y \neq 0$. This is one more advantage of the small-$\beta_0$ approximation. In this approximation it is possible to choose, in all the examples considered above, the variable P such that $\partial(\Delta P)_m/\partial P = 0$, and the mapping of form (8.6) (see below) turns out to be canonical.

Canonicity of the mapping is essential for the use of certain general ergodic-theory theorems and is necessary in numerical simulation, for otherwise the errors can accumulate rapidly because the phase area is not conserved.

For certain magnetic-field configurations, the mapping that describes particle motion in a trap turns out to be the same as (8.6). For example, for a certain trap (see Subsection 2 of Section 3), we introduce the variable $P = \sqrt{\mu}$ [see (7.14) and (7.17)] and obtain

$$\overline{P} = P + (\Delta P)_m \sin\theta_0;\ \bar{\theta}_0 = \theta_0 + G(\overline{P}), \tag{8.7}$$

where

$$(\Delta P)_m = -\frac{3\pi}{8\sqrt{2}} r_0 \sqrt{\omega_0}\, e^{-q} \tag{8.8}$$

and

$$G(P) = \frac{\pi\langle\omega\rangle}{\Omega} = \frac{\pi}{2\sqrt{2}} L\sqrt{\omega_0}\left(\frac{1}{P} + \frac{v^2}{2\omega_0 P^3}\right). \tag{8.9}$$

The last expression can be obtained directly from (8.3) in view of the complete analogy between (3.4), (3.11) and (3.5), (3.12). The adiabaticity parameter is, in this case, $\varepsilon \sim 1/q = 3\rho_m/2L$.

The mapping takes the same form as well for a multimirror trap (see Subsection 4), and also for planar geometry of field-reversed mirrors (Subsection 5). In all these cases there is only one singularity on a half-period of the longitudinal oscillations (in one pass of the particle through the trap), and the motion is symmetric about the field minimum. The mapping for the remaining examples of Section 3 will be constructed later (see Section 14). We note right away that in all the magnetic-trap examples the variable is $P = \sqrt{\mu}$, since we have always used Eq. (4.4) in the calculation of $\Delta\mu$.

In the case of a multimirror trap, we confine ourselves to the region near the separatrix (see Subsection 4 of Section 3). The frequencies $\langle\omega\rangle$ and $\Omega$ are specified here as functions of the parameter w [Eqs. (3.26), (3.29), (3.31)] that depends only on $\mu$ [see (3.25), $H^0$ = const]. As a result we obtain for the function G(P) in the mapping (8.7):

$$\left.\begin{aligned} G(P) &= \frac{\sqrt{2}}{\pi}\,\frac{\lambda}{\sqrt{\lambda-1}}\,\frac{L\sqrt{\omega_0}}{P}\left(\ln\frac{32}{|w|}-2\frac{\lambda-1}{\lambda}\right); \\ w &= \frac{v^2}{(\lambda-1)\,\omega_0 P^2}-\frac{2\lambda}{\lambda-1}. \end{aligned}\right\} \qquad (8.10)$$

The expression for $(\Delta P)_m$, however, remains (8.8) as before, with the parameter

$$\frac{1}{q}=\frac{3\pi}{4}\sqrt{\lambda-1}\,\frac{\rho_m}{L}, \qquad (8.11)$$

where 2L is the distance between neighboring mirrors [see (3.18) and (7.19)].

In the case of field-reversed mirrors we use for the field the asymptotic expression (3.36), which corresponds to large amplitudes of the longitudinal oscillations, since we assume, as before, that $\beta_0$ is small. In our case, the frequencies are given by relations (3.38) and we get

$$G(P)\approx\frac{2}{3}\,\frac{r_0 v^3}{\omega_0 P^4}. \qquad (8.12)$$

The quantity $(\Delta P)_m$, on the other hand, is given by (7.34).

The mappings obtained in this section can be used directly for numerical simulation of prolonged motion of a particle in a magnetic trap of suitable configuration. These mappings are quite simple, and each iteration of a mapping corresponds to one complete pass of the particle through the trap between reflection points in the mirrors.

For analytic study, the mappings obtained can be further simplified and "standardized."

## 9. Standard Mapping

The mapping (8.7) obtained in the preceding section is equivalent to the Hamiltonian (6.9) in the sense that it describes the same system of nonlinear resonances. The resonance condition (in first approximation – see Section 10 below) for the mapping can be written as

$$G(P_r) = 2\pi r, \tag{9.1}$$

where r is an arbitrary integer. This condition determines the resonance values $P = P_r$ and, accordingly, $\mu = \mu_r = P_r^2$, for which the successive values of the phase $\theta_0$ remain unchanged. The latter is correct, of course, if $\mu$ also remains unchanged, i.e., if $\theta_0 = 0$ or $\pi$. It can be easily seen that one of these phase values is unstable, depending on the sign of $(\Delta P)_m$ [at $(\Delta P)_m > 0$, for example, the unstable point is $\theta_0 = 0$].

Since $G(P) = \pi\langle\omega\rangle/\Omega$ (see Section 8), the resonance condition can also be written in the form $\langle\omega\rangle = 2r\Omega$. It differs from the condition (6.2) by a factor of two which is connected with the symmetry of the motion described by the mapping (8.7) about the minimum of the field. In this case, the amplitudes of all the odd harmonics of the longitudinal oscillations vanish in the Fourier expansion (6.1).

We linearize the function G(P) near one of the resonant values of P and introduce a new variable p:

$$G(P) \approx G(P_r) + \left.\frac{dG}{dP}\right|_{P=P_r} (P - P_r) \rightarrow G_r'(P - P_r) \equiv p. \tag{9.2}$$

Here $G_r' \equiv (dG/dP)_{P=P_r}$, and we have discarded the term $G(P_r) = 2\pi r$, which does not change the mapping. In terms of the variables p and $\theta$ (we drop the zero subscript of the phase), the mapping (8.7) takes the form

$$\begin{aligned} \bar{p} &= p + K\sin\theta, \\ \bar{\theta} &= \theta + \bar{p} = \theta + p + K\sin\theta. \end{aligned} \tag{9.3}$$

It was introduced similarly in [8] and named the standard mapping, since many (although by far not all) actual problems of nonlinear dynamics of Hamiltonian systems are reducible to it. In particular, all the magnetic trap types considered in the preceding section reduce to this mapping. Any actual system is distinguished only by a single standard-mapping parameter that can be represented in the form

$$K = G_r'(\Delta P)_m. \tag{9.4}$$

Note that if even the quantity $(\Delta P)_m$ depended on P, it would have to be taken in (9.4) at $P = P_r$ (just as G'). Consequently, K is simply a constant, and the standard mapping turns out to be canonical. The canonicity of the initial mapping (8.7) is therefore immaterial if this mapping is to be analytically investigated by changing to a standard mapping.

The first-approximation resonances for a standard method are defined by the conditions

$$p = p_r = 2\pi r, \tag{9.5}$$

i.e., they are infinite in number, and all are located at equal distances $\delta p = 2\pi$ from one another. From a comparison of the last expression with (9.1) it can be seen that the standard mapping describes the initial system (in particular, its resonant structure) locally with respect to P (and $\mu$). Indeed, the initial system is a particle in a trap, and the mapping (8.7) that describes it has, generally speaking, a finite number of resonances (in view of the limited range of $\mu \lesssim v^2/2\omega_0$), which are not uniformly distributed with respect to $\mu$ (or P). A standard mapping transforms this section of the resonant structure and makes it homogeneous and infinite. For this reason, the standard mapping is also called the homogeneous model of a resonant structure.

For the local model to have meaning, it is necessary that the number of resonances be large, and that their characteristics, particularly K, differ little from neighboring resonances. This is in fact the condition under which linearization of (8.7) with respect to P is admissible. This condition can be written in the form

$$\frac{\delta P}{P} \approx \frac{2\pi}{|PG'|} \sim \frac{2\pi}{|G|} \sim \varepsilon \ll 1, \tag{9.6}$$

where $\delta P = |P_{r+1} - P_r|$ is the distance in P between neighboring resonances.

## 10. Limit of Global Stability

As already noted, the central question of prolonged confinement of a particle in a magnetic trap is whether or not successive resonant changes $\Delta\mu$ of the particle magnetic moment are cumulative. In the language of standard mapping, the same question reads: Is the motion of this system finite, limited in p, or infinite?

The only parameter K of the standard mapping has the meaning of a perturbation parameter. In fact, at $K = 0$ we have $p = \text{const}$ and hence also $\mu = \text{const}$. Since we are dealing in this case with an adiabatic perturbation, K can be regarded as a new adiabaticity parameter. More accurately speaking, the dimensionless adiabaticity parameter should be taken to be $|K|/(2\pi)^2$, the square of the ratio of the perturbation-induced oscillation frequency [$\sim\sqrt{|K|}$; see Eq. (10.2) below] to the perturbation frequency (equal to $2\pi$, with the period of the perturbation equal to 1). The new parameter is connected with the old ($\varepsilon$) by the estimate

$$k = \frac{|K|}{4\pi^2} \sim \frac{e^{-1/\varepsilon}}{\varepsilon}. \tag{10.1}$$

The exponential factor stems here from the expression for $(\Delta P)_m$ (or $\Delta\mu$), and the preexponential one from $G_r' \sim G/P \sim 1/\varepsilon P$. For understandable reasons we shall call k the resonant adiabaticity parameter.

Since the function $k(\varepsilon)$ has a singularity at $\varepsilon = 0$, the new parameter k cannot be obtained, as we have verified also directly, from the asymptotic expansion in $\varepsilon$. Once, however, we have found it by exact integration of the perturbation and have arrived thus ultimately at the standard mapping, we can now use for the analysis an asymptotic expansion in k and, in particular, a very simple averaging method (see, e.g., [12]). Thus, for sufficiently small $k \to 0$ and nonresonant $p \neq 2\pi r$ we can expect the variation of p to be quasiperiodic and bounded. In exactly the same manner, for $k \to 0$ and resonant p (for example, $p \approx 0$) we can neglect all the resonances except the given one ($r = 0$). The standard-mapping difference equations can then be replaced by differential ones:

$$\bar{p} - p \approx \frac{dp}{dt} \approx K \sin\theta; \qquad \bar{\theta} - \theta \approx \frac{d\theta}{dt} = p, \tag{10.2}$$

which turn out to be canonical with a Hamiltonian

$$H_r(\theta, p) = \frac{p^2}{2} + K\cos\theta. \tag{10.3}$$

The time t is measured here in numbers of iterations of the mapping, or in units of $T/2$.

The resonant Hamiltonian (10.3) is equivalent to the Hamiltonian (6.8) – both describe one nonlinear resonance. As already noted in Section 6, motion in this case is finite under all initial conditions. The maximum amplitude of the p oscillations corresponds to motion near the separatrix ($H_r = |K|$). The separatrix has two branches:

$$p_c = \pm 2\sqrt{K}\sin\frac{\theta}{2}. \tag{10.4}$$

This expression is valid at $K > 0$: reversal of the sign of K is equivalent to a shift of the phase $\theta$ by $\pi$: $\theta \to \theta + \pi$ [see (10.3)]. The maximum change $(\Delta p)_m$ of p is equal to the width of the separatrix, i.e., to the largest distance between the branches of the separatrix (at $\theta = \pi$):

$$(\Delta p)_m = 4\sqrt{|K|} = 8\pi\sqrt{k}. \tag{10.5}$$

The dynamics of one nonlinear resonance (10.3) is perfectly similar to the motion of an ordinary pendulum or of a particle in a field with a harmonic potential, particularly in a multimirror trap of type (3.19). The picture of the phase trajectories in all these cases has the form shown schematically in Fig. 1. Consequently, as already noted in Section 6, the oscillations of p, meaning also of $\mu$, are bounded in the case of one resonance. Up to which K is this simple picture of the motion preserved? The usual condition of asymptotic theory requires that the frequency ($\sim|K|^{1/2}$) of the averaged (smoothed) system (10.2) be much lower than the frequency of discarded perturbation in the initial system (9.3), which is equal to $2\pi$ (one iteration is taken as the unit of time), i.e., that $|K| \ll (2\pi)$ or $k \ll 1$ (10.1). The critical value of the perturbation is therefore

$$k_{cr} \sim 1, \quad |K|_{cr} \sim (2\pi)^2 \approx 40. \tag{10.6}$$

A much better estimate can be obtained by using the criterion called the overlap of nonlinear resonances [8]. The simplest variant of this criterion is obtained in the following manner. All integer [$p_r/2\pi = r$] resonances of the standard mapping are identical and are described by the same Hamiltonian (10.3) with shifted momentum $p \to p - p_r$. In particular, each of them has the same width (10.5). The simplest resonance-overlap criterion is determined from the condition that the separatrices of neighboring resonances be tangent. Since the distance between them is $\delta p = 2\pi$ (9.5), the tangency condition is $\delta p = (\Delta p)_m$, whence

$$k_{cr} = 1/16, \quad |K|_{cr} = \pi^2/4 \approx 2.5. \tag{10.7}$$

This is already much closer to the numerical simulation of the standard mapping [8], which yields

$$|K|_{cr} \approx 1, \; k_{cr} \approx 1/40 \tag{10.8}$$

accurate to several percent. Note that at the stability limit the adiabaticity parameter is $k = k_{cr} \ll 1$, thereby justifying the use of an asymptotic expansion in this region.

A very simple criterion thus permits a correct estimate of the order of a critical perturbation. Moreover, the picture of the overlap of the resonance is clear enough to gain a qualitative idea of what happens when the perturbation is larger. Clearly, when the resonances overlap, the trajectory of the system can move over from the region of one resonance to that of a neighboring and the motion becomes infinite, at least for certain initial conditions. The character of this motion will be considered below (Section 11). We note here merely that the overestimate of the critical value by the simple resonance-overlap criterion is due mainly to the fact that in this form the criterion takes into account only resonance of first-order approximation in the adiabaticity parameter k. In the second order in this parameter there appear half-integer resonances $p_r/2\pi = r/2$; in third order we get resonances $p_r/2\pi = r/3$, etc. The complete system of resonances turns out to be everywhere dense in p:

$$p_{rq} = 2\pi r/q, \tag{10.9}$$

where r and q are arbitrary integers. This does not mean, of course, that the resonances always overlap (for any $k \to 0$), since their widths decrease rapidly with increasing denominator q. Indeed, resonances with a denominator q appear only in the q-th (and higher) approximations, i.e., in perturbation terms whose amplitudes are of order of $k^q$, and the resonance width is $\sim k^{q/2}$. Recognizing that the number of q-resonances in a given p interval (for example $2\pi$ between neighboring integer resonances) is proportional to q, we find that the overlap of resonances in all approximations is estimated by the sum

$$S = \sum_{q=1}^{\infty} q\eta^q = \eta \frac{d}{d\eta} \sum_{q=1}^{\infty} \eta^q = \frac{\eta}{(1-\eta)^2} = \frac{1}{4}. \tag{10.10}$$

Here $\eta = \sqrt{k}$, and the overlap condition $S = 1/4$ is obtained from the critical value $\eta_1 = 1/4$ (10.7) with account taken of only integer resonances. The higher-approximation resonances increase S (at a given $\eta$) and accordingly decrease $\eta_{cr}$ to

$$\eta_{cr} = 3 - \sqrt{8} \approx 0.17, \quad k_{cr} \approx \frac{1}{34}, \quad |K|_{cr} \approx 1.16. \qquad (10.11)$$

This is already quite close to the stability limit (10.8), although the estimate (10.10) cannot, of course, be regarded beforehand as reliable, since the terms of the sum in (10.10) contain unknown numerical coefficients. According to the data of [8], the coefficients of the second and third powers of $\eta$ are equal to $\pi/2 \approx 1.57$ and $(2.2)^2$, respectively. At any rate, expression (10.10) demonstrates that a system of resonances that is everywhere dense does not necessarily lead to overlap and instability.

This fundamentally and practically important conclusion can nevertheless not be regarded as sufficiently convincing, since it is based in final analysis on an asymptotic expansion in the parameter k. The result of numerical simulation is also limited, in view of the finite computation time (up to $\sim 10^7$ iterations of the standard mapping). A rigorous proof of existence of a critical perturbation below which the particle remains perpetually (i.e., at $-\infty < t < \infty$) in a magnetic trap was obtained in [10] by constructing converging (and not asymptotic) perturbation-theory series. Unfortunately, technical difficulties prevent us from obtaining, in this manner, an effective estimate of the magnitude (and even of the order) of the critical perturbation. With allowance for second- and third-order perturbation-theory approximations, as well as for higher approximation effects, a value $|K|_{cr} \approx 1.1$ was obtained in [8]. An entirely different method [23] leads to $|K|_{cr} \approx 0.97$ (this method is discussed in [24]). In sum, it can be concluded that the values (10.8) for the critical perturbation are reliable enough and can be used to solve specific nonlinear-dynamics problems that are reducible to standard mapping.

Bounded oscillations of the action $\mu$ for subcritical perturbation are sometimes called superadiabaticity (see [25], where a related problem was solved). In essence, in this case the adiabatic invariant becomes an exact integral of the motion, albeit nonanalytic and even singular in the dynamic variable and in the initial small parameter $\varepsilon$ [10]. It should be noted in this connection that the famous Poincaré theorem that a Hamiltonian system has in the general case no analytic integrals of motion other than the energy is not as significant as heretofore assumed.

The onset of instability of motion for overlapping resonances can be regarded as an interaction of resonances, i.e., as joint action of several resonances; the result of this action differs (radically in this case) from the sum of actions of individual resonances. This situation might be described as interference of resonances, an interference with catastrophic aftereffects. At any rate, resonance overlap is a clear example of the profound difference between linear and nonlinear mechanics.

Since standard mapping describes the initial system locally, the critical value of the parameter K determines the stability boundary in the phase space of the system. We express the position of this boundary in terms of the parameters of the initial system.

For the model (3.1) the standard-mapping parameter is [see (7.8), (8.3), (9.4)]:

$$K \approx \frac{\pi^2}{2}\left(\frac{3}{\sin^3\beta_0} + \frac{1}{\sin\beta_0}\right)\frac{\exp\left[-\frac{4}{3v_0\cos\beta_0}\right]}{v_0}$$

$$\approx \frac{3\pi^2}{2}\,\frac{\exp\left(-\frac{4}{3v_0}\right)}{v_0\beta_0^3}, \quad \beta_0 \ll 1. \qquad (10.12)$$

The critical value $|K| = 1$ (10.8) determines the stability boundary on the plane of the dynamic variables $\beta_0$ and $v_0 = \sqrt{2H^0}$, which is a two-dimensional projection of the four-dimensional phase space of the model under consideration.

This boundary can also be represented in the form

$$\beta_0^{cr} \approx 2.5 v_0^{-1/3}\exp(-4/9v_0). \qquad (10.13)$$

On the velocity plane $(\dot{x}_0, \dot{y}_0)$ the quantity $\beta_0^{cr}$ defines an unstable-motion sector $\beta_0 < \beta_0^{cr}$, in which the amplitude of the longitudinal oscillations increases without limit. The adiabatic regime corresponds here to small $v_0 \ll 1$, since the frequency ratio is $\omega_x/\langle\omega_y\rangle \sim v_0\beta_0^3 \sim \exp(-4/3\,v_0)$ [the last estimate is valid at the stability boundary (10.13)].

For a short magnetic trap, using (8.8) and (8.9), we obtain in analogy with the preceding case

$$K \approx \frac{81\pi^2}{64}\,\frac{r_0}{L}\,\frac{q^2 e^{-q}}{\beta_0^4}, \tag{10.14}$$

where $q = 2L/3\rho_m \gg 1$ and $\beta_0 \ll 1$. The critical angle is therefore

$$\beta_0^{(cr)} \approx \frac{3\sqrt{\pi}}{4}\left(\frac{r_0}{L}\right)^{1/4}\sqrt{q}\, e^{-q/4}. \tag{10.15}$$

It defines in the trap an instability cone in the particle-velocity space at the magnetic-field minimum. In the assumed short-trap model (3.10), the field increases along the magnetic line without limit, and an instability cone exists for all $q \to \infty$, although its width does decrease very rapidly with increasing q. In real traps the field is bounded by some maximum value $\omega_m = \lambda\omega_0$, where $\lambda$ is the mirror ratio. There therefore exists the known adiabatic particle-departure cone $\beta_0^{(a)} \approx \lambda^{-1/2}$ (at $\lambda \gg 1$), even if $\mu = \text{const}$. The instability of the particle motion in a magnetic trap is significant only if $\beta_0^{(cr)} > \beta_0^{(a)}$, or

$$\lambda > \frac{8}{9\pi}\left(\frac{L}{r_0}\right)^{1/2}\frac{e^{q/2}}{q}. \tag{10.16}$$

This estimate can also have another meaning – an upper bound on the amplitude of the stable longitudinal oscillations ($a < a_{cr}$). Actually, the minimum stable $\beta_0^{cr}$ ($\ll 1$) occurs when the particle reaches a field $\omega_{cr} \approx \omega_0\lambda_{cr} \approx \omega_0(\beta^{cr}_0)^{-2}$. Consequently, $\lambda_{cr} = 1 + a_{cr}^2/L^2$ (for our short-trap model) and is equal to the right-hand side of inequality (10.16).

An estimate of the stability is obtained quite similarly as well for opposing mirrors (in planar geometry). From (8.12) and (7.34) we obtain, with the aid of (9.4),

$$K \approx -\frac{64\sqrt{2}}{3}\,\frac{q^{9/8}e^{-q}}{\beta_0^5}. \tag{10.17}$$

Here $q = r_0\omega_0/2v = r_0/2\rho_m$ and $\beta_0 \ll 1$. The width of the unstable sector is

$$\beta_0^{(cr)} \approx 2q^{9/40}e^{-q/5},$$
$$\lambda_{cr} \approx \frac{e^{0.4q}}{4q^{9/20}} \approx \frac{|s|_{cr}}{r_0} \approx \frac{x_{max}}{r_0} \approx \frac{y_{max}}{r_0}, \tag{10.18}$$

where $x_{max} = y_{max} \approx |s|_{max} \gg r_0$ is the maximum deflection of the particle along the coordinate axes (see Fig. 2a), at which its oscillations are still stable.

The location of the stability boundary on the (x, y) intersection plane of the magnetic field can be obtained in the following manner. Since $\omega_0(r_0) = Cr_0$, where C is a certain constant (see Subsection 5 of Section 3), we have

$$\frac{r_0 v}{\omega_0} = r_0 \rho_m = \frac{v}{C} = L^2 = \text{const}, \quad q = \frac{r_0 \omega_0}{2v} = \frac{r_0^2}{2L^2} \tag{10.19}$$

and the length L is a certain physical parameter of the given trap. Substituting these expressions in the second relation of (10.18) and using the magnetic-line equation $2xy = r_0^2$, we obtain the equation of the stability boundary in the form

$$y_b \approx \frac{1}{2\sqrt{2}} q^{1/20} e^{0.4q} \approx \frac{e^{0.4 x_b y_b}}{2\sqrt{2}}$$

or

$$y_b \approx \frac{5}{2} \cdot \frac{1 + \ln x_b}{x_b}. \tag{10.20}$$

Here $x_b$ and $y_b$ are measured in units of L. This expression is valid only at $x_b \gg 1$, since the asymptotic form (3.36) of the field was used in (10.18). Curve (10.20) symmetrized relative to the line x = y is shown in Fig. 5 (the lower hatched line) together with one of the magnetic lines it intersects. Formally, curve (10.20) intersects the line x = y at the point $x_b = y_b \approx 2.08$, where $r_0^{min} = 2.9$ and $q^{min} \approx 4.3$. The unstable region is located below the curve.

In a multimirror trap we take explicitly into account an adiabatic loss cone with aperture angle $\beta_0^{(a)} \approx \lambda^{-1/2} \ll 1$. Therefore, the unstable region takes the form of a conical layer adjacent to the loss cone. We characterize this layer by its width, which we assume to be small (see below):

$$\Delta\beta_0 = \beta_0^{cr} - \beta_0^{(a)} \ll \beta_0^{(a)} \ll 1. \tag{10.21}$$

The main parameter of the function G(P) [Eq. (8.10)] is then equal to

$$w \approx -4\sqrt{\lambda}\, \Delta\beta_0. \tag{10.22}$$

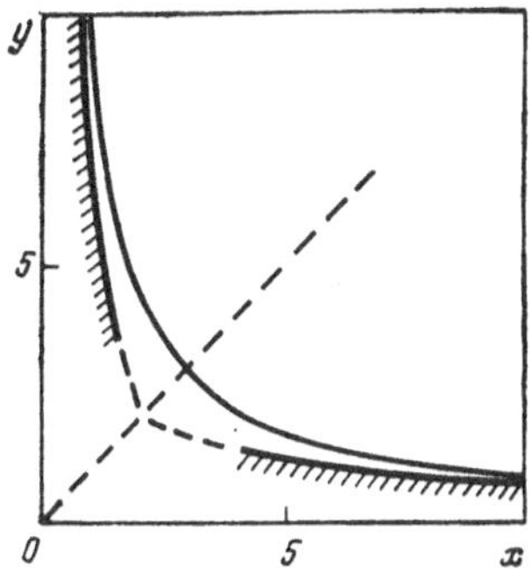


Fig. 5. Motion-stability boundary (lower hatched curve) for planar geometry of opposing mirrors; the instability region is below this curve. Upper curve – one of the magnetic lines. The unit of length is defined by the condition $r_0 \rho_m(r_0) = 1$ (see text).

Since $|w| \ll 1$, it suffices to differentiate in the expression for $G'(p)$ only $w$:

$$G' \approx -\frac{2\sqrt{2}}{\pi}\,\frac{L\lambda\omega_0^{3/2}}{\Delta\beta_0 v^2}. \qquad (10.23)$$

Using for $(\Delta P)_m$ the relation (8.8) with the parameter q from (8.11), we get

$$K \approx \frac{27\pi^2}{64}\,\frac{r_0}{L}\,\frac{\lambda^2}{\Delta\beta_0}\,q^2 e^{-q}. \qquad (10.24)$$

The stable-motion layer therefore is

$$|\Delta\beta_0| \approx 4.2\,\frac{r_0}{L}\,(\lambda q)^2 e^{-q}. \qquad (10.25)$$

In a multimirror trap the layer consists of two halves, each of width $|\Delta\beta_0|$: "upper," in which the particles are untrapped ($w > 0$; $\Delta\beta_0 < 0$), and "lower," with trapped particles ($w < 0$; $\Delta\beta_0 > 0$).

A similar layer is produced, of course, in any other real trap, since the field has maxima in the mirrors. A model of such a "real" trap can be one section of a multimirror trap (two mirrors). In this case the upper half of the layer plays no role, since a particle landing there will leave the trap in one pass. In the lower half, however, the particle can stay for an arbitrarily long time (see Section 11), so that the existence of a layer is perfectly observable. Of course, the presence of this layer is of no importance whatever for prolonged confinement of the particles in a trap, since it hardly increases the adiabatic loss cone.

## 11. Local Diffusion

We consider the character of the motion in the unstable region defined by the condition $|K| > 1$ (see Section 10). We start from the standard mapping (9.3), which describes the motion of the particle at relatively small changes of $\mu$. A characteristic feature of the standard-mapping dynamics is local instability of the motion, i.e., the rapid "scatter" of almost all very close trajectories. This instability must not be confused with the global instability considered in Section 10, which means simply unrestricted motion, i.e., an unbounded variation of p over a sufficiently long time. To investigate the behavior of close standard-mapping trajectories we consider the derivatives [see (9.3)]

$$\frac{d\bar{\theta}}{d\theta} = 1 + K\cos\theta \approx K\cos\theta. \tag{11.1}$$

The last expression is valid at $|K| \gg 1$, i.e., deep in the region of global instability. Exactly the same expression is obtained also for the derivative $(d\bar{p}/dp)$ if it is recognized that the phase $\theta$ in the first equation of (9.3) is connected with the preceding value of $\underline{\theta}$ by the relation $\theta = \underline{\theta} + p$. It is clear, therefore, that the average distance between very close trajectories (in the linear approximation) will increase exponentially with time:

$$|\delta\theta(t)| \approx |\delta\theta|_0 e^{-ht}, \tag{11.2}$$

where the velocity h of the local instability is (per iteration)

$$h = \langle \ln|K\cos\theta| \rangle = \ln\frac{|K|}{2}. \tag{11.3}$$

The last expression is obtained by simply averaging over $\theta$ and assuming a uniform distribution. This simplifying assumption is actually well satisfied at sufficiently large $|K|$ (see [8]). At smaller K, even in the region of global instability, i.e., at $|K| > 1$, considerable regions of stable motion, i.e., of bounded quasiperiodic p oscillations, are preserved on the standard-mapping cylinder (Fig. 6). The

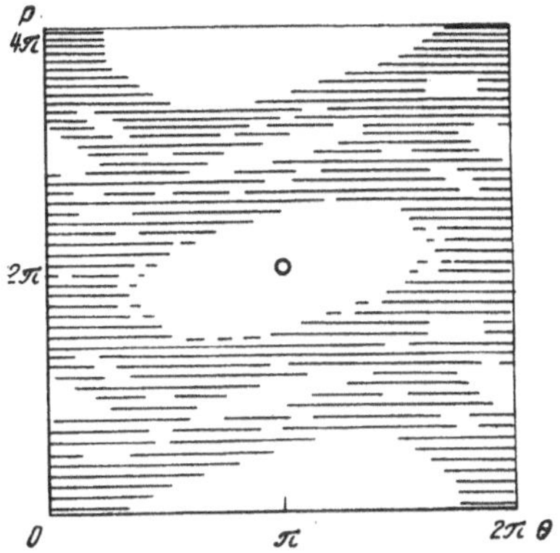


Fig. 6. Section of developed phase cylinder of standard mapping (9.3) at K = 1.13 (according to the data of [8]). The region of unrestricted motion in p (the stochastic component) is shaded. The circle marks a stable immobile point at the center of one of the integer resonances $p_r = 2\pi$.

largest regions surround stable periodic trajectories of the standard mapping with a period 1 (immobile points): $p_1/2\pi = r$, $\theta_1 = \pi$ (r is an arbitrary integer; $K > 0$). These points correspond to centers of integer resonances. An elementary analysis of the linear stability of these points (see, e.g., [8]) shows that the stable regions around them are preserved up to K = 4. At larger K there can also exist several other stable regions, but their area is small and we shall neglect them. This question was investigated in greater detail in [8] (at K = 5, for example, the relative area of stable regions is less than 2%). Accordingly, numerical simulation of the standard mapping shows [8] that the last expression in (11.3) for h is correct to within no more than 5% for $K > 6$.

The quantity h plays a major role in modern theory of dynamic systems and is called metric entropy (see, e.g., [26]) or sometimes KS entropy (the Krylov–Kolmogorov–Sinai entropy) [8]. It turns out that at $h > 0$, i.e., if local instability evolves exponentially, almost all the trajectories of a dynamic system, more accurately of the ergodic (stochastic) component of the motion, are random [27]. In the case considered, at $K > 1$ the stochastic component coincides with the region of global instability.

A more detailed discussion of the extremely interesting question of the physical meaning of such a dynamic fortuity, i.e., random motion of a deterministic dynamic system, is beyond the scope of this article. We confine ourselves only to brief remarks, referring the interested readers to the original papers [27] (a brief popular discussion of these questions can also be found in [28]).

In modern theory of dynamic systems, randomness of trajectories is associated with maximum complexity of the system. The latter means, roughly speaking, that there exists no (simpler) method of describing a given random trajectory other than specifying the trajectory itself. In other words, the equations of motion are utterly useless for the calculation of the trajectory over a sufficiently long time interval, since the entire complexity of the trajectory is contained in its initial conditions. It can be easily seen that this is directly connected with the exponential local instability of the motion, owing to which the trajectory is determined in the course of time by arbitrarily minute details of the initial conditions. A nontrivial fact here is the rigorous result that almost all the initial conditions of the motion correspond here precisely to random trajectories. The dynamic randomness is thus due in the final analysis to the continuity of the phase space, while the role of the dynamic system (of the equations of motion) reduces only to ensuring local stability of motion, which reveals the microscopic structure of the phase space [28].

It is important, within the scope of the considered problem of particle motion in a magnetic trap, that strong local instability makes it impossible to represent the motion in terms of trajectories and forces the use of a statistical description. The basis of the statistical description is the ergodicity of motion in the entire phase space or in part of it. In the latter case one speaks of the ergodic component of the motion. A sufficient albeit not necessary condition of ergodicity of motion is the local instability described above. The average time that an ergodic trajectory stays in any region of phase space is proportional to the invariant measure of this region. For Hamiltonian (canonical) dynamic system such an invariant (i.e., conserved in the course of motion) measure is known to be the phase volume (the Liouville theorem). This permits a quantitative determination of the probability of a state (more accurately, of a region of states) of the dynamic system as a quantity proportional to the phase volume of this region (the proportionality coefficient depends on the normalization used). This is one of the advantages of the canonical (Hamiltonian) equations of motion or of mappings. For the standard mapping, for example, the probability density is $dw = dp d\theta$.

All other statistical properties of a dynamic system depend substantially on the behavior of the correlations. Consider, for example, the correlations of the phase $\theta$ in standard mapping or, more accurately, consider the correlation function

$$C(\tau) = \langle \sin\theta(t) \sin\theta(t+\tau) \rangle, \qquad (11.4)$$

where the averaging is either over t on one almost arbitrary (random) trajectory, or else, since the motion is ergodic, directly over $\theta$. At $h > 0$ all the correlations are damped irreversibly, i.e., $C(\tau) \to 0$ as $\tau \to \infty$, although not necessarily exponentially (see below).

If this damping is fast enough, a simple diffusive description of the motion is possible. Indeed, using the first equation of (9.3), we can write

$$(\Delta p)_t = K \sum_{t'=1}^{t} \sin\theta(t'),$$

where the time t is an integer (the number of the iteration) and $(\Delta p)_t$ is the change of p after t iterations. Neglecting the stable component of the motion (see above), it follows from the ergodicity that $\langle \sin\theta \rangle = 0$, meaning also that

$$\langle (\Delta p)_t \rangle = 0. \qquad (11.5)$$

Let us find $\langle (\Delta p)^2_t \rangle$. We have

$$\langle (\Delta p)^2_t \rangle = K^2 \sum_{t'=1}^{t} \sum_{t''=1}^{t} \langle \sin\theta(t') \sin\theta(t'') \rangle.$$

Recognizing that $\langle \sin^2\theta \rangle = 1/2$, we can represent the last expression as $t \to \infty$ in the form

$$\langle (\Delta p)^2_t \rangle \to tK^2 \left[ \frac{1}{2} + 2 \sum_{\tau=1}^{\infty} C(\tau) \right]. \qquad (11.6)$$

If the last sum is finite, we have a simple statistical diffusion with respect to p at a rate

$$D_p = \lim_{t\to\infty} \frac{\langle(\Delta p)_t^2\rangle}{2t} = \frac{K^2}{4} R(K), \qquad (11.7)$$

where R(K) is a correlation factor that can depend only on K, the only standard mapping parameter that determines all its dynamic and static properties. As $K \to \infty$, the KS entropy $h \to \infty$ and the correlations vanish even for neighboring values of the phase $\theta$. In this limit we have $R \to 1$, and obtain

$$D_p^\infty = K^2/4. \qquad (11.8)$$

Since K is constant for the standard mapping, the diffusion equation for the distribution function f(p, t) takes the simple form

$$\partial f/\partial t = D_p \partial^2 f/\partial p^2. \qquad (11.9)$$

Its particular solution corresponding to the initial conditions $f(p, 0) = \delta(p)$ is the Gaussian distribution

$$f(p, t) = \frac{\exp\left(-\dfrac{p^2}{4D_p t}\right)}{\sqrt{4\pi D_p t}}. \qquad (11.10)$$

Numerical simulation confirms well these simple considerations [8].

Note that for validity of any kinetic equation in general and of the simple diffusion equation (11.9) in particular, it is necessary that the diffusion time scale $T_D \sim p^2/D_p$, i.e., the characteristic evolution time of the distribution function f, be much longer than the dynamic scale $T_h \approx 1/h$, i.e., the correlation damping time [29]. This is necessary because $D_p$ is simultaneously also a local (in time) parameter of the diffusion equation, i.e., in the scale of $T_D$, and asymptotic in the scale of $T_h$ (11.7).* For standard mapping we have

*If sight is lost of this important physical condition, it is easy to arrive at a contradiction (see [30-32] and below).

$$\frac{T_D}{T_h} \sim \left(\frac{p}{K}\right)^2 \Big/ \ln K \sim \left(\frac{p}{K}\right)^2 \gg 1.$$

We now turn from the standard mapping to a mapping of type (8.7), which describes the initial problem more accurately. Since the latter mapping contains the same phase, the correlation properties of the motion remain unchanged, and we can write down right away the rate of diffusion in P:

$$D_P = \frac{(\Delta P)_m^2}{4} R(K) \tag{11.11}$$

with the same correlation factor R(K) (11.7). The last expression is obtained, of course, also from the general relation

$$D_P = D_p \left(\frac{dP}{dp}\right)^2, \tag{11.12}$$

since $K + G_r'(\Delta P)_m^2$ and $dP/dp = 1/G_r'$. In particular, the rate of diffusion in $\mu$, which is exactly the one we need, is

$$D_\mu = \mu (\Delta P)_m^2 R(K), \tag{11.13}$$

where $(\Delta P)_m$ depends on the type of magnetic trap (see Section 8).

We write finally the diffusion rate in ordinary continuous time (and not in terms of the number of iterations of the mapping). Obviously, it suffices for this purpose to divide the expressions derived above by $T/2 = \pi/\Omega$. For example,

$$D_\mu = \frac{\Omega\mu}{\pi} (\Delta P)_m^2 R(K). \tag{11.14}$$

In contrast to the standard mapping, however, the quantity D no longer suffices to set up a diffusion equation for the distribution function $f(\mu, t)$. The reason is that the Fokker–Planck–Kolmogorov (FPK) equation that describes the diffusion process at small changes of the diffusing quantity ($|\Delta\mu| \ll \mu$ in our case) contains, besides $D_\mu$, also another function

$$U_\mu = \lim_{t\to\infty} \frac{\langle(\Delta\mu)_t\rangle}{t}, \tag{11.15}$$

called "drift." This quantity is sometimes called also "friction," but generally without justification, and can be called more readily the average rate (in terms of $\mu$).

With the aid of the quantities $U_\mu$ and $D_\mu$, we can write the FPK equation in the form (see, e.g., [33])

$$\left.\begin{aligned}\frac{\partial f}{\partial t} &= -\frac{\partial Q}{\partial \mu},\\ Q &= -\frac{\partial}{\partial \mu}(D_\mu f) + U_\mu f,\end{aligned}\right\} \quad (11.16)$$

where Q is the probability flux. In the general case $U_\mu \neq 0$. Consider, for example, the mapping (8.7). As shown in Section 8, it is canonical just in the variables P and $\theta_0$, but not $\mu$ and $\theta_0$; i.e., at $|K| \gg 1$ the stochastic trajectory is uniformly distributed just on the P, $\theta_0$ surface. Since $\mu = P^2$, we have $\Delta\mu = 2P\Delta P + (\Delta P)^2$. Hence,

$$U_\mu = \frac{\langle(\Delta\mu)_t\rangle}{t} = \frac{\langle(\Delta P)_t^2\rangle}{t} = 2D_P. \quad (11.17)$$

Comparing this expression with the diffusion rate $D_\mu$ (11.13) we arrive at the relation ($R \approx 1$):

$$U_\mu = \frac{1}{2}\,\frac{dD_\mu}{d\mu}, \quad (11.18)$$

which in fact solves the problem of finding the function $U_\mu$ and with it the complete FPK diffusion equation for the mapping (8.7). The question is, however, how accurately does this mapping describe the initial problem of particle motion in a trap? It is undoubtedly correct in first-order perturbation theory in the adiabaticity parameter $k \sim (\Delta P)_m$ (10.1). But $\mu U_\mu \sim D_\mu \sim (\Delta P)_m^2 \sim k^2$; i.e., the velocity $U_\mu$ is of second order of smallness. As we shall see later (see Section 13), the mapping (8.7) does not ensure such an accuracy, so that relation (11.18) is not correct.

The general method of finding $U_\mu$ will be considered in Section 13, and now we shall show that in the initial diffusion state the rate $U_\mu$ plays no role at all, and can be simply neglected. In fact, the change of $\mu$ consists of two parts: first, the diffusion proper or the scatter $(\delta\mu)_1 \sim \sqrt{D_\mu t}$ and, second, the average displacement $(\delta\mu)_2 \sim U_\mu t$. Their ratio is [see (11.18)]

$$\frac{(\delta\mu)_1}{(\delta\mu)_2} \sim \frac{\sqrt{D_\mu t}}{U_\mu t} \sim \frac{\mu}{(\delta\mu)_1} \gg 1 \quad (11.19)$$

so long as $(\delta\mu)_1 \ll \mu$, i.e., so long as the diffusion has a local character.

By way of example we consider diffusion in a thin stochastic layer in a multimirror trap. In the expression for the diffusion rate (11.13) (discrete t) we assume for simplicity that $R(K) \approx 1$. And since we have in a narrow layer $\mu \approx \mu_a$ (the value of $\mu$ on the adiabatic loss cone), the diffusion coefficient is $D_\mu \approx$ const, and we arrive at the simple diffusion equation (11.9). If the multimirror trap is long enough and the particle flux along the trap can be neglected, the diffusion in the layer leads simply to relaxation of the initial distribution, $f(\mu, 0) \to$ const. To determine the relaxation time, we note that the diffusion flux is zero on both boundaries of the layer, i.e., $\partial f/\partial\mu = 0$. This yields the eigenfunctions of the diffusion equation:

$$\left.\begin{aligned} &f_n(\mu, t) = \exp\left(-\frac{t}{\tau_n}\right)\cos(\pi n\nu), \\ &\nu = (\mu - \mu_{cr})/\Delta_\mu; \quad \tau_n = \Delta_\mu^2/\pi^2 n^2 D_\mu, \end{aligned}\right\} \tag{11.20}$$

where $\mu_{cr}$ is one of the boundaries of the layer, and $\Delta_\mu$ is its total width. The equilibrium distribution corresponds to $n = 0$. The relaxation time is determined by the maximum $\tau_n = \tau_1$. Using the expressions for $D_\mu$ and for the layer width $(\Delta_\mu/\mu = 4|\Delta\beta_0|/\beta_0^{(a)})$, we obtain for the relaxation time

$$\tau_1 \approx \frac{64}{\pi^2}\left(\frac{\lambda L}{\rho_m}\right)^2 \tag{11.21}$$

by iterating the mapping (8.7), i.e., of the passes of the particle between neighboring mirrors. We note that the next eigenfunction with $n = 2$ in (11.20) relaxes at a rate four times faster.

Let us digress somewhat. At equilibrium ($t \to \infty$, $f \to$ const) we have $\langle(\Delta\mu)^2\rangle$ = const. Since Eq. (11.6) remains valid also in this limit, we get $R(K) \to 0$ as $t \to \infty$. Clearly, this is due to the onset of long-range correlations on reflection of the particle from the layer boundaries. But it does not follow at all that the diffusion coefficient (11.7) in Eq. (11.9) is zero, as is sometimes assumed [30-32]. This would be too formal and straightforward an interpretation of the limit in (11.7). In fact, infinity in this limit is infinity in small, i.e., in a dynamic time scale $T_h \ll T_D$. In other words, the limit as $t \to \infty$ in (11.7) means in fact the strong double inequality $T_h \ll t \ll T_D$.

We now consider one section of a multimirror trap. In this case, the particles are confined only in the lower half of the layer. In the solution of the diffusion equation, the second boundary condition must therefore be $f(\mu_a, t) = 0$, i.e., the condition for absorption (emission) of the particles on the adiabatic cone. It is easily seen that the first eigenfunction of (11.20) satisfies this boundary condition (at $\nu = 1/2$). Therefore, expression (11.21) yields also the lifetime of the particles in the lower half of the layer. It is interesting to note that this time increases like the square of the magnetic field, whereas the layer width decreases exponentially with increasing field (10.25). Note also that the next eigenfunction, which satisfies the absorption condition at $\mu = \mu_a$, corresponds now to $n = 3$ and is damped nine times faster than the fundamental ($n = 1$).

## 12. Dynamic Correlations

Let us examine in greater detail the correlation factor R(K) in the expression for the diffusion rate in the case of standard mapping [see (11.7)]. We assume as before that if $|K| \gg 1$ we can neglect the stable component of the motion. Since the standard mapping is canonical in the variables p and $\theta$, in this case the trajectories will fill uniformly the surface of the phase cylinder. Since the structure of the standard mapping is periodic not only in $\theta$ but also in p (with the same period $2\pi$), it suffices in the calculation of the correlations to average over the phase square $2\pi \times 2\pi$ (developed part of the phase cylinder).

Let us find the first few correlations of the function $\sin\theta$ for the standard mapping. The quantity C(1) in (11.4) is obtained directly from the second mapping equation (9.3), and is equal to

$$C(1) = \langle \sin\theta \sin\bar{\theta} \rangle = \langle \sin\theta \sin[\theta + K\sin\theta + p] \rangle = 0 \qquad (12.1)$$

upon averaging over p (for fixed $\theta$). To find C(2), we express the succeeding ($\bar{\theta}$) and preceding ($\underline{\theta}$) phases in terms of $\theta$ and p in the present step:

$$\bar{\theta} = \theta + K\sin\theta + p, \quad \underline{\theta} = \theta - p.$$

Hence

$$C(2) = \langle \sin \underline{\theta} \sin \bar{\theta} \rangle = \frac{1}{2} \langle \cos (2p + K \sin \theta) \rangle - \frac{1}{2} \langle \cos (2\theta + K \sin \theta) \rangle$$

$$= -\frac{1}{2} J_2(|K|) \approx \frac{1}{\sqrt{2\pi |K|}} \cos\left( |K| + \frac{15}{8|K|} - \frac{\pi}{4} \right). \qquad (12.2)$$

In the last expression we used an improved asymptotic representation of the Bessel functions of $|K| \gg 1$ [22]. The entire derivation is apparently valid down to $|K| \approx 4$. At smaller $|K|$ there appears an appreciable stable region (see above) and simple averaging over $\theta$ and p is not valid.

Somewhat more cumbersome calculations yield for the following three correlations:

$$\left. \begin{aligned} C(3) &= \frac{1}{2} [J_3^2(|K|) - J_1^2(|K|)] \approx -\frac{4J_1 J_2}{|K|} \sim |K|^{-2}, \\ C(4) &\approx \frac{1}{2} J_2^2(|K|) \sim |K|^{-1}. \end{aligned} \right\} \qquad (12.3)$$

The approximate equality means here retention of only the principal term of the expansion in the small parameter $|K|^{-1/2}$. Starting with C(4), the exact expressions contain infinite sums of Bessel functions, so that further progress by this method is impossible. Since, however, the correlations are rapidly damped at $|K| \gg 1$ (see below), the obtained values of $C(\tau)$ already approximate sufficiently well the correlation factor

$$R(K) \approx 1 - 2J_2(|K|) + 2J_2^2(|K|), \qquad (12.4)$$

where we have neglected the contribution of C(3). The decisive quantity here is the second term, which leads to characteristic slowly damped oscillations of the diffusion rate with increasing K. Such combinations were observed in numerical simulation of the standard mapping in [8] and in other papers. In [34] was calculated for the first time the correction (12.4) for the asymptotic diffusion rate $D_p^\infty = K^2/4$. The calculation was carried out by an entirely different method (actually, even by two different methods). In particular, ergodicity of the motion was not assumed. The result of [34] can therefore be regarded also as independent proof of the ergodicity of the standard mapping at $|K| \gg 1$.

The numerical values of the factor R(K), taken from [34], are compared in Fig. 7 with the theoretical relation (12.4). The latter differs from the result of [34] (first paper) in that the correlation C(4) is taken into account, thereby improving somewhat the agreement with the numerical data. This correction was introduced in several papers, including the second paper of [34], and also in [35], by a method close to that described above (and postulating likewise ergodicity of the motion).

Thus, the dependence of the first (short-range) correlations on K is explained quite satisfactorily and simply enough. Much less clear is the time dependence of the correlations. Numerical experiments show that this dependence is far from always the simply exponential

$$C(\tau) \propto \exp(-Ah\tau),$$

as was assumed at one time, even if the KS entropy $h > 0$. Figure 8 shows an example of correlation damping for standard mapping at $K = 7$. It can be seen that the function $C(\tau)$ with even $\tau$ can be fairly well fitted with the aid of the function

$$C(\tau) \approx \frac{1}{2} \exp(-\sqrt{\tau}). \qquad (12.5)$$

The nature of this correlation damping remains unclear, although similar results were obtained analytically also for other dynamic systems, at any rate, as an upper bound estimate [36]. On the other hand, for similar mappings the damping of the correlations can also differ significantly from (12.5) (see [37]).

The behavior of the correlations becomes quite complicated as $|K| \sim 1$. A major role is assumed here by the boundary between the stochastic and stable regions with very complicated structure (see, e.g., [38]). We have seen in Section 10 that such a boundary is a distinguishing feature of the phase space of a particle in a hole. The diffusion rate falls off steeply near the boundary. This correlated with the dependence of the diffusion rate for the mapping on K as $K \to 1$. According to numerical data [8], the correlation factor in this region is

$$R(K) \propto (|K| - 1)^s, \quad s \approx 2.6. \qquad (12.6)$$

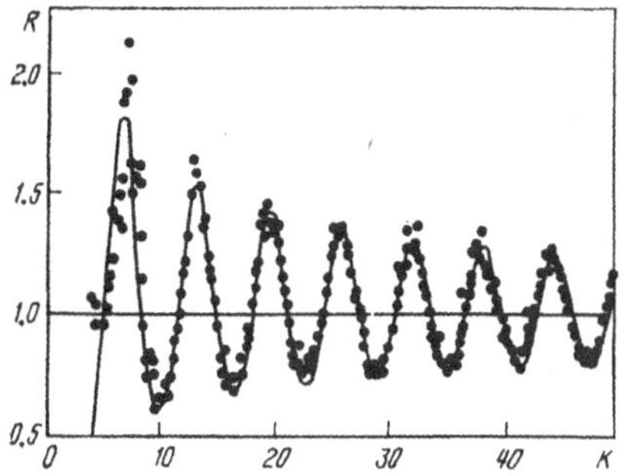


Fig. 7. Relative diffusion rate $R = 4D_p/K^2$ for the standard mapping vs. the mapping parameter K: points – numerical data from [34]; curve – calculated from Eq. (12.4).

By an entirely different method, numerical simulation of the motion in the stochastic layer yielded in [39] $s \approx 4$. Owing to the slow diffusion near the boundary, the trajectory remains stuck in this region for a long time, and this leads, in particular, to a slower decrease of the correlations. According to [39], the qualitative change of the character of the motion near the boundary sets in at $s > 2$. The diffusion to the boundary becomes, in this case, infinite; i.e., the stochastic trajectory does not reach the boundary of the stochastic component. This can be obtained from the following estimate [39]. Let x be the appropriately normalized distance from the boundary of the stochastic component and $d < (\Delta x)^2 > /dt \sim D_x(x) \sim x^3$ [in a thin boundary layer we have $K(x) - 1 \propto x$; $K(0) = 1$ (12.6)]. Since the diffusion rate increases rapidly with x, the time required to leave a layer smaller than x will be determined by the time needed to diffuse over a distance of order x, i.e., $<(\Delta x)^2> \sim x^2$. Hence,

$$t > x^{2-s} \tag{12.7}$$

and for $s > 2$ we have $t \to \infty$ as $x \to 0$.

In the last case ($s > 2$) a qualitative change takes place also in the behavior of the correlation function. Indeed, x near the boundary is proportional to the relative phase volume (area) w of the boundary layer, meaning also to the probability of the trajectory landing in this region.

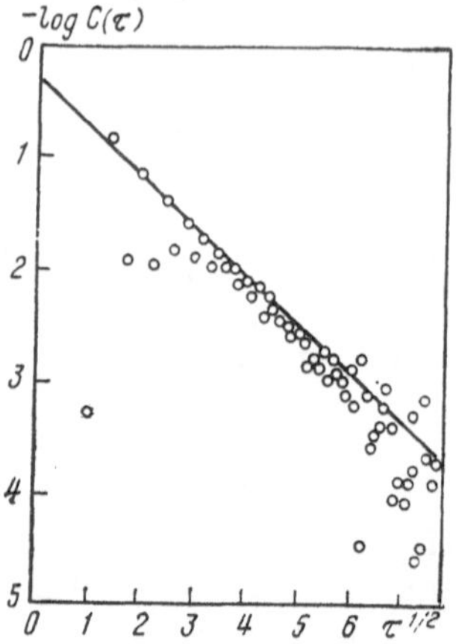


Fig. 8. Damping of the correlations $C(\tau)$ in the standard mapping at $K = 7$: $\tau$ – delay (number of iterations); the logarithm is to base 10. For $\tau = 1$ and $\tau \geq 4$ the numerical values of $C(\tau)$ are determined by the fluctuations of averaging of (11.4) for one trajectory over $10^6$ iterations. Straight line – the function $C(\tau) = 0.5 \exp(-\sqrt{\tau})$.

When expression (11.4) for the correlation function is averaged over the stochastic component, the fraction of trajectories remaining in the layer is less than $x(\tau) \sim \tau^{-1/(s-2)} \propto w(\tau)$, and determines the asymptotic behavior of the correlation as $\tau \to \infty$:

$$C(\tau) \to F(\tau)/\tau^q, \quad q = 1/(s-2). \tag{12.8}$$

This expression is valid only if $s > 2$ $(q > 0)$, for otherwise the correlation is no longer determined by the boundary layer. The function $F(\tau)$ depends, generally speaking, on the behavior, in the boundary layer, of the functions $f(\theta, p)$ to which the correlations $C(\tau)$ pertain [e.g., $f = \sin\theta$ in (11.4)]. In particular, $F(\tau)$ can oscillate, so that $\langle F(\tau) \rangle = 0$ (the average over $\tau$).

As already noted (see Section 11), for the standard mapping at $|K| \sim 1$ there exist sizable stability regions so that a power-law correlation decrease of type (12.8) can be expected. Such a behavior was indeed observed in numerical simulation, and it turned out that $q \approx 1$ $(K = 2.1)$, corresponding to $s \approx 3$ (12.8). For $f = \cos\theta$, the mean value $\langle F(\tau)\rangle \neq 0$, and the integral of $C(\tau)$ diverges. In addition, the average over the stochastic component is $\langle\cos\theta\rangle \neq 0$, in view of the presence of stability. This, however, does not influence greatly the diffusion in p, since the latter is determined by the correlations of the function $f = \sin\theta$ (11.7), i.e., of the very same function that determines $\Delta p$ in the mapping (9.3). In this case, $\langle F(\tau)\rangle = \langle\sin\theta\rangle = 0$ for any form of the stability region around the immobile point $\theta_f$ at which $\sin\theta_f = 0$. Indeed, the average over the stochastic component is

$$\langle\sin\theta\rangle_{sto} = \langle\sin\theta\rangle_0 - \langle\sin\theta\rangle_{stab},$$

where $\langle\sin\theta\rangle_0 = 0$ is the average over the entire phase square, and $\langle\sin\theta\rangle_{stab}$ is the average over the stable comcomponent. The latter, however, is also zero, since it is proportional to the average change of the momentum $\langle\Delta p\rangle = 0$ in the stable region. It can be similarly shown that in this case we also have $\langle F(\tau)\rangle = 0$. At larger $|K|$, namely at $K = 2\pi n$ ($n \neq 0$ an integer), there exist stable regions of another type, called "accelerational," with $(\theta_a)$ at their center, for which $\sin\theta_a \neq 0$, and for which $\langle\Delta p\rangle \neq 0$ [8]. We then have $\langle\sin\theta\rangle_{sto} \neq 0$; $\langle F(\tau)\rangle \neq 0$, and the diffusion becomes anomalously fast. It is possible that just this explains the appreciable spread of the points of Fig. 7 at the minimum of $R(K)$, which corresponds precisely to $K = 2\pi$. In the succeeding maxima the areas of the accelerational regions decrease rapidly, and with them their contribution to the average diffusion rate.

An anomalously slow decrease of the correlations is possibly contained also in the experimental data, described in [40], on electron confinement in a magnetic trap. Figure 9 shows a typical example of the experimental semilog plot of the flux $(-\dot{N})$ of the electrons leaving the trap versus time. The authors expected an exponential decrease of the flux (meaning also of the number N of the electrons remaining in the trap) with time. Such a dependence was indeed observed for some time ($\approx 2.5\tau_e$), but the subsequent decrease of the flow slows down (circles in Fig. 9). The authors fit the "tail" of the $\dot{N}(t)$ plot using a second exponential with larger $\tau_e$.

Generally speaking, a relation in the form of a sum of several exponentials is not surprising, and is, on the con-

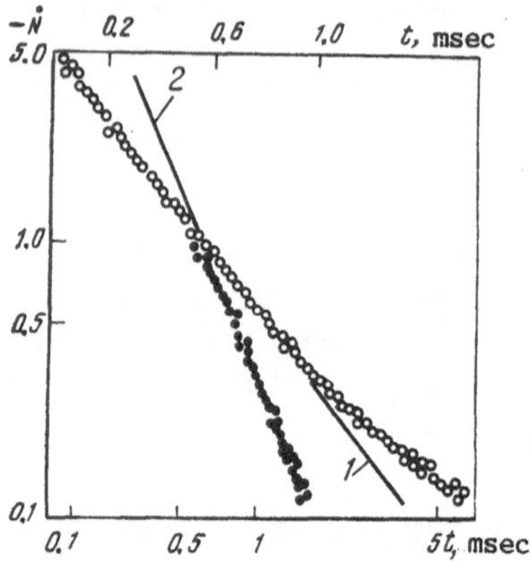


Fig. 9. Time dependence of electron flow from a magnetic trap according to the data of [40]: $(-\dot{N})$ is the flow of outgoing electrons (arbitrary scale). The circles show a semilog $\dot{N}(t)$ dependence (upper scale for t); curve 1 corresponds to the exponential relation $\dot{N} = A\exp(-t/\tau_e)$, $\tau_e \approx 0.36$ msec. The points show the same data (for $t \gtrsim 0.6$ msec) in log–log scale (lower scale for t); curve 2 is the power-law dependence $\dot{N} = Bt^{-\gamma}$; $\gamma \approx 2.3$.

trary, typical of the solution of the linear equation in general and of the diffusion equation in particular (see Section 11). If, however, the data of [40] are plotted in log–log scale (points of Fig. 9), the "tail" of the $\dot{N}(t)$ plot fits even better a power law with exponent $\gamma \approx 2.3$. This is already unusual for a diffusion equation and denotes the absence of (nonsingular) eigenfunctions. The final choice between an exponential and a power-law dependence calls for additional experiments at longer t and in a better vacuum. The latter is needed because electron scattering by the residual gas assumes rapidly an increasing role with time, since the electrons are confined to a boundary layer of ever-decreasing thickness. The volume of this layer is approximately proportional to its thickness and to the number of electrons remaining in the stochastic component, a number that decreases with time like $t^{-\gamma+1} \approx t^{-1.3}$. The same law also governs the decrease of the correlations, i.e., $q \approx 1.3$ (12.8) and $s \approx 2.8$.

Although the exponents s and q given above differ noticeably, the possibility of a universal description of the asymptotic structure of the boundary layer as $x \to 0$, $t \to \infty$ is not excluded. This possibility is based in final analysis on the hierarchic structure of the nonlinear resonances (see [8], Section 4.4). The first successful attempts at a scale-invariant description of several characteristics of the pressure near the stochasticity boundary were undertaken in [23, 41, 42].

## 13. Global Diffusion

To describe prolonged diffusion (in $\mu$ or $\beta_0$) of a particle in a magnetic trap we must know the complete FPK equation (11.16) rather than its local variant with $U_\mu = 0$, $D_\mu$ = const. It is necessary, therefore, first of all to find an expression for the average velocity U. As already noted above (see Section 11), direct calculation of $U_\mu$ in second-order perturbation theory is difficult. More convenient is the following simple procedure used in many papers (see, e.g., [33]). Let the equilibrium distribution function [$f_s(\mu)$ in our case] be known. For closed Hamiltonian systems it is easily obtained from a microcanonical distribution such as

$$F_e(\mu, J)\, d\mu dJ = \delta(H^0(\mu, J) - E)\, d\mu dJ, \tag{13.1}$$

where E is the given value of the energy. Note that we are using here the unperturbed actions $\mu$ and J and, accordingly, an unperturbed Hamiltonian. The perturbation, on the other hand, is assumed small enough to neglect the distortion it produces in the energy surface. In other words, we assume that the perturbation leads only to trajectory distribution over the entire unperturbed energy surface, with ergodic measure (probability density) $F_e(\mu, J)$. When necessary, it is possible (at least in principle) to change to the more precise actions $\mu^{(n)}$, $J^{(n)}$; the Hamiltonian is in general also altered in this case (see Section 5). Note that one more advantage of our choice of the unperturbed system (see Section 2) is the possibility of obtaining an equilibrium distribution function from the unperturbed Hamiltonian.

To obtain an equilibrium distribution in $\mu$ only, we integrate (13.1) over J:

$$f_s(\mu)\,d\mu = d\mu \int \delta(H^0(\mu, J) - E)\left(\frac{\partial J}{\partial H^0}\right)_\mu dH^0$$

$$= d\mu \int \frac{\delta(H^0 - E)\,dH^0}{\Omega(\mu, J)} = \frac{d\mu}{\Omega(\mu, E)}. \qquad (13.2)$$

In the last expression we expressed J in terms of $\mu$ and $H^0$ and assumed $H^0 = E$.

If, for example, $\Omega(\mu) \propto \sqrt{\mu}$, as in the short trap (3.12), the equilibrium density is $df_s \propto d\mu/\sqrt{\mu} \propto d\beta_0$ (at $\beta_0 \ll 1$), i.e., proportional to a planar-angle element rather than a solid-angle one as for a free particle. The reason is that as the angle $\beta_0$ increases, the frequency $\Omega$ of the longitudinal oscillations increases, their amplitude (given the energy E) decreases, and with it the "longitudinal part" of the phase volume. If, however, the magnetic field is uniform, then, $J \to p_{\parallel} = v_{\parallel}$, as well as $\Omega \to v_{\parallel} = v \cos\beta$ and $df_s \propto d\mu/v_{\parallel} \propto \sin\beta\, d\beta$, i.e., simply in proportion to the solid-angle element.

On the other hand, the equilibrium function $f_s(\mu)$ can be obtained from the FPK equation by equating the flux Q to zero. Hence,

$$U_\mu = \frac{1}{f_s}\frac{d}{d\mu}(D_\mu f_s) = \frac{dD_\mu}{d\mu} + D_\mu \frac{d(\ln f_s)}{d\mu}. \qquad (13.3)$$

This is in fact the general formula for the average velocity $U_\mu$. We note that in its derivation it is quite immaterial whether an equilibrium distribution actually exists or not in this system. An equilibrium distribution may also not be reached for electrons in a trap, owing to their escape to the loss cone or to the fact that the function $f_s$ is not normalizable, i.e., the total measure on the energy surface is infinite (see below). None of this influences in any way the limiting function $f_s(\mu)$. Relation (13.2) is simply an invariant measure of an ergodic Hamiltonian system on the basis of the Liouville theorem.

Returning to (13.3), we see that if $f_s$ = const we get

$$U_\mu = dD_\mu/d\mu. \qquad (13.4)$$

This simple relation between the coefficients of the FPK equation was (implicitly) obtained in [43] from the seemingly rather general detailed balancing principle. Nonetheless, it does not always hold [44, 45]. Nor does it hold for our problem if only $d\Omega/d\mu \neq 0$.

The detailed balancing principle is usually associated with dynamic reversibility of motion. The probabilities of direct and reverse transitions in reversible dynamics are, of course, equal, but they must be calculated in full measure (i.e., with all the dimensions of phase space taken into account) of the ergodic component. In the present problem this is not at all the case. First, we disregard completely the phases that are conjugate to the actions. It is shown in [45] that this is permissible in the absence of stable regions. Second, we have excluded one of the actions (J) and, finally and most importantly, the ergodic component of the motion is bounded for a closed system by the energy surface. As a result of all this the density $f_s$ of the probability (of the invariant measure) for the remaining variable $\mu$, whose change we wish to describe by the FPK diffusion equation, is proportional to $\Omega^{-1}(\mu)$ and not to a constant, and relation (13.4) does not hold, nor, therefore, does the detailed balancing "principle" (for the same variable $\mu$). The latter means simply that the transition probability is not proportional in this case to $d\mu$, but $df_s \propto d\mu/\Omega(\mu)$.

It is clear, therefore, that to use the reversibility of the motion in the form of the detailed balancing principle, the latter must be formulated in terms of adequate variables, which could be called ergodic [45]. In the general case they do not coincide with the unperturbed integrals of motion, as is sometimes assumed. In our problem, for example, the ergodic variable is a quantity (we designate it by $\Gamma$) proportional to the equilibrium distribution function, i.e., to the probability on the ergodic component. The variable $\Gamma$ can be defined, for example, via the relation

$$d\Gamma = d\mu/\Omega(\mu). \tag{13.5}$$

The probabilities of the direct and reverse transitions between any two regions with identical $d\Gamma$ will then be equal (the detailed-balancing principle), $f_s(\Gamma) = \text{const}$ and $U_\Gamma = dD_\Gamma/d\Gamma$ (13.3). In the last equation we can now return to the variable $\mu$ with the aid of the equations

$$\left.\begin{aligned} D_\Gamma &= D_\mu (d\Gamma/d\mu)^2, \\ U_\Gamma &= U_\mu \frac{d\Gamma}{d\mu} + D_\mu \frac{d^2\Gamma}{d\mu^2} = \frac{U_\mu}{\Omega} - \frac{D_\mu}{\Omega^2}\frac{d\Omega}{d\mu}. \end{aligned}\right\} \tag{13.6}$$

In the calculation of $U_\Gamma$ we must expand $\Delta\Gamma$ in terms of $\Delta\mu$ up to second order, inclusive. Hence,

$$U_\mu = \Omega \frac{dD_\Gamma}{d\Gamma} + \frac{D_\mu}{\Omega} \frac{d\Omega}{d\mu} = \frac{dD_\mu}{d\mu} - \frac{D_\mu}{\Omega} \frac{d\Omega}{d\mu}. \tag{13.7}$$

This relation can, of course, be obtained also directly from (13.3).

Besides the dynamic variables, we can also transform the time. In particular, a new time $\tau$ can be chosen such that the new frequency $\Omega(\mu) = d\psi/d\tau = \text{const}$, for example, $d\tau = \Omega(\mu)dt$. According to (13.5), $\mu$ becomes, in this case, an ergodic variable, and the coefficients of the FPK equation for the function $f(\mu, \tau)$ satisfy the relation (13.4). This is precisely a property possessed by the discrete mapping time, which is measured in units of $T/2 = \pi/\Omega$. The same result can also be obtained in a more formal manner, by using (13.7). Indeed, putting

$$D_\mu = \frac{\Omega}{\pi} D_\mu^{(0)}, \quad U_\mu = \frac{\Omega}{\pi} U_\mu^{(0)}, \tag{13.8}$$

where the quantities $D_\mu$ and $U_\mu$ pertain to the continuous time, and $D_\mu^{(0)}$ and $U_\mu^{(0)}$ to the discrete one, we get from (13.7)

$$U_\mu = \frac{\Omega}{\pi} \frac{dD_\mu^{(0)}}{d\mu} + \frac{D_\mu^{(0)}}{\pi} \frac{d\Omega}{d\mu} - \frac{D_\mu}{\Omega} \frac{d\Omega}{d\mu} = \frac{\Omega}{\pi} \frac{dD_\mu^{(0)}}{d\mu}$$

or

$$U_\mu^{(0)} = dD_\mu^{(0)}/d\mu. \tag{13.9}$$

We note that (13.9) differs from the relation (11.18) obtained directly from the mapping (8.7). This mapping is thus valid only in first order in the small parameter k. It hence follows, in particular, that it is not suitable for numerical simulation of prolonged (global) diffusion ($\delta\mu \gtrsim \mu$), when the effect of the average velocity $U_\mu$ becomes significant (see Section 11).

The problem of an accurate diffusive description for the analogous problem (the Fermi stochastic acceleration model) was discussed from a variety of viewpoints in [46-48] (see also [9]).

Since (13.9) is known to simplify the FPK equation, it is expedient to solve the problem in discrete time. From (11.16) we get

$$\frac{\partial f}{\partial t} = \frac{\partial}{\partial \mu} \mu \frac{\partial f}{\partial \mu}, \tag{13.10}$$

where we have dropped the zero subscript of $D_\mu$, with the latter given by (11.13).

By way of example we consider a magnetic trap in which the resonant $\Delta\mu$ and the stability parameter K can be estimated in the short-trap approximation (10.14). If $\beta_0^{(cr)} \gg \beta_0^{(a)} \approx \lambda^{-1/2} \ll 1$, we can neglect in first-order approximation the oscillations of the diffusion rates as functions of K and put $R(K) \approx 1$ in (11.13). Introducing a new time (which is likewise discrete)

$$s = t\,(\Delta P)_m^2\,\lambda/4\mu_0, \tag{13.11}$$

where $\mu_0 = v^2/2\omega_0$, and a new variable $x = \beta_0/\beta_0^{(a)}$, we write the equation for the eigenfunctions in the form [see (13.10)]

$$\frac{d}{dx}\left(x\frac{df_\varkappa}{dx}\right) + \varkappa^2 x f_\varkappa = 0 \tag{13.12}$$

with boundary conditions

$$f_\varkappa(1) = 0, \quad \left.\frac{df}{dx}\right|_{x=x_{cr}} = 0. \tag{13.13}$$

The first condition corresponds to the particles going off to the adiabatic cone $\beta_0 = \beta_0^{(cr)}$, and the second to the absense of a particle flux on the stochasticity boundary $\beta_0 = \beta_0^{(cr)}$ (10.15) ($x_{cr} = \beta_0^{cr}/\beta_0^{(a)}$). The smallest eigenvalue $\varkappa^2 \neq 0$ determines the lifetime of the particles in the trap under conditions when the motion is stochastic:

$$f(\mu, s) \to \exp(-\varkappa^2 s). \tag{13.14}$$

The solution of (13.12) is expressed in terms of Bessel functions (see, e.g., [22])

$$f_\varkappa(x) = CJ_0(\varkappa x) + N_0(\varkappa x). \tag{13.15}$$

The eigenvalue and the constant C are determined from the boundary conditions (13.13)

$$\left.\begin{aligned} CJ_0(\varkappa) + N_0(\varkappa) &= 0, \\ CJ_1(\varkappa x_{cr}) + N_1(\varkappa x_{cr}) &= 0. \end{aligned}\right\} \tag{13.16}$$

At $x_{cr} \gg 1$, the approximate solution of these equations can be written in explicit form by using the asymptotic expressions

$$\left.\begin{aligned} N_0(z) &\approx \frac{2}{\pi} J_0(z) \ln\left(\frac{\gamma z}{2}\right), \\ N_1(z) &\approx -\frac{2}{\pi z} + \frac{z}{\pi} \ln\left(\frac{\gamma z}{2}\right), \\ J_1(z) &\approx z/2 \end{aligned}\right\} \tag{13.17}$$

at $z \ll 1$; $\gamma = 1.78...$ is the Euler constant. The small value of $\varkappa$ for $x_{cr} \gg 1$ follows directly from the fact that the first eigenfunction (13.15) must be positive everywhere. Therefore, $\varkappa x_{cr} \lesssim 1$ and $\varkappa \lesssim 1/x_{cr}$. Substituting (13.17) in (13.16), we get from the first of these equations

$$C = -\frac{2}{\pi} \ln\left(\frac{\gamma \varkappa}{2}\right)$$

and, from the second,

$$\frac{1}{\varkappa^2} = \frac{x_{cr}^2}{2} \ln x_{cr}. \tag{13.18}$$

Gathering together all the relations, we obtain for the characteristic lifetime of the particle (the number of passes through the trap):

$$\tau_e \approx \frac{\mu_{cr}}{(\Delta P)_m^2} \ln\left(\lambda \beta_{cr}^2\right) \approx \frac{16}{9\pi} \left(\frac{L}{r_0}\right)^{3/2} \frac{\exp(3q/2)}{q} \ln\left(\lambda \beta_{cr}^2\right), \tag{13.19}$$

where $\beta_{cr} \equiv \beta_0^{(cr)}$. The last expression was written using (8.8) and (10.15), i.e., for a short trap with addition of a narrow loss cone ($\lambda \gg 1$). With increasing mirror ratio $\lambda$, the lifetime of the particles increases slowly, just as, incidentally, in multiple scattering by a gas [1]. A similar relation is obtained also (in planar geometry) for field-reversed mirrors, likewise with addition of a finite loss cone.

The lifetimes of the particles in a trap can be estimated also by a different procedure, used in Budker's first paper [1] on adiabatic traps for multiple scattering of particles. Consider stationary diffusion of particles having sources inside the trap. From (13.10) we have

$$\frac{d}{d\mu} \mu \frac{df_q}{d\mu} + q(\mu) = 0, \tag{13.20}$$

where $q(\mu)$ is the source density, which we choose in the form

$$q(\mu) = \mu^{\alpha} \tag{13.21}$$

with a constant $\alpha > -1$. The particle flux in a trap is

$$Q(\mu) = D_\mu \frac{df_q}{d\mu} = -\int_{\mu_{cr}}^{\mu} q d\mu = \frac{\mu_{cr}^{\alpha+1} - \mu^{\alpha+1}}{\alpha+1}, \tag{13.22}$$

where $\mu_{cr} = \mu_0 x_{cr}^2/\lambda$ corresponds to the stochasticity limit and $Q(\mu_{cr}) = 0$. The total stationary particle flux from the trap is

$$Q_0 = Q(\mu_a) \approx \frac{\mu_{cr}^{\alpha+1}}{\alpha+1} = \frac{\mu_{cr} q(\mu_{cr})}{\alpha+1};$$

$\mu_a = \mu_0/\lambda$ is the value of $\mu$ on the adiabatic loss cone, and we assume that $\mu_a \ll \mu_{cr}$. On the other hand, from (13.22) we obtain the particle density in the trap

$$f_q(\mu) = \int_{\mu_a}^{\mu} \frac{Q d\mu}{D_\mu} \approx \frac{1}{(\Delta P)_m^2} \frac{\mu_{cr}^{\alpha+1}}{\alpha+1} \ln \frac{\mu}{\mu_a}, \tag{13.23}$$

by putting, as before, $D_\mu \approx \mu(\Delta P)_m^2$, $f(\mu_a) = 0$.

The total number of the particles in the trap in the stationary regime is

$$N = \int_{\mu_a}^{\mu_{cr}} f_q(\mu) d\mu \approx \frac{\mu_{cr}^{\alpha+2}}{(\alpha+1)(\Delta P)_m^2} \ln \frac{\mu_{cr}}{\mu_a}. \tag{13.24}$$

We can now introduce the average lifetime of the particle

$$\langle\tau\rangle \equiv \frac{N}{Q_0} \approx \frac{\mu_{cr}}{(\Delta P)_m^2} \ln \frac{\mu_{cr}}{\mu_a}, \tag{13.25}$$

which does not depend on the arbitrary parameter $\alpha$ (at $\alpha > -1$) and agrees exactly with $\tau_e$, (13.19), since

$$\mu_{cr}/\mu_a = (\beta_{cr}/\beta_0^{(a)})^2 = \lambda\beta_{cr}^2.$$

This means that in the approximation considered ($\beta_{cr} \gg \beta_0^{(a)}$) the average particle lifetime $\langle\tau\rangle$ is determined for a large class of source distributions by the first diffusive mode (eigenfunction). Indeed, comparing expressions (13.23) and (13.15) [with allowance for (13.17)] we see that in the approximation considered the two distribution functions coincide:

$$f_{\varkappa} \propto f_q \propto \ln \frac{\mu}{\mu_a}.$$

For $\alpha < -2$ we can obtain similarly

$$\langle \tau \rangle \approx \frac{\mu_{cr}}{|1+\alpha| (\Delta P)_m^2}. \qquad (13.26)$$

The average lifetime now decreases with increasing $|\alpha|$, owing to the concentration of the sources on the adiabatic loss cone. A major role is assumed in this case by diffusion modes with constantly increasing numbers, since the stationary distribution function

$$f_q(\mu) \propto 1 - (\mu_a/\mu)^{|1+\alpha|}$$

has at $\mu \to \mu_a$ an abrupt break whose slope increases with $|\alpha|$.

In the relations obtained above for $\tau_e$ and $\langle \tau \rangle$, the particle lifetime unit is the number of passes through the trap or the number of reflections (from the magnetic mirrors). To transform to ordinary (continuous) time, the expressions obtained must be multiplied by the average half-period of the longitudinal oscillations of the particle

$$\left\langle \frac{T}{2} \right\rangle = \left\langle \frac{\pi}{\Omega(\mu)} \right\rangle = \pi \int \frac{f(\mu)\, d\mu}{\Omega(\mu)}. \qquad (13.27)$$

Here $f(\mu)$ is a normalized distribution function corresponding to the given motion regime, say a stationary regime with courses, or the first diffusion mode.

Near the adiabatic loss cone ($\mu = \mu_a$) we have

$$\frac{1}{\Omega} \propto \ln \left( 16 \frac{\mu_a}{|\mu - \mu_a|} \right)$$

[see (3.26)]. If $f(\mu) \propto \mu - \mu_a$, this region makes no significant contribution to the integral (13.27). We can therefore use $\Omega(\mu)$ without allowance for the loss cone. Let, for example,

$$\Omega(\mu) = \Omega_{cr}\sqrt{\frac{\mu}{\mu_{cr}}}, \tag{13.28}$$

where $\Omega_{cr}$ is the frequency on the stochasticity boundary [see (3.12)]. Furthermore, let

$$f(\mu) = \frac{\ln(\mu/\mu_a)}{\mu_{cr}\ln(\mu_{cr}/\mu_a)} \tag{13.29}$$

be the diffusion distribution function (13.23) normalized to unity ($\mu_{cr} \gg \mu_a$). The integral (13.27) then yields

$$\left\langle \frac{T}{2} \right\rangle \approx \frac{2\pi}{\Omega_{cr}} = 2\,\frac{T_{cr}}{2};$$

i.e., the average period of the longitudinal oscillations of the particle is double the minimum value on the stochasticity boundary.

Now let

$$\Omega(\mu) = \Omega_{cr}\mu/\mu_{cr}$$

[field-reversed mirrors; see (3.37) and (3.38)], and let the distribution function be the same (13.29). In this case,

$$\left\langle \frac{T}{2} \right\rangle \approx \frac{T_{cr}}{4}\ln(\lambda\beta_{cr}^2) \tag{13.30}$$

and the average oscillation period diverges logarithmically as $\lambda \to \infty$. Note that the ergodic measure (13.2) also diverges here and no equilibrium is reached (at $\lambda = \infty$, i.e., in the absence of a loss cone). More accurately speaking, the relaxation is, in this case, nonexponential, and the probability density (the distribution function) tends everywhere to zero. A simple example of such a relaxation is longitudinal diffusion in infinite space. In the one-dimensional case $f(x, t) \propto t^{-1/2} \to 0$ as $t \to \infty$.

## 14. Cohen's Mapping

We have considered all the trap examples described in Section 3, except two: a long trap and opposing mirrors in cylindrical geometry. Particle motion in these last traps does not reduce to standard mapping. We consider below this new situation with cylindrical field-reversed mirrors as the example. Particle dynamics in a long trap was investigated in [49] (see also [4]).

A characteristic feature of particle dynamics in cylindrical field-reversed mirrors is the asymmetry, due to the different asymptotic forms of the effective potential as $s \to \pm\infty$ (3.42), of two half-periods of the longitudinal oscillations of the particle. This leads to a difference in time between successive transits through the field minimum, and hence to two different functions G(P) that characterize the successive changes of the Larmor phase at the field minimum. Comparing (3.36) with (3.42) and using (8.12), we can write

$$G_+(P) = \frac{1}{3}\,\frac{l_0 v^3}{\omega_0 P^4}, \quad G_- = \frac{2}{3}\,\frac{l_0 v^3}{\omega_0 P^4}, \tag{14.1}$$

where $G_\pm$ describes the change of the Larmor phase for motion in the regions of positive and negative s, respectively. These changes differ thus by a factor of two (at the same value of P). There is actually also another difference, because we need the Larmor phase not exactly at the field minimum, but at the point $s = s_1 \approx -\ell_0/19$ (7.38). At longitudinal-oscillation frequencies $a \gg \ell_0$, however, this effect can be neglected (see the end of this section).

Since the resonance $\Delta\mu$ is described in this case, as in the others, by one and the same expression (7.37) during both half-cycles of the longitudinal oscillations, we obtain, in lieu of (8.7), the two-step mapping

$$\overline{P} = P + (\Delta P)_m \sin\theta_1, \quad \overline{\theta}_1 = \theta_1 + G_\pm(\overline{P}). \tag{14.2}$$

Here $\theta_1$ includes in the general case the additional phase shifts (7.37), and the functions $G_\pm$ alternate in succession. Therefore, (14.2) comprises in fact a system of four difference equations and describes the change of the dynamic variables during the entire period of the longitudinal oscillations of the particle. The functions $G_\pm(P)$ can be linearized in P, and the result is a new two-step mapping that is similar but does not fully conform to the standard mapping. It can be expressed in the form

$$\bar{p} = p + K_0 \sin\theta; \quad \bar{\theta} = \theta + \gamma_\pm \bar{p} + \alpha_\pm. \tag{14.3}$$

Here $\gamma_\pm$ are two different constants defined by the condition

$$\frac{G'_{\pm}(P_r)}{\gamma_{\pm}} = G'_0(P_r). \tag{14.4}$$

The resonant value $P_r$ is now obtained from the relation

$$G_+(P_r) + G_-(P_r) = 2\pi r, \tag{14.5}$$

where r is an integer and the additional phase shifts are equal to

$$\left.\begin{aligned} \alpha_+ &= G_+(P_r) \mod 2\pi, \\ \alpha_- &= G_-(P_r) \mod 2\pi, \\ \alpha_+ + \alpha_- &= 0 \mod 2\pi. \end{aligned}\right\} \tag{14.6}$$

The parameter (no longer unique) of the mapping (14.3) is

$$K_0 = (\Delta P)_m G'_0(P_r) \tag{14.7}$$

and the variable is

$$p = G'_0(P_r)(P - P_r). \tag{14.8}$$

In this form (with $\alpha_{\pm} = 0$) the two-step mapping (14.3) was first obtained by Cohen [49] in an analysis of particle motion in a long magnetic trap. Although this trap is symmetric about the z = 0 plane (see Section 3), the cause of the asymmetry of the mapping is that the principal singularities of the magnetic field are shifted from the trap center to the inner edge of the mirrors (see Section 7). Two steps of the mapping correspond in this case to transit of the particle between the singularities and its return to each of them after reflection from the corresponding mirror. Note that if such a trap were also intrinsically asymmetric (e.g., had different mirrors), the mapping would consist of three steps (six difference equations).

In our problem of opposing mirrors we have

$$\left.\begin{aligned} &G_0(P) = l_0 v^3/3\omega_0 P^4, \\ &\gamma_+ = 1, \quad \gamma_- = 2, \\ &\alpha_+ = 2\pi\gamma_+/\gamma, \quad \alpha_- = 2\pi\gamma_-/\gamma, \quad \gamma = \gamma_+ + \gamma_- \end{aligned}\right\} \tag{14.9}$$

and $(\Delta P)_m$ is given by (7.37).

Proceeding to the analysis of the dynamics of the Cohen mapping (14.3), we consider first, following [49], the simpler case when $\gamma_+ \ll \gamma_-$. This situation is typical of long traps. In this case, the mapping (14.3) can lead approximately to a one-step mapping over the total period of the perturbation. To this end, we express the variables $\bar{p}$ and $\bar{\theta}$ in terms of $\underline{p}$ and $\underline{\theta}$, which precede the quantities $\bar{p}$ and $\bar{\theta}$ by one complete period of the mapping (14.3). Writing the four difference equations in explicit form*

$$\left.\begin{aligned} \bar{p} &= p + K_0 \sin\theta, \quad \bar{\theta} = \theta + \gamma_- \bar{p}, \\ p &= \underline{p} + K_0 \sin\underline{\theta}, \quad \theta = \underline{\theta} + \gamma_+ p, \end{aligned}\right\} \tag{14.10}$$

we get

$$\bar{p} = p + K_0 [\sin\theta + \sin(\theta + \gamma_+ p + \gamma_+ K_0 \sin\theta)],$$
$$\bar{\theta} = \theta - \gamma_+ K_0 \sin\theta + \bar{p}(\gamma_+ + \gamma_-).$$

In the last equation we neglect the term $\sim K_0\gamma_+$, and in the first we rewrite the expression in the square brackets in the form

$$2\cos\left(\frac{p\gamma_+}{2} + \frac{\gamma_+ K_0}{2}\sin\theta\right)\cos\left(\theta + \frac{\gamma_+}{2}(p + K_0\sin\theta)\right) \approx 2\cos\left(\frac{p\gamma_+}{2}\right)\cos\theta.$$

The reason for retaining the term $(p\gamma_+/2)$ in the first cosine but dropping it from the second is the following. In the first case the quantity $p\gamma_+$ with large enough p can greatly increase the perturbation $K_0$, whereas in the second it leads only to an insignificant phase shift. Of importance in the dynamics of the phase is not the absolute value of the term $(p\gamma_+/2)$, but its change after one iteration of the mapping. This change is of the order of $K_0\gamma_+$, i.e., the same as of the remaining discarded terms. All these terms lead to the appearance of a second harmonic $\sim K_0\gamma_+ \sin 2\theta$ of the perturbation. For the standard mapping this harmonic appears in second-order perturbation theory, with relative amplitude $\approx K/16$ (see [8, Section 5.1]). This yields a roughly approximate condition for the validity of the considered approximation, viz., $\gamma_+ \lesssim 1/16$ (for $\gamma_- \sim 1$).

*If the phases $\alpha_\pm$ are proportional to $\gamma_\pm$, they can be eliminated by the momentum shift $p \to p + 2\pi/\gamma$.

Introducing the new variable

$$J = (\gamma_+ + \gamma_-)\, p = \gamma p, \tag{14.11}$$

we get the mapping

$$\bar{J} \approx J + K_1 \cos\theta, \quad \bar{\theta} \approx \theta + \bar{J}, \tag{14.12}$$

which is outwardly similar to the standard mapping [cf. (9.3)].

An essential property of the mapping (14.12) is, however, that its parameter

$$K_1 = 2\gamma K_0 \cos\left(\frac{\gamma_+}{\gamma}\,\frac{J}{2}\right) \tag{14.13}$$

depends now on the dynamic variable J. This leads to a qualitative change of the structure of the stochastic component, namely, it is all cut up into isolated strips by narrow gaps with stable motion. It follows from (14.13) that the gaps are located in the vicinity of $J = J_n$, where

$$\cos (J_n \gamma_+/2\gamma) = 0, \tag{14.14}$$

whence

$$J_n = \frac{\pi\gamma}{\gamma_+}(1 + 2n), \tag{14.15}$$

with n an integer. The width of the gap can be estimated from the condition that $|K_1| \approx 1$ on the gap boundary. Expanding the cosine in (14.13) at $J = J_n$ we obtain for the total width of the gap

$$\Delta J \approx \frac{2}{K_0\gamma_+} = \frac{2}{\pi K_m} \mid J_{n+1} - J_n \mid, \tag{14.16}$$

where $K_m = 2\gamma K_0$ is the maximum value of $|K_1|$. At $K_m \gg 1$ the relative gap width becomes small.

Of greater interest is the question of the size of the critical perturbation $|K_0| = K_g$, at which the gaps vanish and the isolated stochastic components of the motion merge into one. In [49] was proposed the criterion

$$K_g^2 \gamma_+ \gamma_- \sim 1 \tag{14.17}$$

based on the following simple consideration. At $|K_0| \sim K_g$ a gap of width $\Delta J \sim 1/K_g\gamma_+$ or $\Delta p \sim 1/K_g\gamma\gamma_+$ is overlapped on one step of the mapping (14.10), i.e., $(\Delta p)_g \sim K_g$; hence the estimate (14.17). This estimate, as well as the remaining simpler properties of the Cohen mapping, are well corroborated by his numerical simulation [49], as well as by our numerical data.

From the standpoint of the criterion for resonance overlap (see Section 10) the estimate (14.17) raises a definite problem. The overlap of the resonances means, in this case, that the gap width $\Delta J$ (14.16) becomes smaller than the distance between the resonances ($\delta J = 2\pi$), or that the condition $|K_1| \gtrsim 1$ holds for all resonances inside a gap. From this we have $K_g\gamma_+ \sim 1$, which is larger than (14.17) by a factor $\gamma/\gamma_+ (\gg 1)$.

The reason why the resonance-overlap criterion leads in this case to a grossly incorrect result is apparently the following. In the vicinity of the gap the amplitudes of the resonances increase quite rapidly (linearly) with increasing distance from the gap center. Under this condition, an important role is assumed by one curious manifestation of interaction between resonances – their mutual repulsion. The physical meaning of the effect and an estimate of its magnitude can be easily followed by means of a simple example.

Consider the motion in the vicinity of one resonance in the pendulum approximation [see the Hamiltonian (10.3)]. Let $p_0 \gg \sqrt{K}$ be the unpertured (at $K = 0$) position of the second resonance, which we consider to be weak enough to neglect its influence on the first resonance. Yet the first resonance distorts the second in such a way that its unperturbed straight phase line $p = p_0 = \text{const}$ turns into the curve $p(\theta) = p_0 + \delta p(\theta)$. The distortion of the resonant trajectory $\delta p(\theta)$ is easily obtained in first order in K from the equation

$$\left.\begin{aligned} &p^2(\theta)/2 + K\cos\theta = p_0^2/2, \\ &\delta p(\theta) = p(\theta) - p_0 \approx -\frac{K}{p_0}\cos\theta. \end{aligned}\right\} \qquad (14.18)$$

Since the maximum of the first-resonance separatrix is at $\theta = \pi$ [$K > 0$, see (10.4)], where $\delta p(\pi) > 0$, the resonance is repelled by an amount $\delta p(\pi) \approx K/p_0$.

Returning to the gap problem, we rewrite the last estimate in the form

$$\delta J_{kl} \sim F_l/(J_k - J_l), \tag{14.19}$$

where $J_k \sim k$ is the distance (in terms of J) from the center of the gap to the displaced resonance number k, while $J_\ell$ is the same for the displacing resonance; $F_\ell$ is the amplitude of the displacing resonance in a scale such that the overlap of the resonances correspond to $F \sim 1$, i.e., $F \sim |K_1|$. But the amplitude of the resonances near the gap is $F_\ell \sim F_0\ell$, where $F_0$ is the amplitude of the weakest resonance near the gap center. According to (14.19), the displacement in each pair is then $\delta J_{k\ell} \sim F_0$. Since the motion is stochastic, the phases of the resonances are random and the total displacement of each of the resonances for a given value of the phase $\theta$ is $\delta J_k \sim F_0\sqrt{N}$, where N is the total number of resonances in the layer. The total displacement of the gap edge is $\delta J$ $\delta J_k\sqrt{N} \sim F_0 N$. But $F_0 N \sim K_m$ are the amplitudes of the resonances at the center of the layer. From the condition $\delta J \sim \Delta J$ [$\Delta J$ is the gap width (14.16)] we obtain the estimate (14.17). Although the arguments presented above are excessively sketchy, it is quite likely that the resonance repulsion plays a major role in the problem considered.

This beautiful phenomenon was first considered qualitatively in [50] and later accurately calculated by the same author and others (see [41]). It should be noted that the statement made by the author in his title, that "primary (i.e., first-order-approximation) resonances do not overlap," has no bearing on this effect. This statement is, of course, correct in the sense that rather strong higher-approximation resonances do set in and ensure overlap prior (i.e., in weaker perturbation) to the tangency of the separatrices of the first-approximation resonances (see Section 10 and [8]). This, incidentally, is just the reason why the repulsion effect is negligible, for example, for two resonances of equal width. In fact, Eq. (14.18) leads, in this case, to the estimate

$$\left(\frac{2\delta p}{p_0}\right)_{cr} \approx \frac{2K_{cr}}{p_0^2} \approx 0.055,$$

where the numerical values of $K_{cr}$ for the two resonances were taken from Section 4.1 of [8]. The same is confirmed by the results of [41], according to which $2K_{cr}/p_0^2 = 0.0612$. But the repulsion becomes substantial when the resonance widths are unequal. Thus, at a resonance-width ratio 3 the repulsion offsets completely the higher-approximation effects, and the critical value of the perturbation agrees with the result obtained by the primary-resonance overlap criterion [41]. For the maximum ratio 5 considered in that paper, the critical perturbation exceeds the last value by an approximate factor 1.7.

Note that in standard mapping there are no integer and half-integer resonances at all, owing to the symmetry of the mapping. The repulsion, however, shifts the resonances of higher (particularly third) order, lowering somewhat $K_{cr}$ in this case. From a comparison of the analytic estimates of [8], where repulsion was not taken into account, with the actual value of $K_{cr}$ it follows that the repulsion effect is, in this case, in the 10% range.

We now turn to particular motion in a trap with field-reversed mirrors. Since $\gamma_+/\gamma_- = 1/2$ in this case, the approximation considered above is not valid. However, at such a "round" ratio $\gamma_+/\gamma_-$ the problem can be differently approached. We consider first the atandard mapping (9.3). The action of the perturbation can be regarded in this case as a periodic sequence of short "bumps" (Fig. 10a). This convenient method was used many times in the analysis of mappings (see, e.g., [8, 9]). In this approach the difference equations can be replaced by the exactly equivalent differential ones

$$dp/dt = K \sin \theta \delta_1(t), \quad d\theta/dt = p \tag{14.20}$$

with the Hamiltonian

$$H(p, \theta, t) = \frac{p^2}{2} + K \cos \theta \delta_1(t) = \frac{p^2}{2} + \frac{K}{2}\left\{\sum_{r=-\infty}^{\infty} \exp[i(\theta - 2\pi r t)] + \text{c.c.}\right\} \tag{14.21}$$

The time is measured here in numbers of mapping iterations; the fundamental frequency of the perturbation is accordingly $2\pi$; $\delta_1(t)$ is a $\delta$ function with period 1 (Fig. 10a) and with a Fourier expansion that leads to the last expression of (14.21). We already know (see Section 9) that the standard mapping has a homogeneous system of resonances $(p_r/2\pi) = r$ (Fig. 10a). The overlap of these resonances (with account of the higher approximations) determines in fact the critical $K_{cr} \approx 1$.

Now consider the Cohen mapping (14.10) for $\gamma_-/\gamma_+ = 2$. In this case the perturbation takes the form of a perio-

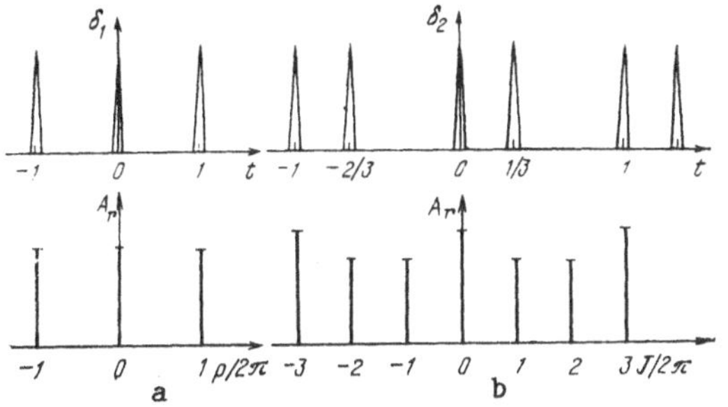


Fig. 10. The perturbation $\delta(t)$ and its spectrum $A_r$: a) for the standard mapping (9.3) – homogeneous system of resonances $r = p/2\pi$; b) for the Cohen mapping (14.10) for a particle in a magnetic trap with field-reversed mirrors. $\gamma_-/\gamma_+ = 2$, $r = J/2\pi$.

dic sequence of pairs of "bumps," and the four equations of (14.10) can be replaced anew by a pair of differential equations of form (14.20). If the unit of time is taken to be the total period of the perturbation, Eqs. (14.10) now become

$$dp/dt = K_0 \sin\theta\delta_2(t), \quad d\theta/dt = 3p, \tag{14.22}$$

since the phase changes after one full period by $(\gamma_+/\gamma_-)p = 3p$, and the function $\delta_2(t)$ is shown schematically in Fig. 10b. Introducing the new momentum $J = 3p$ we arrive at the Hamiltonian

$$H(J,\ \theta,\ t) = \frac{J^2}{2} + 3K_0 \cos\theta\delta_2(t) = \frac{J^2}{2} + \frac{3K_0}{2}\left\{\sum_r A_r \exp[i(\theta - 2\pi r t)] + \text{c.c.}\right\}. \tag{14.23}$$

In contrast to the standard mapping, the resonance amplitudes are now unequal (Fig. 10b):

$$|A_r| = \left|1 + \exp\left(\frac{2\pi i r}{3}\right)\right| = \begin{cases} 2, & r/3 - \text{integer}, \\ \sqrt{3}, & r/3 - \text{fractional}. \end{cases} \tag{14.24}$$

In first-order approximation, the critical value of the perturbation is then determined by overlap of two neighboring resonances of lower amplitude. Since the ratio of the amplitudes of the different resonances ($2/\sqrt{3} \approx 1.15$) differs little from unity, the higher-approximation correction and the resonance repulsion (both decrease somewhat the critical perturbation) are insignificant. But it follows hence that to determine the critical perturbation we can approximately replace the Cohen mapping (14.10) by the standard mapping

$$\bar{J} = J + 3\sqrt{3}\, K_0 \sin\theta,\ \bar{\theta} = \theta + \bar{J}, \tag{14.25}$$

which corresponds to equal amplitudes of all the resonances ($|A_r| = \sqrt{3}$). Note that were we to use the directly mapping (14.12) with the parameter (14.13),

$$K_1 = 6K_0 \cos\left(\frac{\pi r}{3}\right) = \begin{cases} \pm 6K_0, & r/3 - \text{integer}, \\ \pm 3K_0, & r/3 - \text{fractional}, \end{cases}$$

the results would not differ too much from (14.25). This shows that the mapping (14.12) can be used for order-of-magnitude estimates at an arbitrary ratio $\gamma_+/\gamma_-$.

One more difference between the standard mapping (14.25) and the system (14.23) of interest to us should be noted. In the former case the amplitudes as well as phases of all resonances are equal, so that the overlap is determined by the maximum distance between the separatrix branches (see Fig. 1). Now, however, neighboring resonances are separated in phase by $\theta$ on $\Delta\theta = 60°$. Since the distance between the separatrix branches is proportional to $\cos(\theta/2)$ (10.4), its relative decrease by this shift is $\cos 30° = \sqrt{3}/2 \approx 0.87$, which increases the critical perturbation by approximately 13% and offsets in part the aforementioned decrease of the critical perturbation.

Neglecting all these corrections, we get from (14.25) for the critical perturbation in the Cohen mapping with $\gamma_-/\gamma_+ = 2$ the value

$$|K_0|_{cr} \approx \frac{1}{3\sqrt{3}} \tag{14.26}$$

or, in terms of the parameters of a trap with opposing mirrors [see (14.1), (14.7), (7.37)]:

$$\beta_0^{(cr)} \approx 2.6 q^{9/40} \exp(-q/5), \tag{14.27}$$

where $q = \omega_0 \ell_0 / 3v = \ell_0 / 3\rho_m$. This expression differs only by an insignificant numerical factor from relation (10.18) for planar opposing mirrors if the parameter q is suitably chosen (see Section 7).

In the estimate of the diffusion rate we confine ourselves to the case when $3\sqrt{3}|K_0| \gg 1$ and the correlations can be neglected, so that $R(K) \approx 1$ (see Section 12). In the mapping (14.2) we can now assume that both "bumps" are statistically independent. To describe diffusion in discrete time, we choose as the time unit again the half-period of the longitudinal oscillations. Expression (11.3) for $D_\mu$ then remains unchanged, and with it all other relations in Sections 11 and 13.

The correlation factor R(K) will, of course, be different now. It can be obtained in analogy with the procedure used in Section 12 for the standard mapping, but of course with allowance for the correlation of the two "bumps" in the mapping (14.2). We note, finally, that the period of the longitudinal oscillations is given in our case by the relation [see (3.37), (3.38), (3.42)]:

$$T = \frac{T_+ + T_-}{2} = \frac{3vl_0}{\mu\omega_0}, \qquad \Omega = \frac{2\pi}{T} = \frac{2\pi}{3}\frac{\mu\omega_0}{vl_0}. \tag{14.28}$$

In the approximation assumed, no gaps whatever appear in the stochastic component. The reason is that we have considered only longitudinal oscillations with large amplitude $a \gg \ell_0$, for which $\gamma_-/\gamma_+ = 2$. At lower amplitudes this ratio decreases and the resonance amplitude requires a slow dependence on p; this can lead to formation of gaps and to isolation of individual stochastic components of the motion. Generally speaking, gaps can exist also at very large oscillation amplitudes because the ratio $\gamma_-/\gamma_+$ is not exactly equal to 2, but is somewhat smaller because of the additional phase shift in (7.37). From (3.38) and (7.38) we get the estimate

$$\Delta = 2 - \frac{\gamma_-}{\gamma_+} \sim \frac{1}{20}\sqrt{\frac{l_0}{a}}.$$

Since this is a very small quantity, the number of resonances in the isolated part of the stochastic component, i.e., between the gaps ($\sim 1/\Delta$), can be comparable with the total number of resonances, and then the presence or absence of gaps is immaterial.

In a purely dynamic system any gap stops the diffusion completely and thus improves substantially the particle containment in the trap. In a real situation, however, this improvement is doubtful, since the gap is narrow and even insignificant multiple scattering of the particles will cause them to "infiltrate" through the gaps. This question deserves a more detailed investigation, all the more since the asymmetry needed for gap formation can be easily introduced in any trap.

We conclude this section by noting that the effect of a certain special perturbation in a tokamak is also described by Cohen mapping [51].

## 15. Remarks on Adiabatic Invariance

The Budker problem considered in this article, concerning the conditions and accuracy of conservation of the magnetic moment of a charged particle in an adiabatic magnetic trap, is a particular case of a general and, in a certain sense "perpetual," problem of classical mechanics – that of the adiabatic invariance of the action variables. This, as any other invariance, plays an important role in physics, even though it is, generally speaking, approximate. According to the most widely held notions, the basic condition of adiabatic invariance is associated with slowness of the perturbation. This notion dates back to the very beginning of the study of this phenomenon and was subsequently related to the averaging method used to establish adiabatic invariance. It became clear later that slowness of the perturbation does not by itself explain the mechanism whereby adiabatic invariance is violated. A vague idea arose that this mechanism is connected somehow with resonances between an external parametric perturbation and the natural oscillations of the system. During the burgeoning development of quantum mechanics, for which the action variables in general, and their adiabatic invariance in particular, play a special role, Born wrote, for example (as cited in [52]): "We regard as adiabatic such a system change which, first, is not related in any way with the period of the unperturbed system ..." This, however, was only intuition. The role of resonances for adiabatic invariance was first precisely formulated and solved in 1928 [52]. It was sufficient for this purpose to examine attentively, from the standpoint of physics, the well-known Mathieu equation and its solutions. Indeed, instability zone or regions of parametric resonance exist in the vicinity of any half-integer frequency ratio $\omega_0 / \Omega = r/2$, where $\omega_0$ is the unperturbed frequency of the linear oscillator, $\Omega$ is the frequency of the harmonic parametric perturbation, and r is a positive integer. But as $r \to \infty$ we have $\Omega \to 0$ and the perturbation becomes adiabatic in accord with the slowness criterion. Nonetheless, once the oscillator becomes resonant, its energy (and action) can vary arbitrarily strongly with time, i.e., adiabatic invariance is violated no matter how slow the perturbation. This leads to a different concept of adiabatic perturbation as a nonresonant one (see Born's statement above). These two seemingly different concepts are in fact closely related, inasmuch as slowness of the perturbation ensures exponential smallness of the resonant harmonics (see Section 7). This holds true, in particular, also for the Mathieu equation:

$$J = J_0 e^{\gamma t}, \quad \gamma \approx r\Delta\Omega \approx \frac{\Omega}{3}\left(\frac{e^2\varepsilon^2}{8}\right)^r; \tag{15.1}$$

i.e., the rate of exponential growth in the resonance region is itself decreased exponentially with increasing frequency ratio $r = 2\omega_0/\Omega$. The frequency-modulation depth is determined here by the relation $\omega^2(t) = \omega_2{}^0(1 - \varepsilon^2 \cos\Omega t)$. An expression for the instability growth rate $\gamma$ and for the total width $\Delta\Omega$ of the resonance zone is obtained from (7.1) at $r \gg 1$ if $\omega_x = 1 + x^2$ is replaced by $\omega(t)$, and takes the form (15.1) at $\varepsilon \ll 1$.

For a linear oscillator, the decisive of the two adiabaticity conditions is the nonresonance. The resonance zones in this case are very distinct and are determined only by the parameters of the system.

For a nonlinear oscillator, the nonresonance is governed by the initial conditions. Its role differs greatly with the number of degrees of freedom of the system. For a closed system with two degrees of freedom, and also for a linear oscillator with one degree of freedom and an external parametric perturbation, the decisive factor for the adiabatic invariant is the slowness of the perturbation. A result new in principle and at the same time rigorous is here the proof of existence of a finite critical slowness of the perturbation, below which the adiabatic invariant becomes an exact integral of the motion [10]. This result is strongly connected with the special topology of the resonance systems, which can be called ordered or one-dimensional. In this case the frequencies are in only one ratio that depends on one action variable (the second action is excluded in a closed system with the aid of the energy integral; see Section 8). Such an ordered topology is significant because an exact integral exists not for all initial conditions, but only for the "nonresonant" ones. This term has a special meaning in a nonlinear system, since the region at the very center of the resonance, i.e., at an action-variable value for which the resonance conditions are exactly satisfied (see Fig. 6), is here also nonresonant. Moreover, the region in question is as a rule even more stable in the sense that an exact integral is preserved here also when resonance overlap leads to development of global instability. The most unstable, however, is the vicinity of the separatrix of the nonlinear resonance, where a stochastic layer is produced and is preserved at arbitrarily small (and slow) perturbation (see, e.g., [8]). In the case when the nonlinear resonance can be described in the pendulum approximation, the stochastic layer in the vicinity of its separatrix is quite similar to the stochastic layer in a multimirror trap (see Section 10). The width of this layer, and accordingly the share of the "resonant" initial conditions, is exponentially small in terms of the slowness parameter of the perturbation. This region is nonetheless finite and contains no exact integral. Since the resonant structure is one-dimensional, however, the stochastic trajectory is strictly confined to the interior of the layer and collapse of the integral does not alter (with exponential accuracy!) the trajectory of the motion.

The situation changes radically, however, in the case of a multidimensional (even two-dimensional) or disordered topology of the resonance. The trajectory can now go from one stochastic layer to another (i.e., with different resonance), bypassing the nonresonant region with the exact integral of motion. This bypassing of the stable regions is made possible by the intersection of the resonances (and of their stochastic layers) in multidimensional space. In other words, if the dimensionality of the phase space is increased even by unity,* motion becomes possible not only across the layer, whose width is rigorously bounded and small, but also along the layer, which is in general bounded only by the energy integral. This beautiful phenomenon was predicted by Arnol'd, who constructed the first example of such a system [53]. A similar process was subsequently named Arnol'd's diffusion and investigated in great detail in [8] (see also [9, 54]).

*The dimensionality of the phase space of a closed Hamiltonian system is always even. If, however, the system is acted upon by a periodic perturbation, it is said to have one dimension (the phase of the perturbation) and a half-integer number of degrees of freedom.

From the standpoint of the adiabatic invariance discussed here (which is, of course, only a particular case of the general dynamic problem of the stability of motion), in a one-dimensional system the requirement that the initial conditions be nonresonant, is at least just as important as the requirement of slow perturbation. This last requirement, just as in order topology, ensures absence of resonance overlap and the existence of an exact integral for the overwhelming majority of nonresonant initial conditions.

In this situation, however, the probability of the stability of motion in general and adiabatic invariance in particular becomes, in the language of the mathematicians, an incorrect one, or simply physically meaningless. The point is that although the total volume of the stochastic layers in phase space is indeed exponentially small, they form everywhere a dense system. Of course, the same takes place in one-dimensional topology. For example, for standard mapping, any rational value of $p/2\pi$ is resonant (see Section 10). However, in view of the already described ordered structure of the resonances, the system motion is not affected. In the multidimensional case, however, unbounded Arnol'd diffusion sets in.

There are several methods of so-called regularization of the problem, i.e., of formulating it unambiguously and independently of the infinitely small changes of the initial conditions. One can, for example, pose the problem of adiabatic invariance over an arbitrarily large but finite time interval. Since the rate of the Arnol'd diffusion falls off extremely rapidly with increasing order of the resonances, any time limitation transforms right away the infinite and everywhere-dense system of resonances (and of their stochastic layers) into a finite one, and the problem acquires physical meaning. In this case, the nonresonance condition (with respect to the remaining "working" stochastic layers) is essential for adiabatic invariance.

Another regularization method consists of introducing in the problem an additional arbitrarily weak but finite external diffusion [8]. This again leaves a finite number of stochastic layers in which the Arnol'd diffusion is faster than the external diffusion. With the problem so formulated, the initial conditions of the motion are immaterial, since the external diffusion will continuously displace the dynamic trajectory.

## REFERENCES

1. G. I. Budker, "Thermonuclear reactions in a system with magnetic mirrors. Contribution to the problem of direct conversion of nuclear energy into electricity," in: Collected Works [in Russian], G. I. Budker (ed.), Nauka, Moscow (1982), p. 72.
2. B. V. Chirikov, "The problem of stability of motion of a charged magnetic trap," Fiz. Plazmy, 4, No. 3, 521 (1978).
3. R. H. Cohen, "Orbital resonances in nonaxisymmetric mirror machines," Comments Plasma Phys. Controlled Fusion, 4, No. 6, 157 (1979).
4. B. V. Chirikov, "Adiabatic invariants and stochasticity in magnetic confinement systems," Proceedings Intern. Conf. on Plasma Physics, Nagoya (1980), Vol. II, p. 176.
5. R. H. Cohen, G. Rowlands, and J. H. Foote, "Non-adiabaticity in mirror machines," Phys. Fluids, 21, No. 4, 627 (1978).
6. B. V. Chirikov, "Homogeneous model of resonant diffusion of particles in an open magnetic trap," Fiz. Plazmy, 5, No. 4, 880 (1979).
7. G. I. Dimov, V. V. Zakaidakov, and M. E. Kishinevskii, "Thermonuclear trap with tandem mirrors," Fiz. Plazmy, 2, No. 4, 597 (1976); T. K. Fowler and B. G. Logan, "The tandem mirror reactor," Comments Plasma Phys. Controlled Fusion, 2, No. 6, 167 (1977).
8. B. V. Chirikov, "A universal instability of many-dimensional oscillator systems," Phys. Rep., 52, No. 5, 263 (1979).
9. A. J. Lichtenberg and M. A. Lieberman, Regular and Stochastic Motion, Springer, Berlin (1983).
10. V. I. Arnol'd, "Small denominators and the problem of stable motion in classical and celestial mechanics," Usp. Mat. Nauk, 18, No. 6, 91 (1963).
11. G. I. Budker, V. V. Mirnov, and D. D. Ryutov, "Influence of magnetic-field corrugation on expansion and cooling of a dense plasma," Pis'ma Zh. Eksp. Teor. Fiz., 14, No. 5, 320 (1971).
12. N. N. Bogolyubov and Yu. A. Mitropol'skii, Asymptotic Methods in the Theory of Nonlinear Oscillations [in Russian], Nauka, Moscow (1974).
13. R. W. B. Best, "On the motion of charged particles in a slightly damped sinusoidal potential wave," Physica, 40, No. 2, 182 (1968).

14. A. S. Bakai and Yu. P. Stepanovskii, Adiabatic Invariants [in Russian], Naukova Dumka, Kiev (1981).
15. J. E. Howard, "Nonadiabatic particle motion in cusped magnetic fields," Phys. Fluids, 14, No. 11, 2378 (1971).
16. A. M. Dykhne and A. V. Chaplik, "Change of adiabatic invariant of a particle in a magnetic field," Zh. Eksp. Teor. Fiz., 40, No. 2, 666 (1961).
17. M. Kruskal, Adiabatic Invariants [Russian translation], Inostr. Lit., Moscow (1962).
18. C. S. Gardner, "Magnetic moment in second order for axisymmetric static field," Phys. Fluids, 9, No. 10, 1997 (1966).
19. O. B. Firsov, "Repulsion of charged particle from regions with strong magnetic field. (Accuracy of adiabatic invariant)," in: Plasma Physics and the Problem of Controlled Thermonuclear Reactions [in Russian], Izd. Akad. Nauk SSSR, Moscow (1958), Vol. III, p. 259.
20. R. J. Hastie, G. D. Hobbs, and J. B. Taylor, "Nonadiabatic behavior of inhomogeneous magnetic fields," Plasma Physics and Controlled Thermonuclear Fusion Research, IAEA (1969), Vol. I, p. 389.
21. E. M. Krushkal', "Nonadiabatic motion of particles in nonuniform magnetic fields," Zh. Tekh. Fiz., 42, No. 11, 2288 (1972).
22. I. S. Gradshtein and I. M. Ryzhik, Tables of Integrals, Sums, Series, and Products, Academic Press, New York (1965).
23. J. M. Greene, "A method for determining a stochastic transition," J. Math. Phys., 20, No. 6, 1183 (1979).
24. F. M. Izrailev, B. V. Chirikov, and D. L. Shepelyanskii, "Dynamic stochasticity in classical mechanics," Preprint Inst. Nucl Phys., Siberian Branch, Acad. Sci. USSR, No. 80-209, Novosibirsk (1980).
25. M. N. Rosenbluth, "Superadiabaticity in mirror machines," Phys. Rev. Lett., 29, No. 7, 408 (1972).
26. I. P. Kornfel'd, Ya. G. Sinai, and S. V. Fomin, Ergodic Theory [in Russian], Nauka, Moscow (1980).
27. V. M. Alekseev and M. V. Yakobson, "Symbolic dynamics and hyperbolic dynamic systems," Suppl. to Methods of Symbolic Dynamics, R. Bowen (ed.) [Russian translation], Mir, Moscow (1979); A. A. Brudno, "Entropy and algorithmic complexity of trajectories of a dynamic system," Preprint of All-Union Inst. for System Research, Moscow (1980).
28. B. V. Chirikov, "Nature of stochastic laws of classical mechanics," in: Methodological and Philosophical Problems of Physics [in Russian], Nauka, Novosibirsk (1982), p. 181.

29. N. N. Bogolyubov, "Problems of dynamic theory in statistical physics," in: Selected Works [in Russian], Naukova Dumka, Kiev (1970), Vol. 2, p. 99.
30. J. L. Lebowitz, "Hamiltonian flows and rigorous results in nonequilibrium statistical mechanics," in: Statistical Mechanics. New Concepts, New Problems, New Applications, Univ. of Chicago Press, Chicago (1972), p. 41.
31. R. Balescu, Equilibrium and Nonequilibrium Statistical Mechanics, Wiley, New York (1975).
32. G. E. Norman and L. S. Polak, "Irreversibility in classical statistical mechanics," Dokl. Akad. Nauk SSSR, 263, No. 2, 337 (1982).
33. E. M. Lifshitz and L. P. Pitaevskii, Physical Kinetics, Pergamon Press, Oxford (1981).
34. A. B. Rechester and R. B. White, "Calculation of turbulent diffusion for the Chirikov–Taylor model," Phys. Rev. Lett., 44, No. 24, 1586 (1980); A. B. Rechester, M. N. Rosenbluth, and R. B. White, "Fourier-space paths applied to the calculation of diffusion for the Chirikov–Taylor model," Phys. Rev., A23, No. 5, 2264 (1981).
35. J. R. Cary, J. D. Meiss, and A. Bhattacharjee, "Statistical characterization of periodic measure-preserving mappings," Phys. Rev., A23, No. 5, 2744 (1981).
36. L. A. Bunimovich and Ya. G. Sinai, "Statistical properties of Lorentz gas with periodic configuration of scatterers," Commun. Math. Phys., 78, 479 (1981).
37. C. Grebogi and A. N. Kaufman, "Decay of statistical dependence in chaotic orbits of deterministic mappings," Phys. Rev., A24, No. 5, 2829 (1981).
38. G. V. Gadiyak and F. M. Izrailev, "Structure of transition zone of nonlinear resonance," Dokl. Akad. Nauk SSSR, 218, No. 6, 1302 (1974).
39. B. V. Chirikov and D. L. Shepelyanskii, "Statistics of Poincaré returns and structure of stochastic layer of nonlinear resonance," Preprint, Inst. Nucl. Phys, Siberian Branch, Acad. Sci. USSR, No. 81-69, Novosibirsk (1981).
40. D. Bora, P. I. John, Y. C. Saxena, and R. K. Varma, "Multiple lifetimes in the nonadiabatic leakage of particles from magnetic mirror traps," Plasma Physics, 22, No. 7, 653 (1980).
41. D. F. Escande and F. Doveil, "Renormalization method for computing the threshold of the large-scale stochastic instability in two degrees of freedom Hamiltonian systems," J. Stat. Phys., 26, No. 2, 257 (1981).

42. L. P. Kadanoff, "Scaling for a critical Kolmogorov–Arnold–Moser trajectory," Phys. Rev. Lett., 47, No. 23, 1641 (1981); S. J. Shenker and L. P. Kadanoff, "Critical behavior of a KAM surface: empirical results," J. Stat. Phys., 27, No. 4, 631 (1982).
43. L. D. Landau, "Kinetic equation for rarefied gases in strong fields," Zh. Eksp. Teor. Fiz., 7, No. 2, 203 (1937).
44. S. T. Belyaev, "Kinetic equation for rarefied gases in strong fields," in: Plasma Physics and Problem of Controlled Thermonuclear Reactions [in Russian], Izd. Akad. Nauk SSSR, Moscow (1958), Vol. III, p. 50.
45. B. V. Chirikov and D. L. Shepelyanskii, "Diffusion in multiple passage through nonlinear resonance," Preprint Inst. Nucl. Phys., Siberian Branch, Acad. Sci. USSR, No. 80-211, Novosibirsk (1980).
46. M. A. Lieberman and A. J. Lichtenberg, "Stochastic and adiabatic behavior of particles accelerated by periodic forces," Phys. Rev., A5, No. 4, 1852 (1972).
47. A. Brahic, "Numerical study of a simple dynamical system," Astron. Astrophys., 12, 98 (1971).
48. T. A. Zhdanova and F. M. Izrailev, "On Fermi statistical acceleration," Preprint Inst. Nucl. Phys., Siberian Branch, Acad. Sci. USSR, No. 121-74, Novosibirsk (1974).
49. R. H. Cohen, "Stochastic motion of particles in mirror machines," in: Intrinsic Stochasticity in Plasma, G. Laval and D. Gresillon (eds.), Edition de Physique, Orsay (1979).
50. D. F. Escande, "Primary resonances do not overlap," in: Intrinsic Stochasticity in Plasma, G. Laval and D. Gresillon (eds.), Edition de Physique, Orsay (1979).
51. R. J. Goldston, R. B. White, and A. H. Boozer, "Confinement of high-energy particles in tokamaks," Phys. Rev. Lett., 47, No. 9, 647 (1981).
52. L. I. Mandel'shtam, Complete Works [in Russian], Izd. Akad. Nauk SSSR, Moscow (1948), Vol. I, p. 297.
53. V. I. Arnol'd, "Instability of dynamic systems with many degrees of freedom," Dokl. Akad. Nauk SSSR, 156, No. 1, 9 (1964).
54. N. N. Nekhoroshev, "Exponential estimate of the stability time of nearly integrable Hamiltonian systems," Usp. Mat. Nauk, 32, No. 6, 5 (1977).

# TRANSPORT PROCESSES IN AXISYMMETRIC OPEN TRAPS

**D. D. Ryutov and G. V. Stupakov**

## 1. Introduction

One of the most widely used and reliable methods of ensuring magnetohydrodynamic stability in a plasma is to use axisymmetric "minimum-B" magnetic fields. This was, in fact, the stabilization principle used by M. S. Ioffe and co-workers in the early sixties [1], and later successfully employed by the Livermore group in a set of experiments of increasing complexity, culminating in the sensational success with the 2XIIB installation [2]. It is used at present in most traps with ambipolar (tandem) mirrors [3-6]. Stabilization by "average minimum-B" is also used (see [7]) in several variants of multimirror traps.

As noted in [8], the absence of axial symmetry of the magnetic field can cause the transverse transport coefficients to increase above their "classical" values by a mechanism that recalls in certain cases the mechanism of neoclassical diffusion in closed systems (see [9]).

The transport processes considered in the present paper are due mostly to Coulomb collisions, and their lifetime exceeds the particle Coulomb-scattering time (we disregard turbulent transport). Allowance for these processes is accordingly essential only as applied to traps in which the longitudinal lifetime of the plasma is long compared with the lifetime of the ion–ion collisions. This situation is realized, in particular, in the central magnetic bottles of ambipolar and multimirror traps. As for the ordinary mirror trap, the ion lifetime in it does not exceed, roughly speaking, the ion–ion collision time,* and allowance for the transverse losses could, in the worst case, decrease the ion lifetime by 1.5-2 times (compared with purely longitudinal losses). Nonetheless, here, too, the effects of the enhanced transverse losses can be substantial not only for ions but for electrons. Indeed, since the plasma quasineutrality makes the electron lifetime longer than the time of their Coulomb scattering, the problem of electron heat conduction and diffusion by collisions becomes meaningful. Naturally, enhanced electron diffusion and larger thermal conductivity of electrons are also important in the aforementioned systems with improved longitudinal ion confinement.

Typical magnetic-surface shapes of the three types of installation mentioned above are illustrated in Fig. 1.

The enhanced transport is due to certain peculiarities of particle motion in a nonaxisymmetric magnetic field. We assume that the magnetic and electric fields are continuous functions of the coordinates and time, with characteristic scales exceeding the Larmor radius and the cyclotron period, respectively. Under these conditions the quantity

$$\mu \equiv mv_{\perp}^{2}/2B \qquad (1.1)$$

is adiabatically invariant; here m is the particle mass, $v_{\perp}$ is the velocity of rotation along the Larmor circle, and B is the magnetic field strength. Within the context of the present paper, μ can be regarded simply as an integral of the motion.

The particle motion can be described under these conditions in the drift approximation. The Larmor circle oscillates between magnetic (or electrostatic) mirrors and drifts simultaneously along the magnetic axis. If the trap is not very long, the Larmor circle is shifted (by the drift) during one transit from mirror to mirror to a force line close to the initial one (close in the sense that the distance be-

*If account is taken of ion scattering by microfluctuations, a situation is conceivable in which this scattering takes place only in a certain limited ion-velocity-space region that does not intersect the "loss cone." The problem of neoclassical ion diffusion by scattering from microfluctuations can then become important.

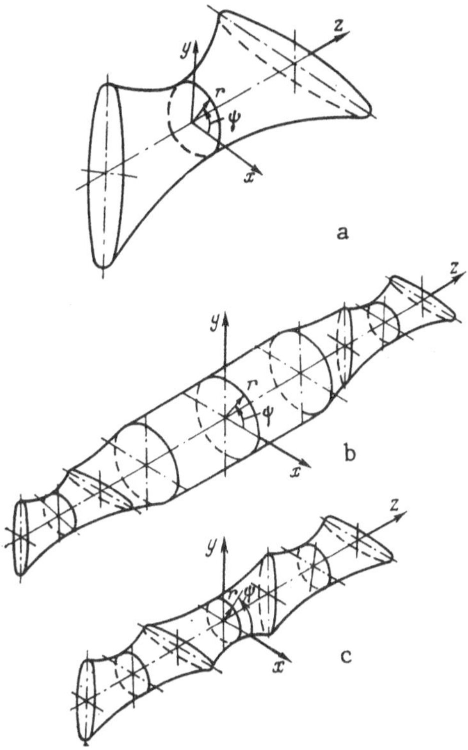


Fig. 1. Shapes of magnetic surfaces: a) in a magnetic bottle with minimum-B; b) in an ambipolar trap; c) in a multimirror trap.

tween these force lines is short compared with the characteristic spatial scale of the problem), and the trajectory of the Larmor circle fills "densely" a certain surface (Fig. 2) called the "drift surface."

The particle motion along the magnetic field constitutes oscillation with slowly varying period, in which is preserved the adiabatic invariant known as longitudinal [10]:

$$I = \sqrt{\frac{2}{m}} \int_{s_1}^{s_2} \sqrt{\varepsilon - \mu B - e\varphi}\, ds, \tag{1.2}$$

where $\varepsilon \equiv (mv^2/2) + e\varphi$ is the total energy of the particle, and the integration is along a magnetic-field force line be-

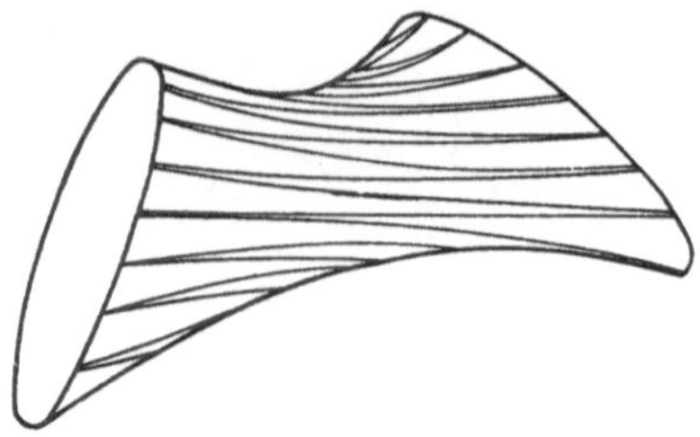

Fig. 2. Drift trajectory that "fills" a drift surface.

tween the stopping points of the particle. In a static field, the particle energy remains unchanged, so that the condition I = const fixes uniquely the drift surface for a particle that has definite values of $\varepsilon$ and $\mu$ and started out from a specified force line.

It is important that, in a nonaxisymmetric magnetic field, the shapes of drift surfaces that pass through one and the same force line depend, generally speaking, on $\varepsilon$ and $\mu$ (Fig. 3), and the maximum excursion of the surfaces can exceed substantially the Larmor radius of the particle. It is this very circumstance that gives rise to the enhanced transport: collisions between particles change the values of $\varepsilon$ and $\mu$ and cause the particles to "jump over" from one drift surface to another, and "spacing" of these random steps can become equal to the maximum excursions of these surfaces [11].

It can be seen from the foregoing that the quality of a magnetic system (in the sense of minimizing transverse losses) depends on how small one can make the excursion of the drift surfaces passing through one force line. It is exactly zero in the trivial case of magnetic fields having a high degree of symmetry – axisymmetric and planar. We say, in this case, that the magnetic field has "embedment"* (omnigenity). For magnetic fields of more complicated symmetry, used in practice to ensure "min B," the tendency is to minimize the omnigenity property at least approximately. In particular, in the paraxial region of magnetic systems of the Yin–Yang type (Fig. 4), which become congruent on reflec-

*The term "omnigenity" is used occasionally, especially in the English language, and was proposed by Hall and McNamara.

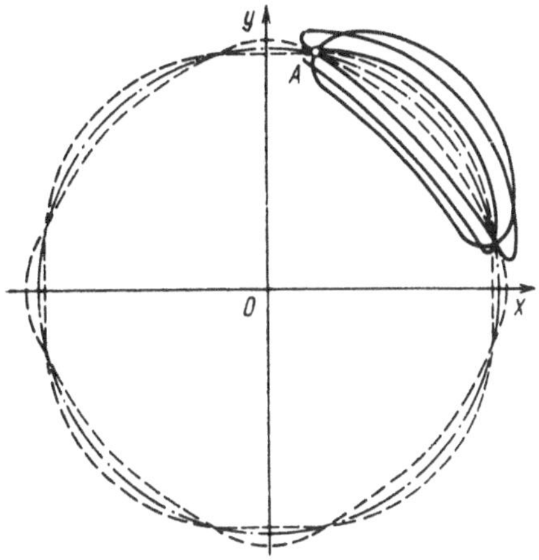


Fig. 3. Characteristic shapes of the intersections of the drift surfaces with the plane z = 0 for the traps shown in Fig. 1. The drift surfaces passing through point A are shown. The shells shown by solid lines lie within one quadrant; those shown dotted encircle the magnetic axis O of the system; the dash-dot curve is a circle with center O and passing through point A.

tion from the median plane and simultaneous 90° rotation about the magnetic axis, this is valid accurate to terms of order $(r/\ell)^2$, where $\ell$ is the longitudinal field-variation scale, and the excursion of the drift surfaces is small. The cause of this smallness in a magnetic field having such a symmetry is that, as can be shown, the radial displacements of the particle on reflection from opposite mirrors cancel one another almost completely. It turns out frequently in this case that the excursion of the drift surfaces that are defined by the condition I = const is small compared with the Larmor radius of the ions. It may be important then to take into account the shell-drift "fine structure" due to the fact that, as on moving "forward" and "backward" along the magnetic field the particle moves "inside" and "outside" the averaged drift surface defined by the condition I = const. The reason for this "splitting" of the drift surface is that the displacements on reflection from the opposite mirrors, as already mentioned, are of opposite sign. The splitting

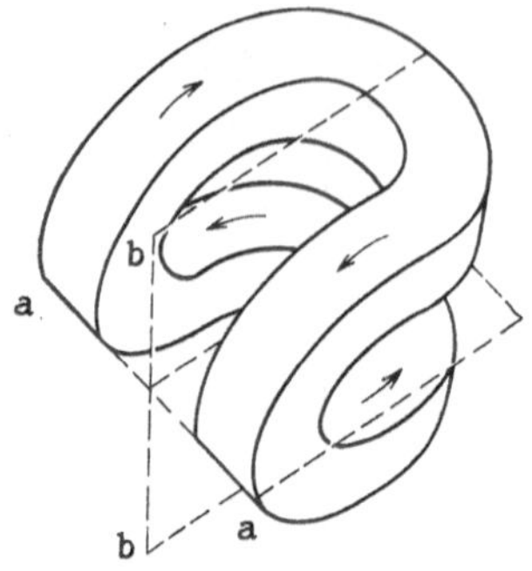


Fig. 4. Magnetic system having quadrupole symmetry: aa and bb – symmetry planes. An arrow marks the current direction.

can be formally taken into account [12, 13] by using a more accurate expression for the longitudinal adiabatic invariant. In some cases the splitting can make a substantial contribution to the transverse transport of the ions [14].

We have dealt so far with the case when the period $t_{\parallel}$ of the longitudinal oscillations is small compared with the time $\tau_D$ to trace the Larmor circle around the magnetic axis. This meant automatically that the particle's azimuthal displacement $\Delta\psi$ around the magnetic axis during the time of one transit from mirror to mirror is short: $\Delta\psi = \pi t_{\parallel}/\tau_D << 1$. For very long traps (real example – the central mirror system of an ambipolar trap) a situation is possible in which $\Delta\psi$ becomes appreciable: $\Delta\psi \gtrsim 1$. The concept of a particle moving gradually through all points of the drift surface becomes meaningless. The adiabatic invariant ceases at the same time to be conserved.

In this situation the largest radial departures are those of the so-called resonant particles, which complete in one transit from mirror to mirror a rational number of turns around the magnetic axis, $\Delta\psi = \pi k/n$, where k and n are integers. In particular, for a magnetic system of the Yin–Yang type large random walks take place at $\Delta\psi = (2p - 1)\pi/2$, p = 10, ±1, ±2, ..., for in this case the radial displacements on reflection from field-reversed mirrors are additive. Since the twist angle $\Delta\psi$ for given $\varepsilon$ and $\mu$ depends, generally speaking, on the radius, the resonance condition is violated for radial displacement of the particle. This is precisely the cause of the finite scale of the radial random walks of the resonant particles. This scale can actually be quite large, so that the Coulomb scattering of the resonant particles can cause an enhanced radial transport [8, 15] named "resonant" in [15].

It is obviously impossible to decrease substantially this type of transport by relative reversed rotation of the field-reversed mirrors around the magnetic axis. It is necessary here to decrease the radial displacement that follows reflection from an individual mirror. This can be accomplished, in principle, by a special choice of the spatial dependence of the field in the mirror.

At very large values of $\Delta\psi$, a new type of loss sets in and is not directly connected with the particle scattering. This is the stochastic diffusion that appears because the correlation between the radial displacements in successive reflections from the mirrors vanishes completely at very large $\Delta\psi$, and the random walks become diffusive even in the absence of collisions [8].

Clearly, in all the aforementioned cases the character of the transport depends essentially on the structure of the magnetic and electric fields in the plasma, since it is they which determine the particle drift. The solution of the transport problem is therefore inseparably linked with the solution of the problem of the plasma MHD equilibrium (only the latter yields the true magnetic field undistorted by the plasma), and with the solution of the problem of determining the electric field in the plasma. Allowance for the magnetic-field corrections necessitated by the plasma pressure does not reduce in many cases merely to obtaining more accurate quantitative results, but leads to strong qualitative changes (for example, at finite $\beta$ there is no cancellation of the displacements from field-reversed mirrors in a Yin–Yang system). Particular magnetic-field corrections are necessitated also by plasma-pressure perturbations connected with radial random walks of the particles.

It follows from the foregoing that in our problem the equilibrium and transport effects are very closely interlinked. This makes a formal solution of the problem quite difficult (apparently more so than for tokamaks).

The plan of the exposition is the following. In Section 2 we report the necessary data from the theory of plasma equilibrium in nonaxisymmetric traps. In Section 3 we analyze the motion of the charged particles. A neoclassical transport theory ($\Delta\psi \ll 1$) is developed in Section 4, and a theory for resonant and stochastic transport ($\Delta\psi \gtrsim 1$) is developed in Section 5.

## 2. Plasma Equilibrium

### 2.1. Vacuum Magnetic Field of Magnetic Bottle in the Paraxial Approximation.

The vacuum magnetic field of nonaxisymmetric magnetic traps can, as a rule, be described with good accuracy in the paraxial approximation, in which it is assumed that the magnetic force lines make a small angle with the system axis. This approximation is valid near the magnetic axis of a trap, at distances small compared with the characteristic scale of the field variation along the axis. The paraxial approximation will hereafter be systematically used both to analyze the plasma equilibrium and to estimate the transport coefficients.

There are no currents in the paraxial region, so that the magnetic field is potential:

$$\mathbf{B} = -\nabla\chi.$$

The function $\chi$ will hereafter be called the magnetic potential. In view of the condition div $\mathbf{B} = 0$, the magnetic potential satisfies the Laplace equation

$$\Delta\chi = 0. \quad (2.1)$$

In practice the magnetic field is usually produced by windings that have a certain spatial symmetry. This means most frequently the presence of two mutually perpendicular planes that pass through the magnetic axis of the system, and that the reflection operation in the plane makes the winding congruent with itself. The shape of one such winding is illustrated in Fig. 4. It can be seen from the figure that reflection in each of the symmetry planes not only makes the winding congruent with itself, but reverses the direction of the currents in the winding.

We choose a cylindrical coordinate frame with the z axis along the magnetic axis and with the polar angle $\psi$ measured from one of the symmetry planes. It can be seen from the indicated symmetry properties of the winding that the magnetic potential satisfies the following symmetry relations:

$$\chi(r, \psi, z) = \chi(r, -\psi, z), \quad \chi(r, \psi, z) = \chi(r, \pi-\psi, z). \tag{2.2}$$

To find the magnetic potential in the paraxial approximation, we expand the function $\chi(r, \psi, z)$ in a Fourier series in $\psi$ and in a Taylor series in r:

$$\chi = \sum_{n,k} A_{nk}(z)\, r^n \cos[k\psi + \psi_{nk}(z)]. \tag{2.3}$$

It follows directly from (2.2) that $\psi_{nk} \equiv 0$, and k can be only even. Stipulating now that the potential (2.3) satisfy Eq. (2.1), we readily find that

$$\chi = A_{00}(z) - \frac{1}{4} A_{00}''(z)\, r^2 - A_{22}(z)\, r^2 \cos 2\psi + O(r^4), \tag{2.4}$$

where $A_{00}$ and $A_{22}$ are arbitrary dependences on z, and the prime denotes differentiation with respect to z.

It is convenient to express the magnetic field corresponding to the potential (2.4) in the form

$$\begin{aligned} B_z &= \mathscr{B} - \frac{1}{4} r^2 \mathscr{B}'' - r^2 b' \cos 2\psi + O(r^4), \\ B_r &= -2rb \cos 2\psi - \frac{1}{2} r \mathscr{B}' + O(r^3), \\ B_\psi &= 2rb \sin 2\psi + O(r^3), \end{aligned} \tag{2.5}$$

where we have introduced the functions

$$\mathscr{B}(z) \equiv -A_{00}'(z), \quad b(z) \equiv -A_{22}(z),$$

which determine, respectively, the field intensity on the axis and the quadrupole component of the magnetic field.

A feature of Yin–Yang systems mentioned in the Introduction is an additional symmetry of the magnetic field. This symmetry manifests itself in the fact that the family of force lines is invariant to a transformation consisting of a 90° rotation around the magnetic axis, followed by reflection in a certain plane perpendicular to this axis (this plane will hereafter be called equatorial). If the coordinate z is measured from the equatorial plane, such a symmetry means that the functions $\mathscr{B}(z)$ and $b(z)$ are even:

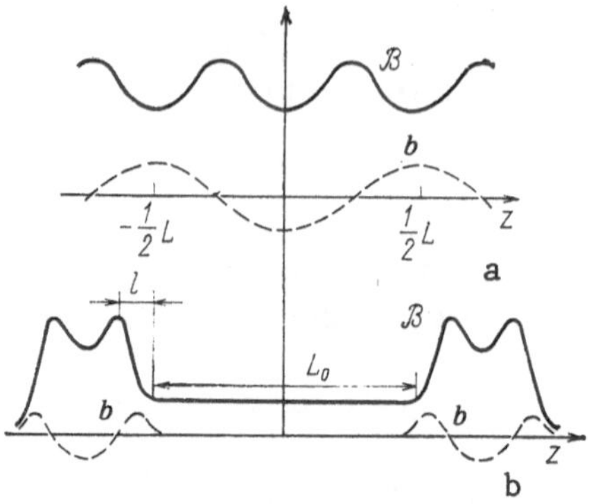


Fig. 5. Forms of the functions $\mathcal{B}(z)$ and $b(z)$ for a multimirror trap (a) and for an ambipolar trap with a long central mirror cell (b).

$$\mathcal{B}(z) = \mathcal{B}(-z), \quad b(z) = b(-z). \tag{2.6}$$

In a multimirror trap (Fig. 5a), the functions $\mathcal{B}(z)$ and $b(z)$, in addition to satisfying (2.6), are periodic. If the period of the field is equal to L (Fig. 5a), the period of the function $\mathcal{B}(z)$ is equal to L/2. The function $b(z)$ vanishes at the maxima of $\mathcal{B}(z)$, and is odd with respect to these maxima.

Simple integration yields readily the equation of the force line that passes in the equatorial plane through the point $r_0$, $\psi_0$:

$$r(r_0, \psi_0, z) = r_0 \sqrt{\frac{\mathcal{B}(0)}{\mathcal{B}(z)}} \left(e^{\Phi} \sin^2\psi_0 + e^{-\Phi}\cos^2\psi_0\right)^{1/2},$$
$$\psi(\psi_0, z) = \arctan\left(e^{\Phi}\tan\psi_0\right), \tag{2.7}$$

where

$$\Phi(z) \equiv 4\int_0^z \frac{b(z')}{\mathcal{B}(z')}\,dz'. \tag{2.8}$$

Equations (2.7) are accurate to $O(r^3/\ell^3)$, where $\ell$ is the scale of variation of the functions $\mathcal{B}(z)$ and $b(z)$. The magnetic surface, whose intersection with the equatorial plane is a circle of radius $r_0$, is deformed on entering the mirror;

its cross section on the coordinate z takes the form of an ellipse given by the equation

$$r = r_0 \sqrt{\frac{\mathcal{B}(0)}{\mathcal{B}(z)}} \left(e^{\Phi}\cos^2\psi + e^{-\Phi}\sin^2\psi\right)^{-1/2}. \tag{2.9}$$

The ratio $r(\psi = 0)/r(\psi = \pi/2)$ is equal to $\exp(-\Phi)$.

We shall need expressions for the modulus of the magnetic field on the force line passing through the point ($r_0$, $\psi_0$) and for the length element ds of this force line. Calculations with the aid of (2.5) and (2.7) yield

$$B(r_0, \psi_0, z) = \mathcal{B}(z) + \delta B(r_0, \psi_0, z),$$

$$\delta B(r_0, \psi_0, z) = \frac{r_0^2\mathcal{B}(0)}{4}\left\{\left(\frac{(\mathcal{B}')^2}{2\mathcal{B}^2} - \frac{\mathcal{B}''}{\mathcal{B}} + \frac{(\Phi')^2}{2}\right)\cosh\Phi + \Phi''\sinh\Phi\right.$$

$$\left. - \cos 2\psi_0\left[\left(\frac{(\mathcal{B}')^2}{2\mathcal{B}^2} - \frac{\mathcal{B}''}{\mathcal{B}} + \frac{(\Phi')^2}{2}\right)\sinh\Phi + \Phi''\cosh\Phi\right]\right\} + O\left(r_0^4\right),$$

$$ds = [1 + \sigma(r_0, \psi_0, z)]\,dz, \tag{2.10}$$

$$\sigma(r_0, \psi_0, z) = \frac{r_0^2\mathcal{B}(0)}{8\mathcal{B}}\left\{\left(\frac{(\mathcal{B}')^2}{\mathcal{B}^2} + (\Phi')^2\right)\cosh\Phi - \frac{2\Phi'\mathcal{B}'}{\mathcal{B}}\sinh\Phi\right.$$

$$\left. - \cos 2\psi_0\left[\left(\frac{(\mathcal{B}')^2}{\mathcal{B}^2} + (\Phi')^2\right)\sinh\Phi - \frac{2\Phi'\mathcal{B}'}{\mathcal{B}}\cosh\Phi\right]\right\} + O\left(r_0^4\right). \tag{2.11}$$

In a Cartesian coordinate system x, y, z with the x and y axes in symmetry planes, the force lines (2.7) are given by

$$x(z) = r\cos\psi = x_0\sqrt{\frac{\mathcal{B}(0)}{\mathcal{B}(z)}}\, e^{-\Phi/2},$$
$$y(z) = r\sin\psi = y_0\sqrt{\frac{\mathcal{B}(0)}{\mathcal{B}(z)}}\, e^{\Phi/2}, \tag{2.12}$$

where $x_0$ and $y_0$ are the coordinates of the force line in the equatorial plane. If we introduce the functions

$$X(z) = \sqrt{\frac{\mathcal{B}(0)}{\mathcal{B}(z)}}\, e^{-\Phi/2}, \quad Y(z) = \sqrt{\frac{\mathcal{B}(0)}{\mathcal{B}(z)}}\, e^{\Phi/2}, \tag{2.13}$$

with the property

$$X(0) = Y(0) = 1, \tag{2.14}$$

relations (2.12) take the rather simple form

$$x(z) = x_0 X(z), \quad y(z) = y_0 Y(z). \tag{2.15}$$

Given the functions X and Y, the paraxial magnetic field can be determined with the aid of (2.13) accurate to a scale factor equal to the value of $\mathcal{B}$ at the point z = 0:

$$\frac{\mathcal{B}(z)}{\mathcal{B}(0)} = \frac{1}{XY}, \quad \frac{b(z)}{\mathcal{B}(0)} = \frac{1}{4XY} \frac{d}{dz} \ln \frac{Y}{X}. \tag{2.16}$$

In some problems it is more convenient to specify the magnetic field in terms of the functions X and Y.

We call a magnetic bottle "short" if its length L (distance between the field maxima) coincides with the characteristic scale $\ell$ of the functions $\mathcal{B}$ and $b$. We shall use below also a long-magnetic-bottle model for which it is assumed that the nonaxisymmetric mirrors are separated by a uniform-field region of length $L_0$ much longer than the mirror lengths $\ell$. The forms of the functions $\mathcal{B}$ and $b$ for this case are shown in Fig. 5b. The condition $L_0 \gg 1$ is satisfied in the central magnetic bottles of ambipolar traps.

We present here also an expression for the integral

$$U \equiv \int \frac{ds}{B} \tag{2.17}$$

that is encountered in the study of the equilibrium and stability of an isotropic plasma in a magnetic field. The integration in (2.17) is either over the period of the corrugation (in the case of a multimirror trap) or between two surfaces that are assumed to block the plasma expansion in the longitudinal direction. We shall assume that the reflecting surfaces are planes $z = \pm L/2$ passing through the maximum points of the magnetic field on the trap axis. The integral (2.17) is a function of the coordinates $r_0$ and $\psi_0$ of the point where the force line passes through the equatorial plane. Using the fact that

$$\int \frac{ds}{B} = \int \frac{dz}{B_z} \tag{2.18}$$

and calculating the variation of $B_z$ along the force line,* we

*This can be easily done with the aid of Eqs. (2.10) and (2.11), recognizing that $B_z = B\,dz/ds$.

obtain after simple integration [with allowance for the symmetry properties (2.6)]

$$U = \int_{-L/2}^{L/2} \frac{dz}{\mathcal{B}} + r_0^2 \mathcal{B}(0) \int_{-L/2}^{L/2} \frac{dz}{\mathcal{B}^3} \left( \frac{1}{4} \mathcal{B}'' \cosh \Phi - b' \sinh \Phi \right) + O(r_0^4). \tag{2.19}$$

It follows from (2.19) that the intersections of the magnetic surfaces U = const with the plane z = 0 are the circles $r_0$ = const [16] [deviation from circularity appears in the next approximation in terms of the parameter $(r/\ell)^2$]. Depending on the sign of the second integral in (2.19), the magnetic field has either an integral minimum ($\partial U/\partial r_0 < 0$) or an integral maximum ($\partial U/\partial r_0 > 0$). Integrating by parts in (2.19) and using the condition $\mathcal{B}'(\pm L/2) = 0$, we easily verify that the axisymmetric paraxial field has a maximum [17]. A minimum of U can be ensured only by a sufficiently large quadrupole component of the field. An exact criterion for its value can be easily obtained from (2.19). Without writing it out, we note only that $b$ should be at least of the order of $\mathcal{B}'$. In this case, the relation between the field components should be

$$B_\psi \sim B_r \sim \frac{r}{l} B_z. \tag{2.20}$$

2.2. Plasma Equilibrium in a Short Magnetic Bottle. We begin with the analysis of the equilibrium in an ordinary short magnetic bottle, regarding the particle motion in the drift approximation. In typical conditions the particle-scattering time is long compared with their drift time around the magnetic axis (and all the more compared with the transit time from mirror to mirror). In this case, the stationary distribution functions of the Larmor circles are functions of the drift integrals of motion

$$f_{e,i} = f_{e,i}(I_{e,i}, \varepsilon, \mu), \tag{2.21}$$

where $I_{e,i}$ is the longitudinal adiabatic invariant. Allowance for the collisions gives rise to a slow time dependence due, in particular, to the neoclassical transport. The dependence on the coordinates enters in $f_{e,i}$ via the invariant $I_{e,i}$.

When considering the equilibrium in general, we shall use the curvilinear coordinates $\xi^1$, $\xi^2$, $\xi^3$, of which $\xi^1$ and $\xi^2$ describe the force line, and the coordinate $\xi^3$ determines the position of a point on the force line ($\xi^1$ and $\xi^2$ are usually chosen to be the polar coordinates $r_0$ and $\psi_0$ of the point where the force line passes through a certain plane perpendicular to the magnetic axis of the system, and $\xi^3$ is the distance s, measured from this plane, along the force line). With this choice of the coordinate system, the invariant I is a function of $\xi^1$, $\xi^2$, $\varepsilon$, $\mu$.

The longitudinal and transverse plasma pressures

$$p_{\parallel} = 4\sqrt{2}\,\pi B \sum_{e,i} m_{e,i}^{-3/2} \int \sqrt{\varepsilon - \mu B \mp e\varphi}\, f_{e,i} d\varepsilon d\mu, \qquad (2.22')$$

$$p_{\perp} = 2\sqrt{2}\,\pi B^2 \sum_{e,i} m_{e,i}^{-3/2} \int \frac{\mu f_{e,i} d\varepsilon d\mu}{\sqrt{\varepsilon - \mu B \mp e\varphi}}, \qquad (2.22'')$$

are expressed in terms of the function f in standard fashion; the functions $p_{\parallel}$ and $p_{\perp}$ satisfy automatically the longitudinal-equilibrium condition

$$\frac{\partial p_{\parallel}}{\partial s} + \frac{1}{B}\frac{\partial B}{\partial s}(p_{\perp} - p_{\parallel}) = 0. \qquad (2.23)$$

It is also possible to express in terms of f the longitudinal ($\mathbf{j}_{\parallel}$) and transverse ($\mathbf{j}_{\perp}$) components of the current in the plasma. When these are determined it must be remembered that f is the distribution function of the Larmor circles and not the true particle distribution function, so that in the calculation of, say, $\mathbf{j}_{\perp}$, one must take into account not only the current connected with the drift of the Larmor circles, but also the diamagnetic current due to their inhomogeneous distribution in space. Methods of finding $\mathbf{j}_{\perp}$ and $\mathbf{j}_{\parallel}$ can be found, respectively, in the review [18] and in paper [13]. The corresponding expressions are

$$\mathbf{j}_{\perp} = \frac{c}{B}\{[\boldsymbol{\tau}\boldsymbol{\varkappa}](p_{\parallel} - p_{\perp}) + [\boldsymbol{\tau}\nabla p_{\perp}]\}, \qquad (2.24)$$

$$\frac{\partial}{\partial s}\left[\frac{1}{B^3} j_{\parallel}\left(B^2 - 4\pi(p_{\perp} - p_{\parallel})\right)\right] = \frac{c}{B^2}\boldsymbol{\tau}[\boldsymbol{\varkappa}\nabla(p_{\parallel} + p_{\perp})], \qquad (2.25)$$

where $\boldsymbol{\tau}$ is the vector tangent to the force line, and $\boldsymbol{\varkappa}$ is the curvature of the force line. The longitudinal current stems from the fact that when the Larmor circle oscillates between the mirrors it moves "forward" and "backward" along lines that are shifted relative to each other (see Section 3.1); if the distribution of the circles varies from one force line to another, a longitudinal current is produced and is thus a certain analog of the diamagnetic current (this illustrative picture is indicated in [13, 19]).

It can be verified that expressions (2.24) and (2.25) lead automatically to the condition div $\mathbf{j} = 0$, which must be satisfied in stationary problems.

Relation (2.22) contains the electrostatic potential. Its distribution along a force line can be determined from the quasineutrality condition $n_e = n_i$. Since $n_e$ and $n_i$, as can be seen from their expressions in terms of the distribution function

$$n_{e,i} = \frac{2\sqrt{2}\,\pi B}{m_{e,i}^{3/2}} \int \frac{f_{e,i} d\varepsilon d\mu}{\sqrt{\varepsilon - \mu B \mp e\varphi}}, \tag{2.26}$$

depend on $\xi^1$, $\xi^2$, B, and $\varphi$, the potential determined from the quasineutrality condition will be a function of the force line and of the modulus of the magnetic field:

$$\varphi = \varphi(\xi^1, \xi^2, B). \tag{2.27}$$

We must make the following reservation: the functions $n_{e,i}$ regarded as functions of B and $\varphi$ can have several branches. This is possible in the case when the so-called Yushmanov potential

$$U_{Yu} \equiv \mu B \pm e\varphi, \tag{2.28}$$

which determines the motion of particles along a force line, has several minima. The particles trapped in each potential well can then each have their own distribution function, and a situation is possible wherein a relation of form (2.27) will be satisfied only piecewise (with respect to s). This is precisely the situation with ion confinement in an ordinary mirror trap, where the functions B(s) and $\varphi$(s) are of the form shown in Fig. 6, and the symmetry of the density (and potential) distributions relative to the z = 0 plane may become distorted. A corresponding asymmetry is realized in the outermost mirror cells of ambipolar traps.

Accurate to the reservation just made, it follows from (2.22) and (2.27) that

$$p_\parallel = p_\parallel(\xi^1, \xi^2, B), \quad p_\perp = p_\perp(\xi^1, \xi^2, B). \tag{2.29}$$

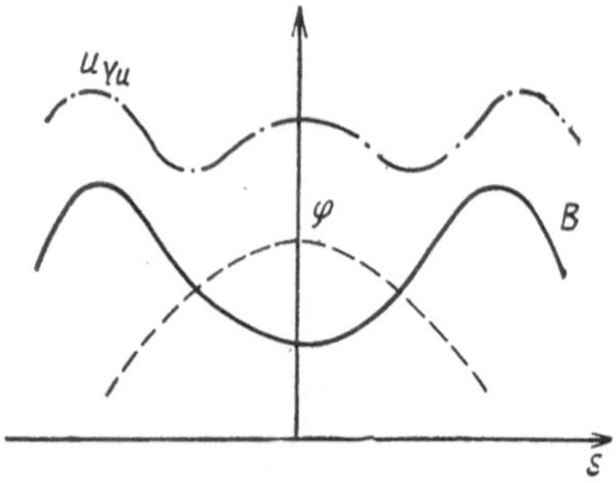


Fig. 6. Distributions of the magnetic field, the electrostatic potential, and the Yushmanov potential $U_{Yu} = \mu B + e\varphi$ along a force line of a mirror trap. The plot of $U_{Yu}$ is shown for a certain fixed value of $\mu$. Independent ion populations can exist near the minima of $U_{Yu}$.

Actually, the longitudinal equilibrium condition (2.23) enables us, knowing one of these functions, say $p_\perp$, to determine the other.

Relations (2.23) and (2.24) can also be obtained by a purely phenomenological approach, by specifying from the outset the plasma-pressure tensor in the form

$$p_{\alpha\beta} = \tau_\alpha \tau_\beta p_{\parallel} + (\delta_{\alpha\beta} - \tau_\alpha \tau_\beta)\, p_\perp$$

and writing the projections of the equilibrium equation

$$\frac{\partial p_{\alpha\beta}}{\partial x_\beta} = \frac{1}{c}\,[\mathbf{jB}]_\alpha$$

on the force line and on a plane perpendicular to the force line. As for relation (2.25), it results in this approach from the requirement that the divergence of the total current be zero:

$$\operatorname{div} \mathbf{j}_{\parallel} = -\operatorname{div} \mathbf{j}_\perp.$$

If no currents flow into the plasma through the end plugs of the trap (this is automatically satisfied, for example, if the plasma borders there on vacuum), relation (2.25) yields the following restriction on the admissible equilibrium configurations:

$$\int \frac{1}{B^2} \tau [\varkappa \nabla (p_{\parallel} + p_{\perp})] \, ds = 0, \tag{2.30}$$

where the integration is between the plasma boundaries. Since $p_{\parallel}$ and $p_{\perp}$ are uniquely related by Eq. (2.23), it can be assumed that only one of these functions, $p_{\perp}$ for the sake of argument, is involved in this restriction.

We have already mentioned that the restriction (2.30) is automatically satisfied if $p_{\perp}(\xi^1, \xi^2, B)$ is obtained by reduction, with the aid of (2.22''), from the microscopic drift equilibrium (2.21). If, however, the macroscopic approach is used as the basis, condition (2.30) becomes significant. The very fact that in the macroscopic approach to equilibrium one must impose the condition (2.30) indicates that by far not for every function $p_{\perp}(\xi^1, \xi^2, B)$ is it possible to find a function $f(I, \varepsilon, \mu)$ connected with $p_{\perp}$ by relation (2.22''). Moreover, it can be shown that even among those functions $p_{\perp}(\xi^1, \xi^2, B)$ that satisfy the condition (2.30) there are some that cannot be obtained from the drift distribution function with the aid of (2.22'').

In the important particular case of isotropic pressure, which is realized in the central magnetic mirror of an ambipolar trap and in a multimirror trap, relations (2.24) and (2.25) can be represented in the form

$$\mathbf{j}_{\perp} = c \frac{[\mathbf{B} \nabla p]}{B^2}, \tag{2.31}$$

$$\frac{\partial}{\partial s} \frac{j_{\parallel}}{B} = \frac{2c\tau}{B^2} [\varkappa \nabla p]. \tag{2.32}$$

It can then be shown (see [20]) that the condition (2.30) is equivalent to the statement that p is constant on the surface U = const (for a definition of U see Section 2.1).

In Section 3, in the analysis of the drift motion of the particle, we shall have to find the distortion, due to the finite plasma pressure, of the vacuum magnetic field in a system of the Yin–Yang type. All the most important singularities of these distortions are already manifested in the case of a low-pressure plasma, to which we confine ourselves now. The analysis is then simplified by the fact that we can first find the distribution of the plasma pressure in a vacuum magnetic field, and then, after calculating the plasma current from Eqs. (2.24) and (2.25), determine the corrections to the vacuum magnetic field.

The problem becomes even simpler when the paraxial approximation is considered. In the zeroth approximation in the parameter $r/\ell$ the equilibrium on each small segment of the plasma pinch can be regarded as if the magnetic field had only a z component, i.e., write the equilibrium equation in the form (we use cylindrical coordinates)

$$p_\perp(r_0, \psi_0, z) + B_z^2/8\pi = \mathcal{B}^2(z)/8\pi, \tag{2.33}$$

where $\mathcal{B}(z)$ is the vacuum field on the system axis. We allow for the possible presence of several branches of the dependence of $p_\perp$ on B, and therefore use, in (2.33), the notation $p_\perp(r_0, \psi_0, z)$ [rather than $p_\perp(r_0, \psi_0, B)$].

In the small-$\beta$ approximation ($\beta$, as usual, is the ratio of the plasma pressure to the magnetic-field pressure), we get from (2.33)

$$B_z = \mathcal{B} + \delta B_z,$$
$$\delta B_z = -\frac{4\pi p_\perp(r_0, \psi_0, z)}{\mathcal{B}(z)}. \tag{2.34}$$

On the whole, with allowance for small deviations from paraxiality, the magnetic-field structure can be obtained by double expansion in the parameters $\beta$ and $r/\ell$. The structure of the result is

$$B_z = \mathcal{B} + O(\beta) + O\left(\frac{r^2}{l^2}\right) + O\left(\beta\frac{r^2}{l^2}\right) + \ldots, \tag{2.35}$$

$$B_r = O\left(\frac{r}{l}\right) + O\left(\beta\frac{r}{l}\right) + \ldots,$$
$$B_\psi = O\left(\frac{r}{l}\right) + O\left(\beta\frac{r}{l}\right) + \ldots. \tag{2.36}$$

The term $O(\beta)$ in (2.35) is determined by relation (2.34), while the terms of order $O\left(\frac{r^2}{l^2}\right)$ in (2.35) and $O\left(\frac{r}{l}\right)$ in (2.36) are determined by relations (2.5). As for the terms of type $O\left(\beta\frac{r^2}{l^2}\right)$ and $O\left(\beta\frac{r}{l}\right)$, to find them we must take into account the longitudinal current ($j_\parallel \sim j_\perp r/\ell$), and the dependence of $j_\perp$ on z. The actual calculation of these terms turns out to be very cumbersome. An idea of the calculation can be gained from [21].

The next important task is to find the possible form of the functions $p_\perp(r_0, \psi_0, z)$ that determine the equilibrium. We consider the question for a low-$\beta$ plasma, and confine ourselves for simplicity to the case of isotropic pressure. It is implied here, of course, that longitudinal confinement is assured, either because the system is periodic along the z axis (multimirror trap) or because the ends of the trap constitute walls that are perpendicular to the magnetic field and reflect the particles ideally.

Under these conditions the constant-pressure surfaces coincide with the surfaces U = const, where U is given by Eq. (2.17). As $\beta \to 0$, as shown in Section 2.1, we have

$$U(r_0, \psi_0) = U_0 + \mathrm{const}\,\frac{r_0^2}{l^2} + O\left(\frac{r_0^4}{l^4}\right). \tag{2.37}$$

A dependence on $\psi_0$ appears here only in the terms $\sim(r_0/\ell)^4$. Accordingly, the lines $U(r_0, \psi_0)$ = const, accurate to terms $\sim(r_0/\ell)^2$, are circles.

To take into account the effect of the finite pressure, it is convenient to express U in the form (2.18) and substitute for $B_z$ expression (2.35). Obviously, U now takes the form

$$U = U_0 + O(\beta) + O\left(\frac{r_0^2}{l^2}\right) + O\left(\beta\frac{r_0^2}{l^2}\right) + O\left(\frac{r_0^4}{l^4}\right) + \ .\ .\ . \tag{2.38}$$

The term $O(\beta)$ can be easily calculated with the aid of Eq. (2.34) with $p_\perp = p(r_0, \psi_0)$, and turns out to be equal to

$$-4\pi p(r_0, \psi_0)\int \frac{dz}{\mathcal{B}^3(z)}, \tag{2.39}$$

while the term $O(r_0^2/\ell^2)$ coincides with const $(r_0^2/\ell)^2$ in (2.37). Cumbersome calculations show that not only the terms $O(r_0^4/\ell^4)$, but also the terms $O(\beta r_0^2/\ell^2)$ are now dependent on $\psi_0$.

The problem of equilibrium with allowance for finite $\beta$ was solved in the following manner. After specifying a certain function $p(r_0, \psi_0)$, one must use (2.38) to calculate $U(r_0, \psi_0)$ and then stipulate that the lines $U(r_0, \psi_0)$ = const coincide with the isobars $p(r_0, \psi_0)$ = const. Since the term $O(\beta)$, given by (2.39), is proportional to $p(r_0, \psi_0)$, it automatically meets the aforementioned condition. The form of the equilibrium is therefore determined by the last three terms. At $\beta \ll 1$, the principal of these three is the axisymmetric term $O(r_0{}^2/\ell^2)$, so that at $\beta < 1$ the isobars are close to circles, i.e., $p \cong p(r_0)$. The deviation of the isobars from circular at $\beta < r_0{}^2/\ell^2$ is determined by the term $O(r_0{}^4/\ell^4)$, and at $\beta > r_0{}^2/\ell^2$ by the term $O(\beta r_0{}^2/\ell^2)$. The value of this deviation can be estimated at $\delta r \sim (r_0/\ell)^2 + \beta$.

2.3. Equilibrium of Isotropic Plasma in a Long Magnetic Bottle. The magnetic-bottle system (probkotron) of an ambipolar trap has some specific features. These features, noted in [22], are due to its large length ($L_0 \gg \ell$). The plasma in the central magnetic bottle can be described in terms of the isotropic pressure p. Without going into structural details of the electrostatic barriers present in the end bottles, we consider below an idealized model in which it is assumed that the longitudinal confinement in the central magnetic bottle is ensured by two planes (1 and 4 in Fig. 7). These are located on the ends of the cell and reflect elastically the plasma particles that reach them. It is also assumed for simplicity that the magnetic-field force lines intersect these planes perpendicularly (in the paraxial approximation this means that the conditions $\mathcal{B}' = 0$ and $b = 0$ are satisfied on these planes).

The problem of determining the equilibrium configurations of an isotropic plasma consists of finding constant-pressure magnetic surfaces. In our formulation of the problem, wherein no currents flow out through the plugs of the magnetic bottle, the form of these surfaces, as discussed above, is determined by the requirement that the integral U be constant, and as $\beta \to 0$, accurate to small $\sim(r/\ell)^2$, the intersections of the magnetic surface with the equatorial plane are circles, $r_0 = \text{const}$.

At finite but low plasma pressure, the initial magnetic field is only slightly distorted. In a short magnetic bottle this leads to a weak change of the shape of the vacuum magnetic surfaces (see Section 2.2), a change of interest primarily in connection with the particle motion in the trap (see Section 3.3). In a long magnetic bottle, on the contrary, considerable distortions of the magnetic surfaces can occur even at $\beta \ll 1$ [22].

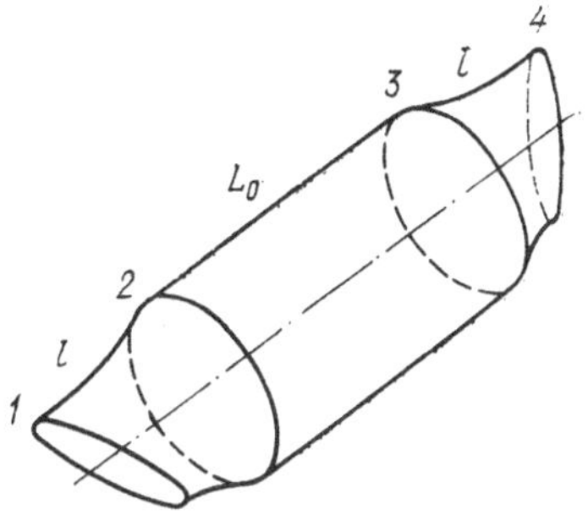


Fig. 7. Illustrating the problem of the equilibrium of an isotropic plasma in a long magnetic bottle.

We now estimate, with the aid of (2.32), the longitudinal current in sections 2 and 3 (Fig. 7), where the vacuum force lines are already straight, i.e., the current that flows into the uniform part of the trap from the mirrors. In a magnetic mirror we have $\varkappa \sim R/\ell^2$, $j_\perp \sim cp/RB$, where R is the characteristic radius of the plasma, and we obtain from (2.32)

$$j_\parallel \sim cp/lB. \tag{2.40}$$

This current, flowing parallel to the long part of the magnetic bottle, produces a transverse magnetic field:

$$B_\perp \sim j_\parallel R/c \sim \beta BR/l, \tag{2.41}$$

which causes the force line to deviate, over a length $L_0$, by a distance x in the radial and azimuthal directions:

$$x \sim \frac{B_\perp}{B} L_0 \sim R\frac{L_0}{L_c}, \quad L_c \equiv l/\beta. \tag{2.42}$$

Thus, if the length $L_0$ of the uniform part is of the order of or larger than $L_c$, one should expect a strong deviation of the constant-pressure surfaces from circular cylinders even at $\beta \ll 1$.

In this case, however, the force-line displacement, over the length $\ell$ of the mirror, from the initial position can be neglected, and by specifying the pressure distributions $p_2(r, \psi)$ and $p_3(r, \psi)$ at the entrances to the mirror in the sections 2 and 3, we can determine the pressure distribution in the mirrors, by assuming simply that p = const along the vacuum force lines. Calculation of $j_{\parallel}$ by Eq. (2.32) yields next the longitudinal-current distributions $j_2(r, \psi)$ and $j_3(r, \psi)$ in sections 2 and 3. The corresponding calculations for a paraxial mirror are given in Appendix 1. The result is

$$j_2 = ar \frac{\partial p_2}{\partial r} \sin 2\psi + (a \cos 2\psi + d) \frac{\partial p_2}{\partial \psi},$$
$$j_3 = ar \frac{\partial p_3}{\partial r} \sin 2\psi + (a \cos 2\psi - d) \frac{\partial p_3}{\partial \psi}, \qquad (2.43)$$

where $a$ and d are constants determined by the geometry of the magnetic field in the mirror. It is easily seen that the total current in sections 2 and 3 is zero, as it should be.

We now consider the variation of the current flowing out of the mirror along the long part of the trap. The curvature of the force lines is due here to the magnetic field of the longitudinal currents themselves, and is estimated from the condition that the displacement of a force line over a length $L_c$ would be equal to R: $\varkappa \sim R/L_c^2$. Now using (2.32), we obtain the change of $j_{\parallel}$ over the length $L_0$:

$$\Delta j_{\parallel} = L_0 \frac{cp}{RB} \frac{R}{L_c^2} \sim j_{\parallel} \frac{L_0}{(L_c/\beta)}. \qquad (2.44)$$

We assume hereafter that the following relation is valid:

$$L_0 \ll L_c/\beta \sim l/\beta^2. \qquad (2.45)$$

The variation of the longitudinal current between planes 2 and 3 can then be neglected, and the condition that $j_{\parallel}$ be constant along the force lines can be written in the form of the equation

$$\mathbf{B} \nabla j_{\parallel} = 0$$

or, in expanded form,

$$\frac{\partial j}{\partial z} = -\frac{B_r}{B_0} \frac{\partial j}{\partial r} - \frac{1}{r} \frac{B_\psi}{B_0} \frac{\partial j}{\partial \psi}, \qquad (2.46)$$

where $B_0$ is the z component of the magnetic field. At the accuracy required, it is simply equal to the vacuum magnetic field in the homogeneous part, so that for $j_{\parallel}$ we can write for brevity simply j. To determine the components of the transverse magnetic field, it suffices to assume that the vector $\mathbf{j}_\parallel$ is directed along the z axis; then,

$$B_r = r^{-1}\partial A_z/\partial\psi, \quad B_\psi = -\partial A_z/\partial r, \tag{2.47}$$

where $A_z(r, \psi, z)$ is the z component of the vector potential and satisfies the equation

$$\frac{1}{r}\frac{\partial}{\partial r} r \frac{\partial A_z}{\partial r} + \frac{1}{r^2}\frac{\partial^2 A_z}{\partial \psi^2} = -\frac{4\pi}{c} j. \tag{2.48}$$

Rewriting (2.46) in the form

$$\frac{\partial j}{\partial z} = \frac{1}{rB_0}\left(\frac{\partial A_z}{\partial r}\frac{\partial j}{\partial \psi} - \frac{\partial A_z}{\partial \psi}\frac{\partial j}{\partial r}\right) \tag{2.49}$$

and adding the condition that the pressure be constant along the force lines

$$\frac{\partial p}{\partial z} = \frac{1}{rB_0}\left(\frac{\partial A_z}{\partial r}\frac{\partial p}{\partial \psi} - \frac{\partial A_z}{d\psi}\frac{\partial p}{\partial r}\right), \tag{2.50}$$

we obtain the closed system of equations (2.48)-(2.50), which describes, together with the boundary conditions (2.43), the equilibrium configurations of a plasma in a long magnetic bottle.

As $\beta \to 0$, this system can be solved very simply. The magnetic-field perturbation can be neglected, putting $A_z = 0$, and Eqs. (2.49) and (2.50) reduce to the requirement that j and p be constant along the z axis. The latter means that $p_2 = p_3$ and $j_2 = j_3$, and it follows from (2.43) that (at $d \neq 0$) the pressure on the homogeneous section of the magnetic bottle does not depend on $\psi$:

$$p = p_0(r), \quad j = j_0 \equiv ar\frac{dp_0}{dr}\sin 2\psi. \tag{2.51}$$

This result signifies that in the paraxial approximation the surfaces on which the integral (2.17) is constant are circular cylinders in the region where the magnetic field is uniform.

In the case of large $\beta$, Eqs. (2.48)-(2.50) can apparently be solved only by numerical methods. No simple analytic solution can be obtained in the limit $L_0 \ll L_c$, when the distortion of the vacuum magnetic surfaces is small. To this end, we linearize the system (2.48)-(2.50) against the background of the solution (2.51), denoting the corrections to the pressure and current by $\delta p$ and $\delta j$:

$$p = p_0(r) + \delta p(r, \psi, z), \quad j = j_0(r, \psi) + \delta j(r, \psi, z).$$

It suffices then to substitute $j_0$ in the right-hand side of (2.48):

$$\frac{1}{r}\frac{\partial}{\partial r} r \frac{\partial A_z}{\partial r} + \frac{1}{r^2}\frac{\partial^2 A_z}{\partial \psi^2} = -\frac{4\pi}{c} ar \frac{dp_0}{dr} \sin 2\psi. \tag{2.52}$$

To solve this equation we must specify the boundary conditions on the wall of the chamber that surrounds the plasma filament. We consider for simplicity the case when the wall is far enough from the plasma boundary. The necessary conditions are then that $A_z$ be finite at zero and at infinity. The corresponding solution is

$$A_z = -\frac{4\pi a}{c} G(r) \sin 2\psi, \tag{2.53}$$

where

$$G(r) = \frac{1}{r^2}\int_0^r (r')^3 p_0(r')\, dr'. \tag{2.54}$$

In this approximation the function $A_z$ does not depend on z. From (2.49) and (2.50) we find that

$$\left.\begin{aligned} \delta j &= \delta j_2 + \frac{z + \frac{1}{2}L_0}{rB_0}\left(\frac{\partial A_z}{\partial r}\frac{\partial j_0}{\partial \psi} - \frac{\partial A_z}{\partial \psi}\frac{\partial j_0}{\partial r}\right), \\ \delta p &= \delta p_2 - \frac{z + \frac{1}{2}L_0}{rB_0}\frac{\partial A_z}{\partial \psi}\frac{dp_0}{\partial r}. \end{aligned}\right\} \tag{2.55}$$

Substituting $z = L_0$ in (2.55), we obtain the connection between $\delta p_2$ and $\delta p_3$ and between $\delta j_2$ and $\delta j_3$. The two equations (2.43) allow us next to determine all four functions $\delta p_2$, $\delta p_3$, $\delta j_2$, $\delta j_3$, while the second equation of (2.55) gives the distribution of the pressure perturbation in the uniform part of the mirror cell. It is easily verified that, in this case, $\delta p$ is obtained accurate to an arbitrary function of r, the choice of which allows us to impose on $\delta p$ the additional

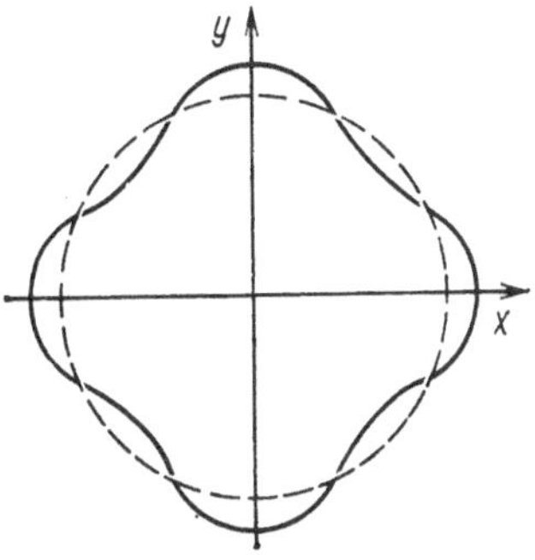


Fig. 8. Form of intersection of the magnetic surface p = const with the equatorial plane at β = 0 (dashed circle) and for finite β (solid line). When the sign of the coefficient of $\cos 4\psi$ in (2.57) is reversed, the entire picture is rotated 45° about the coordinate frame.

condition $\langle \delta p \rangle = 0$, where the angle brackets denote here and elsewhere the operation of averaging over the azimuth:

$$\langle \ldots \rangle = \frac{1}{2\pi} \int_0^{2\pi} \ldots d\psi. \tag{2.56}$$

As a result we have

$$\delta p(r, \psi, z) = \frac{4aL_0}{crB_0} \frac{dp_0}{dr} \left[ \frac{2z}{L_0} G \cos 2\psi + \frac{a}{4d} \left( 2G - r \frac{dG}{dr} \right) \cos 4\psi \right]. \tag{2.57}$$

Taking the smallness of $\delta p$ into account, we can write the equation for the magnetic surface p = const in the region of the uniform magnetic field in the form

$$r(\psi, z) = r_* - \left. \frac{\delta p}{dp_0/dr} \right|_{r=r_*}, \quad r_* = \langle r \rangle.$$

A qualitative form of the intersection of this surface with the equatorial plane of the trap (z = 0) is shown in Fig. 8. The degree of departure of the intersection of the magnetic surface from a circle can be described by the quantity $(r_{max} - r_{min})/r_*$; its order of magnitude is $L_0/L_c$.

## 3. Drift Motion of Charged Particles in Open Traps

3.1. Particle Motion in a Vacuum Magnetic Field. We begin the analysis with a case when the trap is not too long (a more precise criterion is given below), so that during one period of longitudinal oscillations the particle shifts little in azimuth, and the longitudinal adiabatic invariant (1.2) is conserved.

We use the curvilinear coordinates $\xi^1$, $\xi^2$, $\xi^3$ introduced in Section 2.2. In this case, $I = I(\xi^1, \xi^2, \varepsilon, \mu)$ and the drift surface is a magnetic surface defined by the equation $I(\xi^1, \xi^2, \varepsilon, \mu) = \text{const}$. The slow drift from one force line to another is described as a time variation of the coordinates $\xi^1$ and $\xi^2$. The "equations of motion" take the form [23]:

$$\dot{\xi}^1 = \frac{mc}{eBt_{\parallel}} \sqrt{\frac{g_{33}}{g}} \frac{\partial I}{\partial \xi^2}, \quad \dot{\xi}^2 = -\frac{mc}{eBt_{\parallel}} \sqrt{\frac{g_{33}}{g}} \frac{\partial I}{\partial \xi^1}, \tag{3.1}$$

where $g_{33}$ is the component of the metric tensor $g_{\alpha\beta}$; $g$ is the determinant of the matrix $g_{\alpha\beta}$, and the dot denotes differentiation with respect to time. The quantity $t_{\parallel}$ is the time of motion between stopping points (half the period of the longitudinal oscillations of the particle) and is equal to

$$t_{\parallel} = \int_{s_1}^{s_2} ds / |v_{\parallel}|.$$

Note that although B depends on $\xi^3$, the velocities $\dot{\xi}^1$ and $\dot{\xi}^2$ do not depend on $\xi^3$ (as should be the case), inasmuch as, by virtue of the equation div $\mathbf{B} = 0$ we have

$$\frac{\partial}{\partial \xi^3} \left( B \sqrt{\frac{g}{g_{33}}} \right) = 0.$$

In some cases it is important to find the (small) changes of $\xi^1$ and $\xi^2$ during a certain fraction of the period of the longitudinal oscillations of the particle. The corresponding equations are [23]

$$\Delta \xi^1 = \frac{mc}{eB} \sqrt{\frac{g_{33}}{g}} \left( \frac{\partial}{\partial \xi^2} \int_a^b v_{\parallel} ds - v_{\parallel} \frac{B_2}{B} \Big|_a^b \right),$$

$$\Delta \xi^2 = -\frac{mc}{eB}\sqrt{\frac{g_{33}}{g}}\left(\frac{\partial}{\partial \xi^1}\int_a^b v_\parallel ds - v_\parallel \frac{B_1}{B}\Big|_a^b\right), \Bigg\} \quad (3.2)$$

where $B_\alpha$ are the covariant components of the magnetic field, while $a$ and b are the integration limits. If necessary,* the integration with respect to ds is carried out in both directions, with reversal of the sign of $v_\parallel$. Relations (3.1) are obtained from (3.2) by integrating in the latter over the entire period of the longitudinal motion and dividing the result by the period $2t_\parallel$ of the longitudinal oscillations.

Bearing in mind the needs of actual experiments, we consider the forms of the drift surfaces in the case when the magnetic system has a symmetry of the Yin–Yang type (see Section 2.1), and neglecting initially the electric fields† and the magnetic-field perturbation due to the finite pressure of the plasma. In lieu of the coordinates $\xi^1$ and $\xi^2$, we use the polar coordinates $r_0$ and $\psi_0$ of the point of intersection of the force line with the z = 0 plane. We have

$$I = I(r_0, \psi_0, \varepsilon, \mu) = \sqrt{\frac{2}{m}}\int_{s_1}^{s_2}\sqrt{\varepsilon - \mu B}\, ds,$$

where the integration is along the force line between the particle stopping points (i.e., the zeros of the radicand). For untrapped particles (in the case of multimirror traps), the integration should be carried out over the corrugation period.

Since the drift surface becomes congruent with itself under the transformation $\psi \to \psi + \pi/2$, $z \to -z$, its intersection with the plane z = 0 is transformed into itself by the transformation $\psi \to \psi + \pi/2$. It follows hence that the dependence of I on $\psi_0$ should be periodic, with a period $\pi/2$. If $\psi$ is measured from one of the symmetry planes (as shown in Fig. 1a), the dependence will be even. We can accordingly represent I in the form of a series:

$$I(r, \psi, \varepsilon, \mu) = \sum_{k=0}^{\infty} I^{(k)}(r, \varepsilon, \mu)\cos 4k\psi \quad (3.3)$$

*For example, if it is required to find the displacement of a particle reflected from a magnetic mirror.

†That is, putting $\varphi = 0$ in (1.2).

(unless otherwise stipulated, here and below we use, for brevity, the symbols r and $\psi$ in lieu of $r_0$ and $\psi_0$). Clearly, this statement, which follows only from the symmetry properties of the system, remains valid when account is taken of both the plasma electric field and the magnetic-field perturbations by the plasma pressure.*

It is easy to verify that as $\beta \to 0$ (vacuum magnetic field) the expansion of I in powers of r contains only even powers of r. Assuming (in accordance with the actual situation) that r is small compared with the longitudinal scale $\ell$ of the magnetic-field variation, we confine ourselves to calculation of the first two terms of the expansion [O(1) and $O(r^2)$]. We use for this purpose Eq. (2.10) and change over in (1.2), with the aid of (2.11), to integration with respect to z. At the accuracy required, we have

$$I \simeq \sqrt{\frac{2}{m}}\left[\int\limits_{z_1}^{z_2}\sqrt{\varepsilon-\mu\mathcal{B}}\,dz\right.$$

$$\left.+r^2\int\limits_{z_1}^{z_2}\left(\sigma\sqrt{\varepsilon-\mu\mathcal{B}}-\frac{\mu}{2}\frac{\delta B}{\sqrt{\varepsilon-\mu\mathcal{B}}}\right)dz\right], \qquad (3.4)$$

where $z_1$ and $z_2$ are the roots of the solution $\varepsilon - \mu\mathcal{B}(z) = 0$. It can be seen from (2.10) and (2.11) that those terms of $\delta B$ and $\sigma$, which depend on $\psi$, meaning also the corresponding terms of I, are proportional to $\cos 2\psi$. Since, however, I as a function of $\psi$ has, according to the foregoing, a period $\pi/2$, the coefficient of the term $\cos 2\psi$ in the expression for I should vanish [this is readily verified by direct calculation using the even parity of the functions $\mathcal{B}(z)$ and $\Phi(z)$]. Thus, up to second order, inclusive, in the expansion of I in powers of r, the terms do not depend on $\psi$. This means automatically that the intersections of the drift surfaces with the z = 0 plane are circles, and the system has omnigenity properties. The explicit expression for I is

$$I(r,\psi,\varepsilon,\mu)=I_0(\varepsilon,\mu)+r^2 I_1(\varepsilon,\mu),$$

*It is understood that the plasma-production system also has the corresponding symmetry.

where

$$\left.\begin{aligned} I_0(\varepsilon,\ \mu) &= \sqrt{\frac{2}{m}}\int_{z_1}^{z_2}\sqrt{\varepsilon-\mu\mathcal{B}}\ dz, \\ I_1(\varepsilon,\ \mu) &= \int_{z_1}^{z_2}\left(\sigma\sqrt{\varepsilon-\mu\mathcal{K}}-\frac{\mu}{2}\ \frac{\delta B}{\sqrt{\varepsilon-\mu\mathcal{B}}}\right)dz. \end{aligned}\right\} \qquad (3.5)$$

The expression for the correction of fourth order in r is very unwieldy [to find it it is necessary, in particular, to retain fourth-order terms in the representation (2.5) for the vacuum field], and is not very informative. It is important also to note that it is precisely this correction which disrupts the omnigenity properties (since it depends on $\psi$). The general structure of the expression for I, accurate to fourth-order terms, is thus of the form

$$I = I_0(\varepsilon,\ \mu) + r^2 I_1(\varepsilon,\ \mu) + r^4 I_2(\psi,\ \varepsilon,\ \mu).$$

It is convenient here to break up the last term into two parts:

$$I_2' \equiv \langle I_2 \rangle$$

[the meaning of the angle brackets is explained by Eq. (2.56)] and

$$I_2'' = I_2 - \langle I_2 \rangle$$

and represent I in the form

$$I = I_0(\varepsilon,\ \mu) + r^2 I_1(\varepsilon,\ \mu) + r^4 I_2'(\varepsilon,\ \mu) + r^4 I_2''(\psi,\ \varepsilon,\ \mu). \qquad (3.6)$$

It is easy to verify† that

$$I_2'' = \tilde{I}_2''(\varepsilon,\ \mu)\cos 4\psi. \qquad (3.7)$$

The last term in (3.6), which contains the dependence on $\psi$, leads to violation of the omnigenity property: since the form of a drift surface that passes through a certain point ($r_*$, $\psi_*$) is determined by the equation I(r, $\psi$, $\varepsilon$, $\mu$) = I($r_*$, $\psi_*$, $\varepsilon$, $\mu$), it is obvious that at $I_2''$ = 0 the equation for the shell takes the form r = $r_*$, independently of $\varepsilon$ and $\mu$.

†This can be done by writing the general expression for the last term of (3.6) in Cartesian coordinates $r^4 I_2'' = Ax^4 + Bx^3y + Cx^2y^2 + Dxy^3 + Ey^4$, and using the symmetry properties of the drift surface.

At $I_2'' \neq 0$, the deviation of the line I = const from a circle can be easily found by taking the smallness of $I_2''$ into account:

$$\left.\begin{aligned} r(\psi, \varepsilon, \mu) &= r_* + \delta r(\psi, \varepsilon, \mu), \\ \frac{\delta r}{r_*} &= -r_*^2 \cos 4\psi \frac{\tilde{I}_2''(\varepsilon, \mu)}{2I_1(\varepsilon, \mu)}. \end{aligned}\right\} \tag{3.8}$$

The presence in this relation of $\varepsilon$ and $\mu$ as parameters points indeed to absence of omnigenity. The resultant "nonomnigenity" of the drift surfaces is of the order of $(r/\ell)^2$.

The last statement calls for, to be sure, significant refinement: it is valid only when $I_1(\varepsilon, \mu)$ does not vanish in the entire range of values of $\varepsilon$ and $\mu$ of physical interest. If this is not the case, the particles for which $I_1(\varepsilon, \mu) = 0$ will contain large radial random walks. Of course, relation (3.8) is not valid for them and must be replaced by the exact equation $r^4[I'_2(\varepsilon, \mu) + \tilde{I}''_2 \cos 4\psi] = \text{const}$. Since, generally speaking, $\tilde{I}''_2 \sim I'_2$, the radial "swing" of the surfaces for such particles is $\delta r \sim r$ (a situation may even arise in which these drift surfaces are not closed).

The region where the representation (3.8) is violated and $\delta r$ becomes large takes, in the space of the variables $\varepsilon$ and $\mu$, the form of a band of width $\delta\mu/\mu \sim (r/\ell)^2 \ll 1$ around the line on which $I_1(\varepsilon, \mu) = 0$. Narrowness of this band still does not guarantee smallness of the transverse losses, and therefore the magnetic system must be constructed such that the function $I_1(\varepsilon, \mu)$ does not vanish [since reversal of the sign of $I_1$ corresponds to a change in the direction of the particle drift around the magnetic axis, the condition $I_1(\varepsilon, \mu) \neq 0$ means that all particles drift in the same direction].

Everything said above concerning the form of the drift surfaces is applicable to the case of untrapped particles (in multimirror systems), the only exception being that the integration over z in the expression for I must in their case be carried out over the period of the magnetic system.

If the trap is so constructed that $I_1(\varepsilon, \mu) \neq 0$, the displacements from opposite mirrors, as shown above, are cancelled out with high accuracy, and the intersections of the drift surfaces differ little from circles. In this case it is quite feasible for the radial displacement on reflection from each of the mirrors to exceed $\delta r$.

In fact, it follows from (3.2) that on moving from the equatorial plane to the mirror and back the guiding center of the particle is displaced radially and in azimuth by

$$\Delta r_{\pm} \simeq \frac{2mc}{e\mathcal{B}r} \frac{\partial}{\partial\psi} I_{\pm}(r, \psi, \varepsilon, \mu) \tag{3.9}$$

and

$$\Delta\psi_{\pm} \approx \frac{2mc}{e\mathcal{B}r} \frac{\partial}{\partial r} I_{\pm}(r, \psi, \varepsilon, \mu),$$

where $I_{\pm}$ corresponds to the contribution made to the I from the "half" $z \gtrless 0$ of the mirror cell (so that $I = I_{+} + I_{-}$). We note that, owing to the symmetry of the trap, the functions $I_{+}$ and $I_{-}$ are invariant to the transformations $\psi \to -\psi$ and $\psi \to \pi - \psi$ (meaning that they are periodic with period $\pi$); furthermore,

$$I_{+}(r, \psi, \varepsilon, \mu) = I_{-}\left(r, \psi + \frac{\pi}{2}, \varepsilon, \mu\right). \tag{3.10}$$

Calculations similar to the foregoing show that the expansion of $I_{\pm}$ in powers of r contains only even powers; accurate to terms of second order, inclusive,

$$I_{\pm} = \frac{1}{2} I_0(\varepsilon, \mu) + \frac{r^2}{2} I_1(\varepsilon, \mu) \pm \cos 2\psi \frac{r^2}{2} I_1^{*}(\varepsilon, \mu), \tag{3.11}$$

where

$$I_1^{*} = \frac{\mathcal{B}(0)}{\sqrt{2m}} \int_0^{z_1} dz\, (\varepsilon - \mu\mathcal{B})^{-1/2} \mathcal{B}^{-3} [(\varepsilon F_{+} - \mu\mathcal{B}G_{+}) e^{-\Phi} - (eF_{-} - \mu\mathcal{B}G_{-}) e^{\Phi}], \tag{3.12}$$

while the functions F and G are given by

$$F_{\pm} = 4b^2 + \frac{1}{4}(\mathcal{B}')^2 \pm 2b\mathcal{B}',$$

$$G_{\pm} = 6b^2 + \frac{3}{8}(\mathcal{B}')^2 - \frac{1}{4}\mathcal{B}\mathcal{B}'' \pm (3b\mathcal{B}' - \mathcal{B}b').$$

In order of magnitude we have

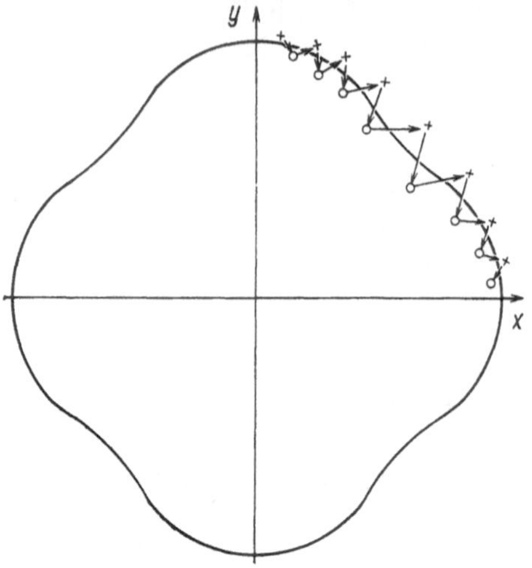


Fig. 9. "Splitting" of drift surface. The points and crosses (shown in the first quadrant only) indicate the crossing of equatorial plane of the trap by the guiding center of the particle as it moves in the positive and negative directions; solid line denotes section of drift surface for I = const.

$$\Delta r_{\pm} \sim r r_H / l, \quad \Delta \psi \sim r_H / l, \tag{3.13}$$

where $r_H$ is the Larmor radius of the particle. It can be seen that $\Delta r_{\pm} > \delta r$ at

$$r^2 < l r_H.$$

When this condition is satisfied, the radial random walks of the particles are due not to the difference between the intersection of the averaged drift surface and a circle, but to successive (opposite in sign) displacements on reflection from the mirrors. The shape of the curve formed by the points of successive crossings of the plane z = 0 by the particle is shown in Fig. 9. We shall refer below to the deviation of the particles from the averaged drift shell as "splitting of the surface" (see [14]).

It can be seen from the estimate (3.13) that the splitting is automatically small compared with the Larmor radius of the particle. This might seem to mean that the contribution of the splitting effect to the transverse transport is automatically small compared with the classical one. It will be shown below, however, that in those cases when the splitting has a narrow maximum as a function of $\varepsilon$ and $\mu$ it can lead to large radial losses (see Section 5.3).

Splitting of the drift surfaces occurs also for untrapped particles (in multimirror systems); a particle passing through successive mirrors is radially shifted in alternating directions.

Surface splitting is that microscopic mechanism that leads to the appearance of the longitudinal closure currents defined by the condition (2.25).

3.2. Effect of Electric Field on Particle Motion in Short Traps. We have dealt so far with the simplest situation, when the particles drift in a vacuum magnetic field. We consider now in succession the effects of an electric field and of finite plasma pressure. We begin with consideration of the role of an electric field. One reservation must be made right away. It can be seen from the estimate (3.13) that the longer the trap the smaller the azimuthal displacement of the particle in one transit and the better the condition of adiabatic invariance of the quantity I is formally satisfied. The electric field, however, gives rise to an electric drift around the magnetic axis, and the angle change $\Delta\psi$ due to this effect increases linearly with $\ell$. At sufficiently large $\ell$ the value of $\Delta\psi$ becomes comparable with unity, and the adiabatic invariance of I is violated. An analysis based on invariance of I is therefore valid only for sufficiently short traps (such that $\Delta\psi \ll 1$).

Let us investigate the influence of the potential on the form of the drift surfaces for a plasma confined in a magnetic system of Yin–Yang type, and confine ourselves to the paraxial approximation. If omnigenity were exactly satisfied, the potential would be, as shown in Section 2.2, a function of r and B [the possible presence of several branches of the function $\varphi(B)$ is not considered as yet].* Allowance for (small) drift surface nonomnigenity, however, will obviously cause $\varphi$ to take the form

*Recall that we are using for brevity the symbols r and $\psi$ in place of $r_0$ and $\psi_0$.

$$\varphi = \varphi_0(r, B) + \delta\varphi(r, \psi, B), \tag{3.14}$$

where $\delta\varphi$ is a small quantity [it will be shown below that $\delta\varphi \sim (r/\ell)^4$].

In the absence of $\delta\varphi$, the expression for I would take the form

$$I(r, \psi, \varepsilon, \mu) = \sqrt{\frac{2}{m}} \int_{z_1}^{z_2} \sqrt{\varepsilon - \mu B - e\varphi_0(r, B)} \frac{ds}{dz} dz \tag{3.15}$$

(we confine ourselves for brevity to positively charged particles). The dependence of I on $\psi$, which leads to deviation from omnigenity, enters here via the small quantities $\delta B$ and $\sigma$, which are defined by relations (2.10) and (2.11) and are of the order of $O(r^2)$. We have therefore, accurate to terms $r^2$,

$$I = I_0(r, \varepsilon, \mu) + \sqrt{\frac{2}{m}} \int_{z_1}^{z_2} \left[ \sigma \sqrt{\varepsilon - \mu\mathcal{B} - e\varphi_0(\mathcal{B}, r)} - \frac{\delta B}{2} \frac{\mu - e\dfrac{\partial\varphi_0}{\partial\mathcal{B}}}{\sqrt{\varepsilon - \mu\mathcal{B} - e\varphi_0(\mathcal{B}, r)}} \right] dz, \tag{3.16}$$

where

$$I_0 = \sqrt{\frac{2}{m}} \int_{z_1}^{z_2} \sqrt{\varepsilon - \mu\mathcal{B} - e\varphi_0(\mathcal{B}, r)}\, dz, \tag{3.17}$$

and $\mathcal{B} = \mathcal{B}(z)$. Repeating the arguments that follow Eq. (3.4), we find that, in the present case, I becomes dependent on $\psi$ only in terms of fourth order in r.

In view of its smallness, the contribution of $\delta\varphi$ to I need be taken into account only in the zeroth-order paraxial approximation; this leads to the following correction to I:

$$\delta I_\varphi(r, \psi, \varepsilon, \mu) = \frac{-1}{\sqrt{2m}} \int_{z_1}^{z_2} \frac{e\delta\varphi(r, \psi, \mathcal{B})\, dz}{\sqrt{\varepsilon - \mu\mathcal{B} - e\varphi_0(\mathcal{B}, r)}}.$$

We thus have the following approximate representation of I:

$$I = I_0(r, \varepsilon, \mu) + r^2 I_1(r, \varepsilon, \mu) + r^4 I_2(r, \psi, \varepsilon, \mu) + \delta I_\varphi(r, \psi, \varepsilon, \mu). \quad (3.18)$$

To assess $\delta\varphi$, we use the quasineutrality condition ($n_e = n_i$), in which we substitute expressions (2.26) for $n_e$ and $n_i$. Obviously, for $\varphi$ in the form (3.14) and I in the form (3.18) the equation for $n_{e,i}$ can be schematically represented as

$$n_{e,i} = n_{e,i}^{(0)}(r, B, \varphi_0) + \left(\frac{r}{l}\right)^4 F_{e,i}(r, \psi, B) + \hat{L}_{e,i}(\delta\varphi),$$

where $F_{e,i}$ are certain functions of the order of unity, and $\hat{L}_{e,i}(\delta\varphi)$ are linear functionals of $\delta\varphi$, and are of zeroth order in the parameter $r/\ell$. The potential $\varphi_0$ is determined from the condition $n_e^{(0)}(r, B, \varphi_0) = n_i^{(0)}(r, B, \varphi_0)$, and the correction $\delta\varphi$ from the condition $(r/\ell)^4(F_i - F_e) = \hat{L}_e(\delta\varphi) - \hat{L}_i(\delta\varphi)$. It is therefore clear that

$$\delta\varphi \sim (r/l)^4\varphi_0; \quad (3.19)$$

i.e., according to (3.18), the $\psi$-dependent correction in the expression for I remains of the order of $(r/\ell)^4$ also when the electric field is taken into account.

An important new factor is that now the term $I_0$ also becomes dependent on r [via the term $e\varphi_0(\mathscr{B}, r)$ in the radicand of (3.17)]. This leads to a change in the estimate (3.8) for $\delta r$. We now have

$$\delta r \simeq -\left.\frac{r^4(I_2 - \langle I_2\rangle) + \delta I_\varphi - \langle \delta I_\varphi\rangle}{\frac{\partial I_0}{\partial r} + 2rI_1}\right|_{r=r_*}. \quad (3.20)$$

The term $\partial I_0/\partial r$ in the denominator of (3.20) becomes predominant at

$$re\left|\frac{\partial\varphi_0}{\partial r}\right| \gg \varepsilon\left(\frac{r}{l}\right)^2.$$

If, as is frequently the case, $e|\partial\varphi_0/\partial r| \sim T/R$, where T and R are, respectively, the plasma temperature and radius, this condition is automatically satisfied for particles with $\varepsilon \sim T$. Using the rough estimates $\partial I_0/\partial r \sim I_0/R$, $I_2'' \sim I_0/\ell^4$, and $\delta I_\varphi \sim I_0(r/\ell)^4$, we find that the order of magnitude of $\delta r$ is

$$\frac{\delta r}{r_*} \sim \left(\frac{r_*}{l}\right)^4; \tag{3.21}$$

i.e., an electric field decreases the "nonomnigenity" of the drift shells.

The physical reason for the decrease of the radial deflections of the particle is that an increase of the transverse electric field increases the electric drift velocity in the azimuthal direction and, accordingly, "smoothens" the drift surfaces.

We now take into account the possible appearance of several branches of $\varphi(B)$. This means actually that $\varphi(B)$, in all the equations, is replaced by a function $\varphi(z)$ having, generally speaking, a symmetry different from that of $B(z)$. A situation with a multiply valued function $\varphi(B)$ is encountered, in particular, in the end magnetic bottles of an ambipolar trap, which contain "central plasma" electrons and ions trapped between the electrostatic potential barrier on the one hand, and the internal mirror of the outermost bottle, on the other (see Figs. 5 and 6). These particles oscillate within the confines of "half" the bottle, and the aforementioned cancellations of the $\psi$-dependent contributions [of order $(r/\ell)^2$] to I from field-reversed mirrors does not take place in them. This is why the nonomnigenity of the drift shells of such particles becomes, generally speaking, appreciable for such particles.

3.3. Allowance for the Finite Plasma Pressure. Plasma pressure distorts the vacuum magnetic field and, as a consequence, alters the shape of the drift surfaces. The perturbation of the magnetic field is determined in the principal (zeroth) paraxiality approximation by relation (2.34):

$$\delta B_z(r, z) = -\frac{4\pi p_\perp(r, z)}{\mathscr{B}(z)}$$

(we write r in lieu of $r_0$). We note that $\delta B_z$, expressed in the form (2.34), is a correction to the magnetic field on the unperturbed magnetic surface.

The perturbations, due to the finite $\beta$, of the r and $\psi$ components of the magnetic field, are of the order of $\beta r/\ell$, and allowance for the next higher paraxial approximation in the expression for $\delta B_z$ leads to the appearance of a correction of the order of $\beta(r/\ell)^2$ [see (2.35) and (2.36)]. It can therefore be stated, starting with Eq. (1.2) for I, that the correction to I for the finite $\beta$, which we designate $\delta I_\beta$, has the following structure:

$$\delta I_\beta \sim \beta[1 + O(r^2/l^2)]. \tag{3.22}$$

The first term of this expansion is very easy to find. It suffices to replace B in (1.2) by $\mathscr{B}(z) + \delta B_z(r, z)$ and ds by dz, and expand in powers of $\delta B_z$. The result is

$$\delta I_\beta(r, \varepsilon, \mu) = \frac{2\sqrt{2}\,\pi\mu}{\sqrt{m}} \int_{z_1}^{z_2} \frac{p_\perp(r, z)\,dz}{\mathscr{B}(z)\sqrt{\varepsilon - \mu\mathscr{B}(z) - e\varphi(r, z)}}. \tag{3.23}$$

In the case of isotropic pressure, $p_\perp$ is replaced by p(r),* and the latter can be taken outside the integral sign, so that the correction $\delta I_\beta$ acquires a particularly simple and universal form.

The correction (3.23) does not depend on the angle $\psi$, and as a function of the radius its variation scale is of the order of the plasma radius. The derivative $\partial \delta I_\beta / \partial r$ is proportional to the velocity of the gradient drift due to the magnetic-field inhomogeneity brought about by the presence of the plasma. The value of $\delta I_\beta$ exceeds already at $\beta > (r/\ell)^2$ the axisymmetric term $r^2 I_1(\varepsilon, \mu)$ [see (3.5)] contained in the expression for I in an unperturbed magnetic field.

Analysis of the second-order approximation in the parameter $r/\ell$ reveals the following important circumstance: even if the function $p_\perp$ can be written in the form $p_\perp = p_\perp(r, B)$, the $\psi$-dependent contributions from field-reversed mirrors do not cancel one another in the terms of order $\beta(r/\ell)^2$ in the entire confinement region. The relevant calculations are very cumbersome and will not be presented here.

The appearance of an angle-dependent term $\sim\beta(r/\ell)^2$ in I leads, generally speaking, to a change of the quantity $\delta\varphi$ introduced with the aid of Eq. (3.14). Arguments perfectly analogous to those set forth in connection with (3.19) show that at $\beta \gtrsim (r/\ell)^2$ we have

*To avoid confusion, we recall that r denotes here the point of intersection of a force line with the equatorial plane of the trap ($r = r_0$).

$$\delta\varphi \sim \beta (r/l)^2.$$

3.4. Drift Trajectories. Summing the results of the three preceding sections and introducing a more convenient notation, we express I in the form

$$I(r, \psi, \varepsilon, \mu) = I_0(r, \varepsilon, \mu) + \delta I(r, \psi, \varepsilon, \mu), \tag{3.24}$$

and assume that

$$\langle \delta I \rangle = 0$$

(this can always be achieved by redefining $I_0$). In contrast to Eqs. (3.5) and (3.16), $I_0$ stands here for all the terms independent of $\psi$, regardless of their origin. The quantity $I_0$ can be represented as a sum of three terms

$$I_0 = I_{0V} + I_{0E} + I_{0P},$$

where $I_{0V}$ describes the particle drift in a vacuum magnetic field and is determined by a formula that follows from (3.16):

$$I_{0V} = \sqrt{\frac{2}{m}} \int_{z_1}^{z_2} \left[ \sigma \sqrt{\varepsilon - \mu \mathcal{B} - e\varphi_0(\mathcal{B}, r)} - \frac{\delta B}{2} \frac{\mu - e\dfrac{\partial \varphi_0}{\partial \mathcal{B}}}{\sqrt{\varepsilon - \mu \mathcal{B} - e\varphi_0(\mathcal{B}, r)}} \right] dz; \tag{3.25}$$

$I_{0E}$ describes the electric drift and is determined by (3.17):

$$I_{0E} = \sqrt{\frac{2}{m}} \int_{z_1}^{z_2} \sqrt{\varepsilon - \mu \mathcal{B} - e\varphi_0(\mathcal{B}, r)}\, dz, \tag{3.26}$$

while $I_{0P}$ describes the gradient drift due to the distortion of the vacuum magnetic field by the plasma pressure,

$$I_{0P} = \frac{2\sqrt{2}\,\pi\mu}{\sqrt{m}} \int_{z_1}^{z_2} \frac{p_\perp(r, z)\, dz}{\mathcal{B}^2(z)\sqrt{\varepsilon - \mu \mathcal{B}(z) - e\varphi(r, z)}}. \tag{3.27}$$

Expressions (3.25)-(3.27) are for ions. For electrons it suffices to replace e in them by $-e$.

Even in a very weak electrostatic field, $|\partial\varphi/\partial r| \gtrsim (T/e) \times (R/\ell)^2$, and for a very small $\beta$, $\beta \gtrsim (R/\ell)^2$, each of the terms of $I_0E$ and $I_0P$ becomes larger than $I_0V$, and the radial dependence of $I_0$ is determined by the sum of the terms $I_0E$ and $I_0P$.

The correction $\delta I$ also receives contributions from several terms. At $\beta < (r/\ell)^2$, the most important of these terms is given by (3.7), while at $\beta > (r/\ell)^2$ the most important is the term $\sim\beta(r/\ell)^2$ from (3.22). Both these terms have a period $\pi/2$ as a function of the angle $\psi$. If, however, the system does not have exactly the symmetry (2.6), or if it contains "Yushmanov" particles (see Section 2.2), then $\delta I$ acquires a term with a period $\pi$. Incidentally, we shall not consider this possibility below.

Bearing in mind that $\delta I$ is small, we introduce a parameter $\alpha$ that characterizes this smallness:

$$\delta I \sim \alpha I_0. \tag{3.28}$$

We proceed to consider the drifting trajectories proper, by which we mean the intersections of the drift shells with the plane $z = 0$, i.e., the curves $r = r(\psi, \varepsilon, \mu)$, defined by the equation $I(r, \psi, \varepsilon, \mu) = \text{const}$. The form of the trajectories can be found by expanding $I_0$ in (3.24) in powers of the small deviation $\delta r = r_* - r$ (where $r_* = \langle r \rangle$) accurate to first-order terms:

$$r(\psi, \varepsilon, \mu) = r_* - \left.\frac{\delta I}{\partial I_0/\partial r}\right|_{r=r_*}. \tag{3.29}$$

It follows from (3.29) that the trajectories of most particles deviate from circles by $\delta r \sim \alpha r$.

A case requiring special attention is when $\partial I_0/\partial r = 0$ and Eq. (3.29) no longer holds. The relation $\partial I_0/\partial r = 0$ defines at each point $r$ a certain curve on the $(\varepsilon, \mu)$ plane. Let the particle be described at the initial instant by coordinates $r^{(1)}$ and $\psi^{(1)}$ and by values of $\varepsilon$ and $\mu$ close to one of the points of the curve mentioned above, plotted for $r = r^{(1)}$. Introducing the deviation $x = r - r^{(1)}$ and expanding $I_0$ in terms of $\Delta\varepsilon$, $\Delta\mu$, and $x$ ($\Delta\varepsilon$ and $\Delta\mu$ are measured from the nearest point on the curve) accurate to terms of second order of smallness, we get

$$\frac{1}{2}\frac{\partial^2 I_0}{\partial r^2}x^2 + \left[\frac{\partial^2 I_0}{\partial r\partial\mu}\Delta\mu + \frac{\partial^2 I_0}{\partial r\partial\varepsilon}\Delta\varepsilon\right]x + \delta I(\psi) - \delta I(\psi^{(1)}) = 0 \tag{3.30}$$

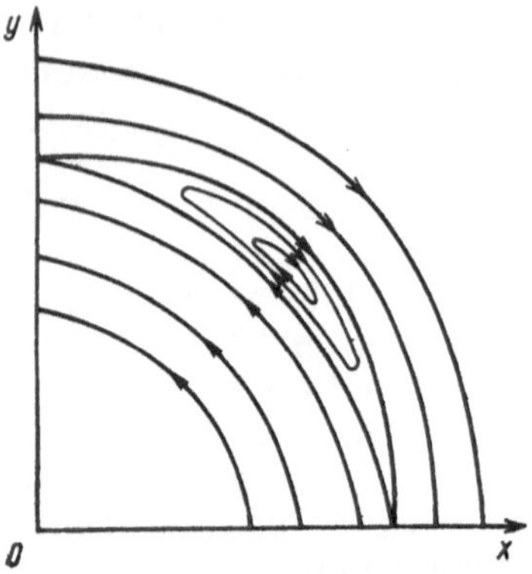


Fig. 10. Intersections of the drift surfaces and the z = 0 plane. Each curve corresponds to the same values of ε and μ, but to different values of I. The arrows point in the direction of the motion. The picture is periodic with a period π/2.

(we omit for brevity the arguments $r^{(1)}$, ε, μ in $I_0$, δ I, and in the derivatives of $I_0$). The form of the trajectory (3.30) varies significantly, depending on the expression in the square brackets of (3.30). If this expression is small, the range of angles in which the particles can move is limited ("trapped" particles), and the trajectories have a distinctive banana shape (Fig. 10). If it is large, however, the particles pass around the magnetic axis of the system ("untrapped" particles).

Numerical calculations of the drift trajectories for a number of nonaxisymmetric open traps are given in [14] and in [24, 25].

Let us discuss separately the important case when a long section of a uniform magnetic field is present between two mirrors (the model of the central magnetic bottle of an ambipolar trap). The main contribution to $I_0$ is made then just by the uniform section, and the following relations are approximately valid:

$$I_{0E} \simeq L_0 \sqrt{\frac{2}{m}} \sqrt{\varepsilon - \mu B_0 - e\varphi_0(r)}, \qquad (3.31')$$

$$I_{0P} \simeq L_0 \frac{2\sqrt{2}\,\pi\mu p(r)}{\sqrt{m}\,B_0 \sqrt{\varepsilon - \mu B_0 - e\varphi_0(r)}}, \tag{3.31''}$$

where $L_0$ is the length of the uniform section, and $B_0$ is the vacuum magnetic field strength in this section. The condition $\partial I_0/\partial r = 0$ then reduces to

$$\mu \frac{4\pi}{B_0} \frac{\partial p}{\partial r} - e\frac{\partial \varphi_0}{\partial r}\left(1 + \frac{\mu B_0}{\varepsilon - \mu B_0 - e\varphi_0} \frac{2\pi p}{B_0^2}\right) = 0.$$

This condition is satisfied for "thermal" particles, i.e., particles with $\varepsilon \sim \mu B_0 \sim T$, at $e|\partial\varphi_0/\partial r| \sim \beta T/R$. At $\beta \ll 1$ the second term in the round brackets can therefore be neglected, and the condition $\partial I_0/\partial r = 0$ takes the particularly simple form

$$\mu = \mu_*(r) \equiv \frac{e}{4\pi} \frac{B_0^2 (\partial\varphi_0/\partial r)}{\partial p/\partial r}. \tag{3.32}$$

This equation is for ions, and is valid for electrons if e is replaced in it by −e. It can be seen from (3.32) that "banana" trajectories can appear simultaneously only for one particle species [depending on the sign of the ratio $(\partial\varphi_0/\partial r)/(\partial p/\partial r)$].

The boundary between the untrapped and trapped particles is determined in this case by the deviation of $\mu$ from $\mu^*$. The particles are untrapped for

$$|\Delta\mu| = |\mu - \mu_*| > \Delta\mu_0$$
$$\equiv \frac{1}{|\partial^2 I_0/\partial r\partial\mu|}\left\{2\frac{\partial^2 I_0}{\partial r^2}\left[\max_{\psi} \delta I(\psi) - \min_{\psi} \delta I(\psi)\right]\right\}^{1/2}$$

(we assume $\partial^2 I_0/\partial r^2 > 0$, for the sake of argument) and trapped for the opposite inequality.

In order of magnitude we have

$$\Delta\mu_0 \sim \alpha^{1/2}\mu_*.$$

From this estimate we obtain easily the characteristic width of the banana

$$\Delta x_b \sim \alpha^{1/2} R.$$

We denote the time of encircling of the bulk of the particles (i.e., of particles with $\Delta\mu/\mu \sim 1$) around the magnetic axis by $\tau_D$ (the subscript D indicates that this motion is due to drift). With decreasing $\Delta\mu$ the time of motion increases in proportion to $(\partial I_0/\partial r)^{-1}$, and becomes infinite on the separatrix between the untrapped and trapped particles. For trapped particles not too close to the separatrix the period of motion along a banana trajectory is of the order of $\alpha^{-1/2}\tau_D$. Outside the separatrix, at $\Delta\mu \gg \Delta\mu_0$, the period of particle motion is $\tau_D\mu/\Delta\mu$, and the maximum deviation of the trajectory from a circle is $\alpha R\mu/\Delta\mu$.

In the presence of a long section, the following expression is valid for the particle transit time between mirrors:

$$t_{\parallel} \simeq \frac{L_0\sqrt{\frac{m}{2}}}{\sqrt{\varepsilon - \mu B_0 - e\varphi_0}}. \qquad (3.33)$$

3.5. Particle Motion in Long Traps. As noted in Section 3.2, the presence of electric drift causes the angle $\Delta\psi$ through which the guiding center of the particle is rotated during one transit from mirror to mirror to increase linearly with the length L of the trap. Varying likewise linearly with L is the contribution made to $\Delta\psi$ by the gradient drift due to the magnetic-field distortion by the plasma pressure. A situation with $\Delta\psi$ larger than unity is therefore possible if the trap is long, and description of the particle motion in terms of the adiabatic invariant I becomes unsuitable. This situation obtains realistically for ions in the central magnetic bottle of an ambipolar trap. For electrons, as a result of the much smaller value of $t_{\parallel}$ (whereas the azimuthal drift velocity is comparable with that of the ions), the condition $\Delta\psi \ll 1$ is usually satisfied.

A characteristic property of the central magnetic bottle is the presence of a long section of uniform magnetic field between the mirrors (see Fig. 5b). We assume that the electrostatic potential and the plasma pressure have axisymmetric distributions in this section:

$$\varphi = \varphi_0(r), \quad p = p(r). \qquad (3.34)$$

It is convenient to describe the particle motion in this case by fixing the coordinates $r_m$, $\psi_m$, m = 0, 1, 2, ... of the points of successive crossings of the equatorial plane of the trap by the particle leading center. The law of motion comprises, in this case, the transformation that determines the values of $r_m$ and $\psi_m$ in terms of their values $r_{m-1}$ and $\psi_{m-1}$ in the preceding step.

The change of the coordinate r between two crossings of the equatorial plane is due to passage of the particle through a region of nonaxisymmetric mirrors. The radial displacement $\Delta r$ is expressed in terms of the adiabatic invariant calculated for the "half" of the magnetic bottle in accordance with Eq. (3.9). The contribution made to $I_{\pm}$ by the uniform-magnetic-field region in which the potential and pressure distributions are axisymmetric does not depend on $\psi$, so that to calculate $\Delta r_{\pm}$ it suffices to take into account in $I_{\pm}$ only the contribution from the region of the magnetic mirror. In this case, it is necessary to replace $\varepsilon$ by $\varepsilon - e\varphi_0(r)$ in expression (3.11), written without allowance for the electric field. On the other hand, allowance for the distribution of the vacuum magnetic field of the mirrors at $\beta \ll 1$ introduced only a small ($\sim\beta$) correction to $r_{\pm}$, and can therefore be neglected. As a result we find that $\Delta r_{\pm} = \pm\Delta r$, where

$$\Delta r \equiv -\frac{2mcr}{eB_0} I_1^*(r, \varepsilon, \mu)\sin 2\psi, \tag{3.35}$$

and $I_1^*(r, \varepsilon, \mu)$ is given by Eq. (3.12), in which the integration is carried out over the region of the inhomogeneous magnetic field (and $\varepsilon$ is replaced by $\varepsilon - e\varphi_0$). The magnetic field in the equatorial plane is replaced in Eq. (3.35) by the value of $B_0$ on the uniform section of the magnetic bottle. The use of (3.35) implies that the mirrors themselves are not very long (as is always the case in real installations), so that the change of the angle $\psi$ on reflection from a mirror is small compared with unity.

In the case of nonparaxial magnetic field, in view of the symmetry (3.10) of the functions $I_+$ and $I_-$, the expressions for $\Delta r_{\pm}$ can be written in the following form:

$$\Delta r_{\pm} = \sum_{k=1}^{\infty} (\pm 1)^k a_k(r, \varepsilon, \mu)\sin 2k\psi. \tag{3.36}$$

The coefficient $a_1$, equal to the factor preceding $\sin 2\psi$ in (3.35), will hereafter be called the reflection coefficient; we omit its subscript, $a_1 \to a$ ; we shall use for $\Delta r_{\pm}$ the paraxial approximation (3.35) and write

$$\Delta r_{\pm} = \pm a \sin 2\psi. \tag{3.37}$$

In order of magnitude we have [see (3.13)]

$$a \sim r_H \frac{R}{l}.$$

We now derive the transformation, referred to above, for $r_m$ and $\psi_m$. It is actually convenient here to use only those transits in which the particle crosses the equatorial plane as it moves in a positive direction along the z axis (i.e., over a time $2t_{\|}$).

Let a particle with $v_z > 0$ be located at the initial instant at point $r_1$, $\psi_1$ of the equatorial plane. After passing through half the magnetic bottle and having a phase $\psi_1 + {}^1/_2\Delta\psi$, it then enters the first mirror and undergoes a radial displacement equal to $a \sin(2\psi_1 + \Delta\psi)$. Next, moving along the uniform part, it enters the second mirror with a phase $\psi_1 + {}^3/_2\Delta\psi$, and is displaced radially by an amount $-a \sin(2\psi_1 + 3\Delta\psi)$. Finally, after overcoming also half the magnetic-bottle length, it reaches the equatorial plane, in which case its phase is increased by ${}^1/_2\Delta\psi$. As a result, the coordinates $r_2$, $\psi_2$ of the new intersection point are

$$r_2 = r_1 + a(r_1)\sin[2\psi_1 + \Delta\psi(r_1)] - a(r_1)\sin[2\psi_1 + 3\Delta\psi(r_1)] + O(a^2), \tag{3.38}$$

$$\psi_2 = \psi_1 + 2\Delta\psi(r_1) + O(a) \tag{3.39}$$

(the arguments $\varepsilon$ and $\mu$ in $a$ and $\Delta\psi$ are left out for brevity).

If $\Delta\psi$ does not satisfy any special conditions, the particle undergoes, after successive reflections from the mirrors, more-or-less random displacements along the radius,* by ~ $a$ in one direction or another, and the resultant radial displacement is small (of the order of $a$). This statement is not valid only for special groups of particles, called "resonant" [15], for which $\Delta\psi$ is close to an odd number of right angles:

$$\Delta\psi(r, \varepsilon, \mu) = (2k+1)\frac{\pi}{2} + \alpha, \quad |\alpha| \ll 1, \tag{3.40}$$

where k is an integer. According to condition (3.40), a particle reflected from one mirror reaches a geometrically

*A more accurate description is given below.

equivalent point on the other mirror and the radial displacements due to successive reflection are additive.*

Since $\Delta\psi$ depends on r, it follows that, given $\varepsilon$ and $\mu$, the condition (3.40) can be satisfied only near definite discrete values $r = R_k(\varepsilon, \mu)$, determined from the condition

$$\Delta\psi(R_k, \varepsilon, \mu) = \frac{\pi}{2}(2k+1). \tag{3.41}$$

It therefore does not follow from the foregoing that a resonant particle moves continuously along the radius in one direction: departure from the resonant value of r leads to violation of condition (3.40) and, generally speaking, to a change in the direction of the motion.

It can be seen from condition (3.40) that if the points of intersection of the resonant particle and the equatorial plane are marked at intervals $4t_{\|}$, a quasicontinuous trajectory is produced. The transformation that determines the position of the neighboring points on this trajectory is obviously of the form

$$r_{m+2} - r_m = 4(-1)^k a \cos 2\psi_m, \tag{3.42}$$

$$\psi_{m+2} - \psi_m = 4\left[\Delta\psi - \frac{\pi}{2}(2k+1)\right]. \tag{3.43}$$

An inessential term $2\pi$ is left out of the right-hand side of the last equation. We shall find it convenient to introduce a (small) departure of the particle from the resonant surface (3.41):

$$x = r - R_k(\varepsilon, \mu).$$

Equation (3.40) then takes the form

$$\Delta\psi - \frac{\pi}{2}(2k+1) = x\frac{\partial\Delta\psi}{\partial r}.$$

The left-hand sides of (3.42) and (3.43) determine the change of the particle position in the equatorial plane after a time $4t_{\|}$. Dividing relations (3.42) and (3.43) by $4t_{\|}$ and transforming in the left-hand sides from finite differences

*In the more general case, when the displacement is given by (3.36), the resonance condition takes the form $\ell\Delta\psi = (2k+1)\pi/2$, where k and $\ell$ are integers.

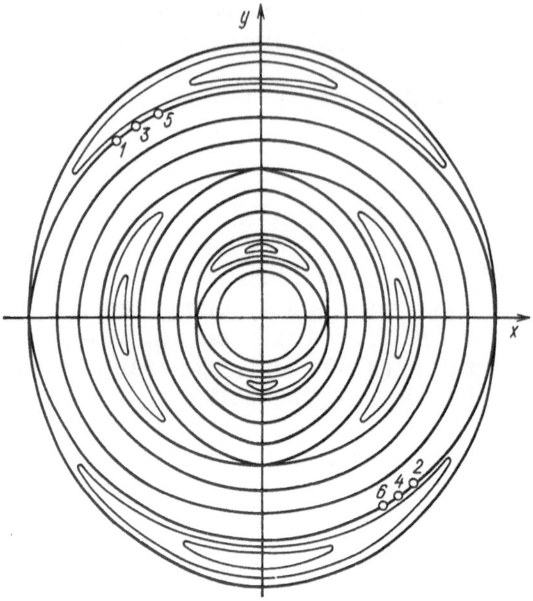


Fig. 11. Family of resonant trajectories in the equatorial plane. It shows the successive points of intersection of a drift trajectory of a particle with the equatorial plane at time intervals $2t_{\parallel}$.

to derivatives, we obtain the equations of motion of the particle along a resonant trajectory:

$$\frac{dx}{dt} = \frac{a}{t_{\parallel}}(-1)^k \cos 2\psi, \quad \frac{d\psi}{dt} = \frac{x}{t_{\parallel}} \frac{\partial \Delta\psi}{\partial r}. \qquad (3.44)$$

Integrating, we find that the shape of the trajectory is given by the equation

$$\mathcal{Y} \equiv (-1)^k \frac{x^2}{a(R_k)} \frac{\partial \Lambda\psi}{\partial r}\Big|_{r=R_k} - \sin 2\psi = \text{const}. \qquad (3.45)$$

Values $|\mathcal{Y}| < 1$ correspond to particles that follow "banana" trajectories, and $|\mathcal{Y}| > 1$ to particles whose trajectories encircle the magnetic axis of the system (Fig. 11).

With the aid of (3.44) it is easy to estimate the lengths $\Delta x_r$ of the radial random walks of resonant particles (i.e., of particles for which $|\mathcal{Y}| \sim 1$):

$$\Delta x_r \sim |a/(\partial\Delta\psi/\partial r)|^{1/2}, \tag{3.46}$$

and their period of motion along the trajectory

$$\tau_r \sim t_{\parallel}\left|\frac{\Delta x_r}{a}\right| \sim t_{\parallel}\left|a\frac{\partial\Delta\psi}{\partial r}\right|^{-1/2}. \tag{3.47}$$

Obviously, the distance between neighboring resonant surfaces is equal to $\pi|\partial\Delta\psi/\partial r|^{-1}$ (we assume that $\Delta\psi \gtrsim 1$). It was implied above that this distance is large compared with the width $\Delta x_r$ of the resonant bananas (see Fig. 11), i.e., that the following condition holds:

$$a\,|\partial\Delta\psi/\partial r| \ll 1. \tag{3.48}$$

If the inverse inequality holds, the resonances overlap and a qualitatively new phenomenon sets in – stochastic diffusion. It is considered in Section 5.5.

We now examine in greater detail the motion of nonresonant particles for which $|\Delta\psi - (2k+1)\pi/2| \sim 1$ [validity of (3.48) is assumed as before]. We shall show that it is likewise possible to introduce for them the concept of a smooth trajectory, on which representative points lie in the $r, \psi$ plane. It will be seen from the sequel that the radial random walks of these particles are of the order of $a$, so that the change of $\Delta\psi(r, \varepsilon, \mu)$ due to the fact that $r$ is not constant during the course of motion can be neglected.

We carry out in succession the transformations (3.38) and (3.39) and track the variation of the distance between the starting point $r_1$, $\psi_1$ and the terminal point $r_n$, $\psi_n$ as a function of the number $n$ of the iterations. As already noted, the quantity $r_n - r_1 \sim a$ is small compared with $r$. The difference $\psi_n - \psi_1$ can, accurate to a term that is a multiple of $2\pi$, always be made smaller than the preselected value of $\delta\psi_0$ if $n$ is taken to be large enough. To this end we find a value of $m$ such that

$$\delta\psi \equiv |2m\Delta\psi - 2l\pi| < \delta\psi_0, \tag{3.49}$$

where $\ell$ is an integer,* and put $n = m$.

*It is easy to show that at least one such $m$ exists even among the numbers smaller than $2\pi/\delta\psi_0$.

As a result, the points $(r_1, \psi_1)$, $(r_{m+1}, \psi_{m+1})$, $(r_{2m+1}, \psi_{2m+1})$, ... will lie alongside one another in the equatorial plane, and it is possible to draw through them a smooth curve – the trajectory of the nonresonant particle.

Proceeding to construct the trajectory determined above, we carry out m times the transformations (3.38) and (3.39):

$$\left.\begin{aligned} r_{m+1} &= r_1 - a(r_1)\cos 2(\psi_1 + m\Delta\psi)\frac{\sin 2m\Delta\psi}{\cos\Delta\psi}, \\ \psi_{m+1} &= \psi_1 + 2m\Delta\psi. \end{aligned}\right\} \quad (3.50)$$

The differential equations for the motion along the trajectory are obtained by using (3.49) and taking account of the fact that a transition from one point to the other corresponds to 2m transits through the magnetic bottle, i.e., it occurs within a time $2mt_{\|}$:

$$\left.\begin{aligned} \frac{dr}{dt} &\simeq \frac{r_{m+1} - r_1}{2mt_{\|}} = \frac{-a\cos(2\psi_1 + \delta\psi)\sin\delta\psi}{2mt_{\|}\cos\Delta\psi} \simeq \frac{-\delta\psi a\cos 2\psi_1}{2mt_{\|}\cos\Delta\psi}, \\ \frac{d\psi}{dt} &\simeq \frac{\psi_{m+1} - \psi_1 - 2l\pi}{2mt_{\|}} = \frac{\delta\psi}{2mt_{\|}}. \end{aligned}\right\} \quad (3.51)$$

The form of the trajectory is obtained from the equation

$$\frac{dr}{d\psi} = \frac{\dot{r}}{\dot{\psi}} = \frac{-a}{\cos\Delta\psi}\cos 2\psi$$

and is given by

$$r(\psi) = -\frac{a}{2\cos\Delta\psi}\sin 2\psi + \text{const.} \quad (3.52)$$

It can be seen that the auxiliary quantities m and $\delta\psi$, introduced above, have dropped out of the final result (3.52).

We have defined above the trajectory of nonresonant particles as a line passing through the points $(r_1, \psi_1)$, $(r_{m+1}, \psi_{m+1})$, $(r_{2m+1}, \psi_{2m+1})$.... It is easy to check, however, that the intermediate points (with $n \neq pm$, where p is an integer) also lie on this trajectory. To this end it suffices to verify that if a point $(r_1, \psi_1)$ lies on the curve (3.52), then a point $(r_2, \psi_2)$ connected with $(r_1, \psi_1)$ by the transformation (3.50) with m = 1 also lies on this curve. The latter is really satisfied since, as follows from (3.52),

$$r(\psi_1 + 2\Delta\psi) = r_1 - 2a\cos 2(\psi_1 + \Delta\psi)\sin\Delta\psi.$$

We note finally that in the transition region between resonant and nonresonant particles, when $\alpha$ of Eq. (3.40) lies in the interval

$$\sqrt{\left| a \frac{\partial \Delta\psi}{\partial r} \right|} \ll \alpha \ll 1,$$

both Eqs. (3.45) and (3.52) give the same result. To check on this we must first recognize that, since $|\mathcal{Y}| \gg 1$ in this region, the value of x changes little along the trajectory:

$$x(\psi) = x_0 + \delta x(\psi), \quad \delta x \ll x_0,$$

so that Eq. (3.45) takes the form

$$(-1)^k \frac{2x_0}{a} \frac{\partial \Delta\psi}{\partial r} \delta x - \sin 2\psi = \text{const.} \tag{3.53}$$

Second, using the smallness of $\alpha$, we can replace $\cos\Delta\psi$ in (3.52) by $(-1)^{k+1}\alpha$ and take into account the fact that

$$\alpha \approx x \frac{\partial \Delta\psi}{\partial r}.$$

Equation (3.52) goes over then into (3.53).

3.6. Decrease of Radial Random Walks of Particles by Special Choice of the Magnetic Field. As stated in Section 1, one of the methods of decreasing the transverse losses is to use in the magnetic bottle a magnetic field that, preserving the min-B property, leads to a minimum divergence of the drift surfaces (a magnetic field in which the forms of the drift surfaces do not depend at all on the energy and on the magnetic moment of the particle was referred to above as having omnigenity). Such a formulation of the problem, however, is adequate only for short central mirrors, where drift surfaces can be defined as constant-I surfaces. At $\Delta\psi \gtrsim 1$ this definition becomes generally speaking meaningless, since the adiabatic invariant I is no longer a conserved quantity. It is therefore natural to generalize the omnigenity concept to include the transverse drift transport also in the case $\Delta\psi \gtrsim 1$. It obviously suffices for this purpose to require that the guiding centers of the particles located at the initial instant on a certain force line remain during the entire time of the drift on a surface that is the same for all particles, regardless of their energy or magnetic moment. Such surfaces (which, by definition, cannot intersect), will be constant-plasma-pressure surfaces (in the case of an isotropic distribution function), and it is the inability of a particle to go from one surface to another which denotes the absence of the effects of neoclassical and resonant transport. We agree to call these also drift surfaces, although now their definition is in no way connected with the conservation of I and, of course, pertains only to a magnetic field with the omnigenity property (in the generalized sense indicated above).

Let us derive the equation that must be satisfied by such a field [19, 26, 27]. We assume everywhere below in this section that rot $\mathbf{B} = 0$. It is known that in a potential magnetic field a particle drifts in the direction of the binormal of the force line:

$$\mathbf{v}_D \backsim [\mathbf{B}\nabla B], \tag{3.54}$$

The omnigenity condition means, obviously, that the drift velocity $\mathbf{v}_D$ and the longitudinal velocity component $\mathbf{v}_\parallel$ must lie in a plane tangent to the drift surface at the given point. In this case, neither drift nor longitudinal motion leads the particle out of the surface. We introduce the force-line curvature vector $\boldsymbol{\varkappa}$. By virtue of its definition,

$$\boldsymbol{\varkappa} = B^{-3}[[\mathbf{B}\nabla B]\mathbf{B}]\,; \tag{3.55}$$

this vector is perpendicular both to $\mathbf{v}_D$ and to $\mathbf{v}_\parallel$. In the sought-for magnetic field the vector $\boldsymbol{\varkappa}$ should be directed along the normal to the drift surface. If the latter is defined by the equation

$$H(\mathbf{r}) = \text{const},$$

where $H(\mathbf{r})$ is a certain function of the coordinates, the normal to the drift surface is collinear with the vector $\nabla H$ and the omnigenity condition takes the form

$$\nabla H = \boldsymbol{\varkappa}\frac{|\nabla H|}{|\boldsymbol{\varkappa}|}. \tag{3.56}$$

Finally, taking the curl of both sides of this equation and taking the scalar product of the result and $\boldsymbol{\varkappa}$, we arrive at

$$\boldsymbol{\varkappa}\,\text{rot}\,\boldsymbol{\varkappa} = 0. \tag{3.57}$$

This is in fact the necessary omnigenity condition. The proof that this is simultaneously a sufficient condition is given in Appendix 2. Substituting in (3.57) the expression (3.55) for $\varkappa$, we find that it takes, in terms of the magnetic field **B**, the form

$$[\mathbf{B}\nabla B]\nabla(\mathbf{B}\nabla B)=0. \tag{3.58}$$

This equation was obtained by another method in [26].

The high degree of nonlinearity of Eq. (3.58) makes it impossible to find for it a general solution. We confine ourselves, therefore, below to magnetic fields that have the omnigenity property in the paraxial approximation. It is convenient in this case to start from Eq. (3.57).

Using the representation (2.15), we easily find the components of the force-line curvature vector, accurate to the terms $\sim(x/\ell)^2$, $(y/\ell)^2$:

$$\left.\begin{aligned}\varkappa_x&=x_0X''=x\frac{X''}{X},\\ \varkappa_y&=y_0Y''=y\frac{Y''}{Y},\\ \varkappa_z&=-y_0^2Y'Y''-x_0^2X'X''=-x^2\frac{X'X''}{X^2}-y^2\frac{Y'Y''}{Y^2}.\end{aligned}\right\} \tag{3.59}$$

Substituting (3.59) in (3.57) we obtain

$$\varkappa\,\mathrm{rot}\,\varkappa=\frac{xy}{X^2Y^2}\left[YY''\frac{d}{dz}(XX'')-XX''\frac{d}{dz}(YY'')\right]=0, \tag{3.60}$$

from which it follows right away that

$$XX''=cYY'', \tag{3.61}$$

where c is a constant. This is in fact the equation that defines the class of quadrupole-symmetry magnetic fields having the omnigenity property in the paraxial approximation. It can be seen that this class is quite large: once we choose an arbitrary function X(z) [satisfying only the requirement X(0) = 1], the constant c, and the initial condition Y'(0), the function Y(z) is obtained uniquely from Eq. (3.61) [with account taken of the second initial condition Y(0) = 1].

The sign of the constant c in (3.61) is very important: it determines whether or not the drift surfaces will or will not envelop the magnetic axis. To prove this statement, let us find the intersection of the drift surface with the plane z = const. Since the vector normal to the drift surface is parallel to $\varkappa$, the form of the intersection is determined from the equation

$$\frac{dx}{dy}=\frac{[\varkappa\mathbf{e}_z]_x}{[\varkappa\mathbf{e}_z]_y}=-\frac{y}{x}\,\frac{Y''X}{X''Y}, \tag{3.62}$$

where $\mathbf{e}_z$ is a unit vector along the z axis. Integrating (3.62) and using (3.61), we obtain

$$c\frac{x^2}{X^2}+\frac{y^2}{Y^2}=\text{const.} \tag{3.63}$$

Thus, the intersections of the drift surfaces form a family of embedded ellipses at $c > 0$ and a family of hyperbolas at $c < 0$. Of course, only the case $c > 0$ can be used if the particles are to be confined in the proximity of the trap axis.

We also present a paraxial-field omnigenity condition written for the functions $\mathcal{B}$ and $b$, confining ourselves to the case $c > 0$. Substituting in (3.61) expressions (2.13) for X and Y, we obtain, after simple transformations:

$$\left(-\frac{1}{2}\mathcal{B}''\mathcal{B}+\frac{3}{4}(\mathcal{B}')^2+4b^2\right)\sinh(\Phi+c_1)$$
$$+2(\mathcal{B}b'-2b\mathcal{B}')\cosh(\Phi+c_1)=0, \tag{3.64}$$

where $c_1 = {}^1/_2 \ln c$. For $c_1 = 0$ (which correspond to circular intersections of the drift surfaces with the plane z = 0) this equation was derived in [28].

Obviously, at $c_1 = 0$ (circular intersections) the term proportional to $\cos 2\psi_0$ in expressions (3.11) for $I_\pm$ vanishes identically. Accordingly, the splitting of the surfaces I = const decreases from the value given by (3.13) to

$$\Delta r_\pm \sim r_H (r/l)^3. \tag{3.65}$$

At the same time, since the noncircularity of the intersections of these surfaces is determined by a term proportional to $r^4$ in I, and this term does not, generally speaking, vanish in the field (3.64), the order of magnitude of their nonomnigenity remains equal to $(r/\ell)^2$.

An example of a numerical calculation of the magnetic configuration (for an ambipolar trap) with the aid of Eq. (3.64) can be found in [29].

In conclusion we note one more advantage of magnetic fields that satisfy Eq. (3.58): No longitudinal currents are generated in a plasma placed in such a field. This can be of practical significance in long magnetic bottles in connection with the strong distortion, discussed in Section 2.3, of the magnetic surfaces in such mirrors. To prove this statement we must check on the satisfaction of the condition

$$\operatorname{div} \mathbf{j}_\perp = 0. \tag{3.66}$$

This is easiest to do for an isotropic plasma (although the statement itself is valid also in the case of anisotropic pressure). Since we have, according to (2.31),

$$\operatorname{div} \mathbf{j}_\perp = c\nabla p \operatorname{rot} \frac{\mathbf{B}}{B^2} = -\nabla \frac{1}{B^2} [\nabla p \mathbf{B}] = B^2 \mathbf{j}_\perp \nabla \frac{1}{B^2} = -2\mathbf{j}_\perp \boldsymbol{\varkappa},$$

and in the magnetic field (3.58) the vector $\varkappa$ is perpendicular to the magnetic surface along the tangent to which $\mathbf{j}_\perp$ is directed, it is obvious that (3.66) is satisfied.

All the results reported in this section pertained to the case of zero plasma pressure. It can be shown that if a vacuum magnetic field satisfying the condition (3.61) is distorted by plasma currents, the omnigenity property is lost, and at $\beta \gtrsim (r/\ell)^2$ the estimate (3.65) must be replaced by

$$\Delta r_\pm \sim \beta r_H \frac{r}{l}.$$

## 4. Neoclassical Transport in Open Traps

4.1. Derivation of the Boltzmann Kinetic Equation and of the System of Transport Equations. We consider in this section a situation in which both plasma components (electrons and ions) in an open trap are in the neoclassical transport regime. We agree to label quantities pertaining to the different components by the subscripts e and i, respectively. Many equations of this section hold for both electrons and ions; the subscripts e and i will be omitted in such cases.

We assume explicitly hereafter, as in the preceding sections, a low-$\beta$ plasma.

As a rule, in open traps the plasma-particle mean-free path $\lambda$ exceeds substantially the longitudinal dimension of the central mirror. With this circumstance taken into account, the fast motion of particles along a force line leads in the neoclassical regime to a phase mixing of the distribution function with respect to the longitudinal coordinate, and permits f to be regarded as independent of $\xi^3$:

$$f = f(\xi^1, \xi^2, \varepsilon, \mu, t).$$

We assume below, for simplicity, that the distribution of the Yushmanov potential (2.28) along the force line has only one minimum, so that f is a single-valued function of its arguments. In the general case, as discussed in Section 2.2, the distribution function can have several branches, each corresponding to particle distribution in a local potential well.

The function f satisfies a kinetic equation which we now proceed to derive. We note first, however, that the equations for the drift velocity (3.1) averaged along a force line, which were derived in [23] for stationary fields, remain in force if the electric field varies slowly with time (compared with $t_{\parallel}$) and if, at the same time, the magnetic field remains constant [30]. This is precisely the situation in our case, when the plasma redistribution along the radius by diffusion alters the electric potential, but the time variation of the magnetic field can be neglected in view of the small $\beta$ of the plasma.

We introduce a function $g(\xi^1, \xi^2, \varepsilon, \mu, t)$ such that $g\,d\xi^1 d\xi^2 d\varepsilon\, d\mu$ is equal to the number of particles in the interval $d\varepsilon d\mu$, located in a force tube specified by the differentials $d\xi^1$, $d\xi^2$. The area $d\Sigma$ of the normal cross-section of this tube (as a function of the coordinate $\xi^3$) is

$$d\Sigma = \sqrt{\frac{g}{g_{33}}}\, d\xi^1 d\xi^2. \tag{4.1}$$

The continuity equation for g is

$$\frac{\partial g}{\partial t} + \frac{\partial}{\partial \xi^1}(\dot{\xi}^1 g) + \frac{\partial}{\partial \xi^2}(\dot{\xi}^2 g) + \frac{\partial}{\partial \varepsilon}(\bar{\dot{\varepsilon}} g) = \left(\frac{\partial g}{\partial t}\right)_{St}. \tag{4.2}$$

Here $(\partial g/\partial t)_{St}$ denotes the change of g, due to Coulomb collisions, while $\bar{\dot{\varepsilon}}$ denotes the rate of change, averaged over the period, of the total particle energy:

$$\bar{\dot{\varepsilon}} = e\frac{\overline{\partial \varphi}}{\partial t}, \tag{4.3}$$

where

$$\frac{\overline{\partial\varphi}}{\partial t}=\frac{1}{t_{\parallel}}\int_{s_1}^{s_2}\frac{ds}{|v_{\parallel}|}\frac{\partial\varphi}{\partial t}, \quad (4.4)$$

and e denotes the charge of the particles of the considered component. We point out that the quantity $\overline{\partial\varphi/\partial t}$ depends not only on the coordinates and time, but also on the parameters $\varepsilon$ and $\mu$ of the particular particle along whose trajectory the averaging is carried out.

We now express g in terms of f. The number of particles per unit volume in the interval $d\varepsilon d\mu$ is

$$dn=2\,\frac{2\pi B}{m^2}f\,\frac{d\varepsilon d\mu}{|v_{\parallel}|}. \quad (4.5)$$

The additional factor 2 in (4.5) is due to the fact that any one set of values $\varepsilon$, $\mu$ corresponds to two points in the $v_{\perp}$, $v_{\parallel}$ plane, with opposite signs of $v_{\parallel}$. To establish the connection between g and f, we integrate (4.5) over the volume of the force tube, using (4.1). The result is

$$g=\frac{4\pi}{m^2}\int_{s_1}^{s_2}B\sqrt{\frac{g}{g_{33}}}f\,\frac{ds}{|v_{\parallel}|}=\frac{4\pi}{m^2}B\sqrt{\frac{g}{g_{33}}}ft_{\parallel}. \quad (4.6)$$

In accord with its meaning $(\partial g/\partial t)_{St}d\xi^1 d\xi^2 d\varepsilon d\mu$ is equal to the collision-induced change of the number of particles in the interval $d\varepsilon d\mu$ within the force tube $d\xi^1 d\xi^2$:

$$\left(\frac{\partial g}{\partial t}\right)_{St}d\xi^1 d\xi^2 d\varepsilon d\mu=\int\frac{4\pi B}{m^2}\frac{d\varepsilon d\mu}{|v_{\parallel}|}d\Sigma ds\,\mathrm{St}\,f$$

$$=d\xi^1 d\xi^2 d\varepsilon d\mu\,\frac{4\pi B}{m^2}\sqrt{\frac{g}{g_{33}}}\int_{s_1}^{s_2}\frac{ds}{|v_{\parallel}|}\mathrm{St}\,f. \quad (4.7)$$

Expressing now g and $(\partial g/\partial t)_{St}$ in terms of f in accordance with (4.6) and (4.7), and substituting the expressions for $\dot{\xi}^1$, $\dot{\xi}^2$, $\dot{\varepsilon}$ in (4.2), we find that

$$\frac{\partial f}{\partial t}+\dot{\xi}^1\frac{\partial f}{\partial\xi^1}+\dot{\xi}^2\frac{\partial f}{\partial\xi^2}+e\frac{\overline{\partial\varphi}}{\partial t}\frac{\partial f}{\partial t}=\overline{\mathrm{St}\,f}, \quad (4.8)$$

where the bar over St f denotes, as in (4.4), averaging along the force line. In the derivation of (4.8) it is necessary to take into account the easily verified relation

$$\frac{\partial t_{\parallel}}{\partial t} = -\frac{\partial\left(t_{\parallel}\dot{\bar{\varepsilon}}\right)}{\partial \varepsilon}.$$

We now take explicit account of the fact that we are considering a magnetic configuration for which the nonomnigenity of the drift surfaces is small. Obviously, this circumstance is mathematically manifested by the fact that in some chosen curvilinear coordinates $\xi^1$, $\xi^2$ the longitudinal adiabatic invariant takes the form

$$I(\xi^1, \xi^2, \varepsilon, \mu) = I_0(\xi^1, \varepsilon, \mu) + \delta I(\xi^1, \xi^2, \varepsilon, \mu), \quad \delta I \ll I_0. \tag{4.9}$$

It follows from the results of Section 3 that $\xi^1$ can be chosen to be the distance from the force line to the axis in the equatorial plane; in the derivation of the general relations, however, it is convenient not to choose the coordinates $\xi^1$, $\xi^2$ beforehand. We agree only to let $\xi^2$ play the role of the angle variable and then have it change by $2\pi$ after encircling the magnetic axis. The subdivision (4.9) can then be made unique (in the given coordinates) by stipulating:

$$I_0 = \langle I \rangle, \quad \delta I = I - I_0, \tag{4.10}$$

where the angle brackets denote averaging over $\xi^2$:

$$\langle \ldots \rangle = \frac{1}{2\pi}\int_0^{2\pi} d\xi^2 \ \ldots . \tag{4.11}$$

We shall refer to the surfaces $\xi^1$ = const as magnetic surfaces; by virtue of (4.9), there is little difference between the drift surfaces and the magnetic surfaces. We note right away that the choice of magnetic surfaces defined in this manner is not unique [for example, we can carry out a transformation $\xi^1 \to \xi^1(1 + \zeta(\xi^1, \xi^2))$ that does not change the order of magnitude of $\delta I$ if $\zeta \sim \delta I/I_0$]. In an actual calculation, this leeway can be used to simplify the calculations.

We introduce a quantity N equal to the ratio of the number of particles included between two magnetic surfaces $\xi^1$ and $\xi^1 + d\xi^1$ to the differential $d\xi^1$:

$$N = 2\pi \left\langle \int_0^\infty d\mu \int_{\varepsilon_{\min}}^\infty d\varepsilon g \right\rangle = \frac{8\pi^2}{m^2} \left\langle B \sqrt{\frac{g}{g_{33}}} \int d\varepsilon d\mu f t_\parallel \right\rangle, \quad (4.12)$$

where $\varepsilon_{\min} = \varepsilon_{\min}(\xi^1, \xi^2)$ is equal to the minimum value of the potential (2.28) on the force line (hereafter, as in the last integral, we shall as a rule not indicate the limits of integration with respect to $\varepsilon$ and $\mu$). The kinetic energy W of these particles, referred to $d\xi^1$, is expressed in terms of the function f as follows:

$$W = \frac{8\pi^2}{m^2} \left\langle B \sqrt{\frac{g}{g_{33}}} \int d\varepsilon d\mu f t_\parallel (\varepsilon - e\bar{\varphi}) \right\rangle. \quad (4.13)$$

From the kinetic equation (4.8) [with account taken of the relations (3.1)] we obtain the following balance equations for N and W (see Appendix 3):

$$\left.\begin{aligned} &\frac{\partial N}{\partial t} + \frac{\partial}{\partial \xi^1} q = 0, \\ &\frac{\partial W}{\partial t} + \frac{\partial}{\partial \xi^1} Q = S + \dot{W}_{\mathrm{ad}} + \dot{W}_{\mathrm{St}} + \Gamma. \end{aligned}\right\} \quad (4.14)$$

Here

$$q \equiv \frac{8\pi^2 c}{me} \left\langle \int d\varepsilon d\mu f \frac{\partial I}{\partial \xi^2} \right\rangle \quad (4.15)$$

and

$$Q \equiv \frac{8\pi^2 c}{me} \left\langle \int d\varepsilon d\mu \frac{\partial I}{\partial \xi^2} f (\varepsilon - e\bar{\varphi}) \right\rangle \quad (4.16)$$

have the meaning of the particle and energy fluxes, and

$$S = -\frac{8\pi^2 e}{m^2} \left\langle B \sqrt{\frac{g}{g_{33}}} \int d\varepsilon d\mu f t_\parallel \left( \frac{\partial \bar{\varphi}}{\partial \xi^1} \dot{\xi}^1 + \frac{\partial \bar{\varphi}}{\partial \xi^2} \dot{\xi}^2 \right) \right\rangle \quad (4.17)$$

is a source connected with the work performed by the electric field on the particles; $\Gamma$ denotes the rate of change of the internal energy of the considered plasma component because of collisions with particles of the other component;

and the meaning of the terms $W_{ad}$ and $W_{St}$ is explained in Appendix 3.*

We shall need hereafter the assertion that any distribution function that depends only on the integrals of motion I, ε, μ, f = F(I, ε, μ) causes the particle flux (4.15) to vanish. This assertion is easy to prove by noting that

$$\left\langle F \frac{\partial I}{\partial \xi^2} \right\rangle = \left\langle \frac{\partial}{\partial \xi^2} \int F dI \right\rangle = 0.$$

As for the flux (4.16), it vanishes in this case if the averaged potential $\bar{\varphi}$ is constant on the magnetic surfaces, $\bar{\varphi} = \varphi(\xi^1, \varepsilon, \mu)$; this is true, generally speaking, only if the nonomnigenity of the drift surfaces is neglected (see below). We note also that it can be verified directly that, for a function of the integrals of motion, the source S coincides with the quantity $\partial Q / \partial \xi^1$. The last statement is quite obvious and means that, in the absence of collisions ($\Gamma = \dot{W}_{St} = 0$) and in a stationary potential ($\dot{W}_{ad} = 0$), any function of the integrals of motion is a solution of the kinetic equation and corresponds to a constant value of W.

Since the plasma diffusion time in systems with small drift-shell nonomnigenity is long compared with the collision time, the kinetic equation can be solved by using locally a Maxwellian approximation, i.e., by assuming that

$$f = f_M + \delta f, \tag{4.18}$$

where

$$f_M = n_*(\xi^1, t) \left[ \frac{2\pi}{m} T(\xi^1, t) \right]^{-3/2} \exp\left[ -\frac{\varepsilon}{T(\xi^1, t)} \right], \tag{4.19}$$

and $\delta f \ll f_M$.

*The following remark is in order here. Whereas the flux q is the number of particles crossing the magnetic surface per unit time, in the case when the potential φ varies along a force line the quantity Q is not the kinetic energy flowing through the magnetic surface per second. Since S and Q enter in the energy balance equation as the combination $S - \partial Q / \partial \xi^1$, the breakup of this combination into two terms is rather arbitrary: some of the terms from Q can always be combined with S and, conversely, divergence-type terms in S can be transferred to Q.

We note right away that the assumption that the plasma has a Maxwellian distribution function (4.19) is, strictly speaking, not valid for an ambipolar trap, where the end mirrors contain hot anisotropic ions that ensure longitudinal confinement of the magnetic-bottle plasma. To be able to disregard the peculiarities of longitudinal containment in this case, we shall assume that the magnetic-bottle plasma is bounded in the longitudinal direction by two surfaces located in the mirrors and reflecting elastically the particles that reach them. We also assume that the force lines intersect these surfaces in a perpendicular direction.

To make the breakup (4.18) unique, we stipulate that the correction $\delta f$ makes no contribution to the values of N and W:

$$\left\langle \sqrt{\frac{g}{g_{33}}} B \int d\varepsilon d\mu \delta f t_{\parallel} \right\rangle = \left\langle B \sqrt{\frac{g}{g_{33}}} \int d\varepsilon d\mu \delta f t_{\parallel} (\varepsilon - e\bar{\varphi}) \right\rangle = 0. \tag{4.20}$$

The potential $\varphi$, according to Section 3.2, should be determined from the quasineutrality condition $n_e = n_i$. Neglecting the difference between f and $f_M$, we have

$$n_{e,i} = n_{*e,i}(\xi^1) \exp[\pm e\varphi/T_{e,i}(\xi^1)] \tag{4.21}$$

(we do not indicate here and below the explicit dependence on the time), and the quasineutrality condition leads to the requirement that the potential on the magnetic surface be constant,

$$\varphi = \varphi_0(\xi^1).$$

In the next approximation, account must be taken of the corrections $\delta f_{e,i}$ to the electron and ion densities:

$$\delta n_{e,i} = \frac{4\pi B}{m_{e,i}^2} \int \frac{d\varepsilon d\mu}{|v_{\parallel}|} \delta f_{e,i}, \tag{4.22}$$

and we then get, from the quasineutrality condition,

$$\varphi = \varphi_0(\xi^1) + \delta\varphi(\xi^1, \xi^2, \xi^3), \tag{4.23}$$

where

$$\delta\varphi = \frac{T_i T_e}{ne(T_e + T_i)} (\delta n_i - \delta n_e). \tag{4.24}$$

It is important to stress here that, unlike in tokamaks [9], diffusion in nonaxisymmetric open traps is not automatically ambipolar. The ambipolarity conditions are satisfied only for a definite distribution of the potential $\varphi_0$ and serves as the equation from which the dependence of $\varphi_0$ on $\xi^1$ should be obtained.

The small difference, which follows from the representation (4.23), between the magnetic surfaces and equipotentials means that the term $\dot{W}_{ad}$ in the energy balance equation can be set equal to zero. It is easy to verify this also by formal calculation, substituting for $\varphi$ in (A3.4) the function $\varphi_0$ that depends only on $\xi^1$.

The potential perturbation $\delta\varphi$ makes a contribution $\delta I_\varphi$ to the term $\delta I$ in (4.9), a term responsible for the divergence of the drift surfaces (see Sections 3.2 and 3.3). The expression for $\delta I_\varphi$ can be written in the form

$$\delta I_\varphi = \int ds \sqrt{\frac{2}{m}(\varepsilon - \mu B - e\varphi_0 - e\delta\varphi)}$$
$$- \int ds \sqrt{\frac{2}{m}(\varepsilon - \mu B - e\varphi_0)} = -\frac{e}{m} t_\parallel \overline{\delta\varphi}. \tag{4.25}$$

We emphasize that one task in the solution of the transport problem is to find the perturbation $\delta\varphi$ of the potential and to calculate the corrections (4.25) to the adiabatic invariant.

We now substitute relations (4.18) and (4.22) in expressions (4.15)-(4.17). Since $\delta f$, $\delta I$, and $\delta\varphi$ are small, it suffices to retain terms of second order (the linear terms vanish):

$$q = \frac{8\pi^2 c}{me}\left\langle \int d\varepsilon d\mu \delta f \frac{\partial \delta I}{\partial \xi^2} \right\rangle, \tag{4.26}$$

$$Q = \frac{8\pi^2 c}{me}\left\{\left\langle \int d\varepsilon d\mu \delta f \frac{\partial \delta I}{\partial \xi^2}(\varepsilon - e\varphi_0)\right\rangle + \left\langle e \int d\varepsilon d\mu \delta I f_M \frac{\partial \overline{\delta\varphi}}{\partial \xi^2}\right\rangle\right\}, \tag{4.27}$$

$$S = -\frac{8\pi^2 c}{m}\left\{\left\langle \int d\varepsilon d\mu \delta f \left(\frac{\partial \varphi_0}{\partial \xi^1}\frac{\partial \delta I}{\partial \xi^2} - \frac{\partial \overline{\delta\varphi}}{\partial \xi^2}\frac{\partial I_0}{\partial \xi^1}\right)\right\rangle\right.$$
$$\left. + \left\langle \int d\varepsilon d\mu f_M \left(\frac{\partial \delta I}{\partial \xi^2}\frac{\partial \overline{\delta\varphi}}{\partial \xi^1} - \frac{\partial \delta I}{\partial \xi^1}\frac{\partial \overline{\delta\varphi}}{\partial \xi^2}\right)\right\rangle\right\}. \tag{4.28}$$

It is now convenient to redefine the quantity* Q, retaining only the first term in the curly brackets of (4.27):

$$Q = \frac{8\pi^2 c}{me}\left\langle \int d\varepsilon d\mu \delta f \frac{\partial \delta I}{\partial \xi^2}(\varepsilon - e\varphi_0)\right\rangle \tag{4.29}$$

and transferring to S the derivative of the second term with respect to $\xi^1$. After integration by parts, the expression for S then takes the form

$$S = -\frac{8\pi^2 c}{m}\left\{\left\langle \int d\varepsilon d\mu \delta f \left(\frac{\partial \varphi_0}{\partial \xi^1}\frac{\partial \delta I}{\partial \xi^2} - \frac{\partial \overline{\delta \varphi}}{\partial \xi^2}\frac{\partial I_0}{\partial \xi^1}\right)\right\rangle + \left\langle \int d\varepsilon d\mu \delta I \frac{\partial f_M}{\partial \xi^1}\frac{\partial \overline{\delta \varphi}}{\partial \xi^2}\right\rangle\right\}. \tag{4.30}$$

The correction to the distribution function δf, which enters in the expressions (4.26), (4.29), and (4.30), is determined from the kinetic equation obtained from (4.8), neglecting the terms proportional to the time derivatives, with allowance for expressions (3.1), (4.9), and (4.18):

$$\delta \dot{\xi}^1 \frac{\partial}{\partial \xi^1}(f_M + \delta f) + \dot{\xi}^2_0 \frac{\partial \delta f}{\partial \xi^2} = \overline{\mathrm{St}\, \delta f}, \tag{4.31}$$

where

$$\delta \dot{\xi}^1 = \frac{mc}{eBt_{\parallel}}\sqrt{\frac{g_{33}}{g}}\frac{\partial \delta I}{\partial \xi^2}, \tag{4.32}$$

$$\dot{\xi}^2_0 = -\frac{mc}{eBt_{\parallel}}\sqrt{\frac{g_{33}}{g}}\frac{\partial I_0}{\partial \xi^1}, \tag{4.33}$$

and St δf denotes the linearized collision term.

To conclude this subsection, we note that the foregoing derivation of the transport equations, as well as the subsequent calculations in this section, can be used with practically no change for the problem of neoclassical diffusion in systems of the BUMPY TORUS type. An idea of the character of the calculations in the latter case can be gained from [31-33].

*See the footnote following Eq. (4.17).

4.2. Transport Problem in the Long-Magnetic-Bottle Model. The system of transport equations derived in the preceding section and the kinetic equation for the correction to the distribution function become noticeably simplified for the long-magnetic-bottle problem. We note only that we shall assume the plasma pressure low enough to be able to neglect the magnetic-surface distortion effects described in Section 2.3.

We agree first on the choice of the curvilinear coordinates $\xi^1$, $\xi^2$ that specify the position of the force line. We choose $\xi^2$ to be the azimuthal angle in the equatorial plane of the trap, measured from one of the symmetry planes. As for the coordinate $\xi^1$, it is convenient to choose it equal to $\pi r^2$ (r is the radius in the equatorial plane). The quantities N and W will then be, respectively, the azimuth-averaged number of particles and the energy is a force tube of unit cross-section area. Bearing in mind that the main contributions to N and W are made by a straight section of length $L_0$, we can write

$$N = nL_0, \quad W = \frac{3}{2} nTL_0,$$

where T is the temperature in the Maxwell function (4.19), and the density is connected with $n_*$ by the relation $n = n_* \exp(-e\varphi_0/T)$.

Since the correction to the distribution function $\delta f$ does not depend on the coordinate $\xi^3$, the electron and ion density perturbations do not vary along the force line on the long section of the magnetic bottle, where the magnetic field is uniform, and, accordingly, $\delta\varphi$ is independent here of the coordinate z:

$$\varphi = \varphi_0(r) + \delta\varphi(r, \psi) \tag{4.34}$$

[recall that $\varphi_0$ is a function of the magnetic surface, $\varphi_0 = \varphi_0(r)$].

We note now that on averaging along a force line [see Eq. (4.4)] the main contribution to the integral is made by the long region of the uniform field, where the force lines are straight. This means, in particular, that $\overline{\delta\varphi} \cong \delta\varphi$, and $\overline{\mathrm{St}\,\delta f} \cong \mathrm{St}\,\delta f$. The latter enables us to neglect the term $\dot{W}_{St}$ in the energy balance equation (4.14) [see (A3.6)] and write the transport equation in the form

$$\left.\begin{aligned}&\frac{\partial n}{\partial t}+\frac{1}{r}\frac{\partial}{\partial r}r\left(\frac{q}{2\pi rL_0}\right)=0,\\&\frac{3}{2}\frac{\partial}{\partial t}nT+\frac{1}{r}\frac{\partial}{\partial r}r\left(\frac{Q}{2\pi rL_0}\right)=\frac{S}{L_0}+\frac{\Gamma}{L_0}.\end{aligned}\right\}\qquad(4.35)$$

The expressions in the parentheses are fluxes per unit area.

In terms of the introduced coordinates, the kinetic equation (4.31) takes the form

$$u_r\frac{\partial}{\partial r}(f_M+\delta f)+u_\psi\frac{\partial\delta f}{\partial\psi}=\mathrm{St}\,\delta f,\qquad(4.36)$$

where $u_r$ and $u_\psi$ denote the rates of change of the coordinates r and $\psi$:

$$u_r=\frac{mc}{reB_0t_\parallel}\frac{\partial\delta I}{\partial\psi},\qquad(4.37')$$

$$u_\psi=-\frac{mc}{reB_0t_\parallel}\frac{\partial I_0}{\partial r},\qquad(4.37'')$$

and $B_0$ is the magnetic-field strength on the uniform sections. An expression for the function $I_0$ in (4.37") is given in Section 3.4. The supplementary conditions (4.20) imposed on the correction $\delta f$ are transformed into

$$\left\langle\int\frac{d\varepsilon d\mu}{|v_\parallel|}\delta f\right\rangle=\left\langle\int\frac{d\varepsilon d\mu}{|v_\parallel|}\delta f(\varepsilon-e\varphi_0)\right\rangle=0.\qquad(4.38)$$

The first of them means that the azimuth-averaged perturbation of the density on the uniform part of the magnetic bottle is zero.

Examples of the solution of the kinetic equation in various regimes with different collision frequencies will be found in the sections that follow. For simplicity, we present everywhere only expressions for the diffusion flux q of the particles.

4.3. Qualitative Treatment. We classify in this section the various transport regimes and estimate the diffusion coefficient as a function of the collision frequency $\nu$.

We consider first the case when the trap contains trapped particles that move on banana trajectories (see Section 3.4) and denote by $\Delta\mu$ the difference $\mu-\mu_*(r)$, where r is the radius at which the diffusion coefficient is estimated. Let the collision frequency be so small that the time $\alpha^{-1/2}\tau_D$ in which the trapped particles complete their circuit is too short for them to experience a collision that transforms them into untrapped ones. This situation corresponds to the banana regime in the theory of transport processes in toroidal machines [34]. Recognizing that the effective frequency of such collisions is $\nu_{ef} \sim \nu(\mu_*/\Delta\mu_0)^2 \sim \alpha^{-1}\nu$, we find that the banana regime is realized at

$$\nu\tau_D \lesssim \alpha^{3/2}. \tag{4.39}$$

The contribution of the trapped particles to the diffusion coefficient is then estimated at

$$\mathcal{D} \sim \Delta x_b^2 \nu_{ef} \frac{\Delta\mu_0}{\mu_*} \sim \alpha^{1/2}\nu R^2, \tag{4.40}$$

where the factor $\Delta\mu_0\mu_*$ takes into account the small fraction of these particles. The contribution made to the diffusion coefficient by untrapped particles with $\Delta\mu \sim \mu_*$ is

$$\mathcal{D} \sim (\delta r)^2 \nu \sim \alpha^2 \nu R^2 \tag{4.41}$$

and is small compared with (4.40). Thus, in the banana regime, the rates of the transport processes are determined by the particles whose magnetic moments lie in a narrow vicinity of width $\Delta\mu \sim \alpha^{1/2}\mu_*$ near the value $\mu_*$.

The region of collision frequencies in which the inequalities

$$\alpha^{3/2} \lesssim \nu\tau_D \lesssim 1 \tag{4.42}$$

are satisfied is the analog of the "plateau" regime in toroidal installations (see [34]). In this case the diffusion is also determined by the particles with small $\Delta\mu$. These are particles for which the reciprocal time of encircling the axis $\tau^{-1} \sim \tau_D^{-1}(\Delta\mu/\mu_*)$ becomes comparable with the effective collision frequency $\nu(\mu_*/\Delta\mu)^2$, i.e., $\Delta\mu/\mu_* \sim (\nu\tau_D)^{1/3}$. The divergence of the drift surfaces for these particles is of the order of $\alpha R\mu_*/\Delta\mu$ and the diffusion coefficient is estimated as

$$\mathcal{D} \sim \left(\alpha R \frac{\mu_*}{\Delta\mu}\right)^2 \nu \left(\frac{\mu_*}{\Delta\mu}\right)^2 \frac{\Delta\mu}{\mu_*} \sim \alpha^2 \frac{R^2}{\tau_D}. \tag{4.43}$$

The diffusion coefficient (4.43) is independent of the collision frequency. The contribution made to the diffusion by

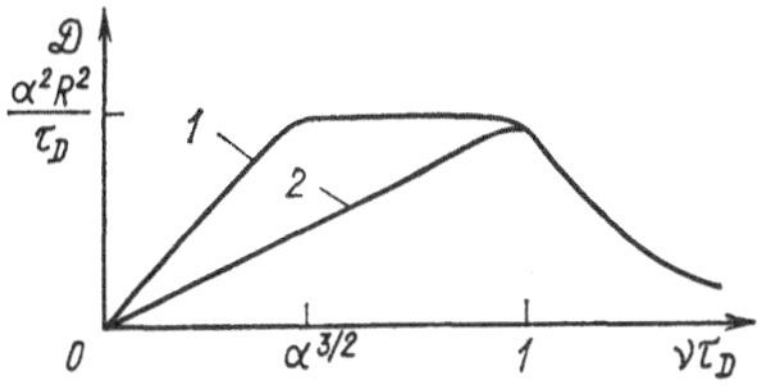


Fig. 12. Qualitative dependence of the diffusion coefficient in the neoclassical regime on the collision frequency in the presence (1) and absence (2) of trapped particles.

particles with large $\Delta\mu$ is therefore estimated from Eq. (4.41) and is small compared with (4.43).

Finally, at $\nu\tau_D \gtrsim 1$, the diffusion velocity is determined by the particles with $\Delta\mu \sim \mu_*$. During the time between collisions such a particle is displaced in azimuth through an angle $\sim (\nu\tau_D)^{-1}$, and its radial displacement is therefore of the order of $\delta r(\nu\tau_D)^{-1}$. The diffusion coefficient is therefore estimated here as

$$\mathscr{D} \sim \left(\frac{\delta r}{\nu\tau_D}\right)^2 \nu \sim \alpha^2 R^2 \frac{1}{\nu\tau_D^2}. \tag{4.44}$$

The picture described above presupposes that the plasma contains trapped particles. Generally speaking, this may also not be the case: the derivative $\partial I_0/\partial r$ can differ from zero at all values of $\varepsilon$ and $\mu$, or can vanish on the "tail" of the Maxwellian distribution, where the number of particles is exponentially small. In this situation, the diffusion coefficient at $\nu\tau_D \lesssim 1$ must be estimated from Eq. (4.41). We call this the rotational-diffusion regime (since the main contribution to $\mathscr{D}$ is made by particles that rotate rapidly around the magnetic axis).

A qualitative plot of the diffusion coefficient vs. the collision frequency, plotted on the basis of the estimates above, is shown in Fig. 12.

4.4. Rotational Diffusion Regime. The fluxes are easiest to calculate in the rotational diffusion regime. Since, by virtue of the condition $\nu\tau_D \ll 1$, the collision term in the kinetic equation (4.36) is small compared with the second term in the left-hand side, we can seek the correction $\delta f$ to the distribution function by using perturbation theory in the small parameter $\nu\tau_D$:

$$\delta f = \delta f^{(0)} + \delta f^{(1)} + \ldots .$$

In the zeroth approximation, with collisions neglected, the kinetic equation is satisfied by any function of the particle's integrals of motion. Recognizing that this function should differ little from the Maxwellian distribution, we set it equal to

$$f^{(0)} = n(r_*)\left[\frac{m}{2\pi T(r_*)}\right]^{3/2} \exp\left[-\frac{\varepsilon - e\varphi_0(r_*)}{T(r_*)}\right], \tag{4.45}$$

where $r_*$ is the radius averaged over the particle trajectories, and is expressed in terms of $r$, $\psi$, $\varepsilon$, and $\mu$ as follows [see (3.29)]:

$$r_* = r + \frac{\delta I}{\partial I_0/\partial r}. \tag{4.46}$$

In this approximation, the correction $\delta f$ is given by the difference $f^{(0)} - f_M$, which can be written, by expanding $f^{(0)}$ in powers of the small difference $r_* - r$, in the following form:

$$\delta f^{(0)} = \frac{\delta I}{\partial I_0/\partial r}\frac{\partial f_M}{\partial r}, \tag{4.47}$$

where

$$\frac{\partial f_M}{\partial r} = \left[\frac{1}{n}\frac{\partial n}{\partial r} + \frac{e}{T}\frac{\partial \varphi_0}{\partial r} + \frac{1}{T}\frac{\partial T}{\partial r}\left(\frac{\varepsilon - e\varphi_0}{T} - \frac{3}{2}\right)\right] f_M. \tag{4.48}$$

The correction $\delta f^{(1)}$, linear in the collision frequency, is obtained from the kinetic equation (4.36), in which we can neglect the small term $u_r \partial \delta f^{(1)}/\partial r$:

$$u_\psi \frac{\partial \delta f^{(1)}}{\partial \psi} = \mathrm{St}\, \delta f^{(0)}. \tag{4.49}$$

Integrating this equation with respect to the variable $\psi$, we get

$$\delta f^{(1)} = \int_0^{\psi} \frac{d\psi'}{u_\psi} \operatorname{St}\left(\frac{\delta I(\psi')}{\partial I_0/\partial r} \frac{\partial f_M}{\partial r}\right). \tag{4.50}$$

A property of this solution is that the function $\delta f$ satisfies the necessary conditions (4.38).

We confine ourselves to the simplest case, when the azimuthal particle drift is determined predominantly by the electric field, so that we can use expression (3.31') for $I_0$. We then find, in accordance with (4.37"),

$$u_\psi = -c\frac{E}{rB_0}, \tag{4.51}$$

where $E \equiv \partial\varphi_0/\partial r$. Substituting (4.50), (3.31'), and (4.51) in (4.26), we obtain an expression for the particle diffusion flux in the rotational regime:

$$q = \frac{8\pi^2 rB_0}{e^2E^2L_0}\Big\langle\left(\frac{1}{n}\frac{\partial n}{\partial r} + \frac{e}{T}\frac{\partial\varphi_0}{\partial r}\right)\int d\varepsilon d\mu \delta I \operatorname{St}(\delta I \mid v_\parallel \mid f_M)$$
$$+\frac{1}{T}\frac{\partial T}{\partial r}\int d\varepsilon d\mu \delta I \operatorname{St}\left[\delta I f_M \mid v_\parallel \mid \left(\frac{\varepsilon - e\varphi_0}{T} - \frac{3}{2}\right)\right]\Big\rangle. \tag{4.52}$$

We now proceed to calculate the perturbation $\delta\varphi$ of the potential. To find the perturbation of the density on the magnetic surface, it suffices to use the zeroth approximation $\delta f^{(0)}$:

$$\delta n = \frac{4\pi B_0}{m^2}\int \frac{d\varepsilon d\mu}{|v_\parallel|}\frac{\delta I}{\partial I_0/\partial r}\frac{\partial f_M}{\partial r}. \tag{4.53}$$

Substituting here (3.31'), we find that, in the region of the uniform magnetic field,

$$\delta n = \frac{4\pi B_0}{emEL_0}\int d\varepsilon d\mu \delta I \frac{\partial f_M}{\partial r}. \tag{4.54}$$

Calculations given in Appendix 4 yield

$$\delta n = \frac{Tn}{eE}\left\{\left(\frac{1}{n}\frac{\partial n}{\partial r} + \frac{e}{T}\frac{\partial\varphi_0}{\partial r}\right)\left(\frac{\delta U}{U_0} - \frac{2e\delta\varphi}{T}\right)\right.$$
$$\left.+\frac{1}{T}\frac{\partial T}{\partial r}\left(\frac{\delta U}{U_0} + \frac{3}{2}\frac{e\delta\varphi}{T}\right)\right\}, \tag{4.55}$$

where $\delta U$ is the difference $U - \langle U \rangle$, and $U_0 = L_0/B_0$. With the aid of (4.24) we now obtain the perturbation $\delta\varphi$ of the potential. We present an expression for it in the case $T_e = T_i = T$:

$$\delta\varphi = -\frac{T^2}{e^2 E}\left(\frac{1}{n}\frac{\partial n}{\partial r} + \frac{1}{T}\frac{\partial T}{\partial r}\right)\frac{\delta U}{U_0}. \qquad (4.56)$$

The obtained potential perturbation in the term $\delta I$ of the adiabatic invariant completes the calculation of the flux (4.52) in the rotational regime.

4.5. Banana Regime. Recognizing that, in an open trap, the particles that describe banana trajectories are usually ions, we carry out the analysis in the next two sections for the ion component, and neglect electron–ion collisions.

As noted in Section 4.3, the main contribution to transport in the banana regime is made by particles whose magnetic moment is close to $\mu_*$. We therefore seek first the correction $\delta f$ to the distribution function of the ions in a phase-space region in the form of a narrow (relative to $\mu$) strip elongated along the $\varepsilon$ axis; the line $\mu = \mu_*$ is located in this strip. In the vicinity of this line, $\delta f$ changes by a factor 2 when $\mu$ changes by a small amount of the order of $\alpha^{1/2}\mu_*$, and has at the same time a smooth dependence on $\varepsilon$, $|\partial \ln \delta f / \partial \ln \varepsilon| \sim 1$. The last circumstance makes it possible to simplify the collision integral for the considered phase-space region, expressing it in terms of the variables $\varepsilon$ and $\mu$ and retaining only terms containing two differentiations with respect to $\mu$ (cf. the corresponding approximations in [35, 36]). We can then put $\mu = \mu_*$ in the coefficient preceding the second derivative with respect to $\mu$. The result takes the form (see Appendix 5)

$$\mathrm{St}\,\delta f \simeq \nu(\varepsilon, r)\,\mu_*^2 \frac{\partial^2 \delta f}{\partial \mu^2}. \qquad (4.57)$$

Here

$$\nu(\varepsilon, r) = \nu_0(r)\, G_1(\zeta, \chi), \qquad (4.58)$$

with the collision frequency $\nu_0$ defined as

$$\nu_0(r) = \frac{\sqrt{2}\,\pi n e^4}{m^{1/2}\,T^{3/2}}\,\Lambda,$$

and the function $G_1$ as

$$G_1 = \frac{1}{\zeta \chi^{3/2}}\left[\left(1 - \frac{1}{2\chi}\right)h + \frac{dh}{d\chi}\right] + \frac{h}{\chi^{5/2}},$$

where

$$h(\chi) = \frac{2}{\sqrt{\pi}}\int_0^{\chi}\sqrt{t}\,e^{-t}\,dt, \qquad \chi = \frac{\varepsilon - e\varphi}{T}, \qquad \zeta = \frac{\mu B}{\varepsilon - e\varphi - \mu B}.$$

Taking (4.58) into account, we rewrite (4.36) in the form

$$u_r\frac{\partial f_M}{\partial r} + u_\psi\frac{\partial \delta f}{\partial \psi} + u_r\frac{\partial \delta f}{\partial r} = \nu\mu_*^2\frac{\partial^2 \delta f}{\partial \mu^2}. \tag{4.59}$$

Note that the last two terms in the left-hand side of (4.59) are of equal order of magnitude: even though $u_r \sim \alpha^{1/2} r u_\psi$ for trapped particles, the scale of radial variation of $\delta f$ is equal for these particles to the "banana" width $\alpha^{1/2}R$, so that $r(\partial\delta f/\partial r)_\mu \sim \alpha^{-1/2}\partial f \sim \alpha^{-1/2}\partial\delta f/\partial\psi$. It is convenient to introduce, in place of $\mu$, a new variable $\tilde{\mathcal{F}}(r, \psi, \varepsilon, \mu)$, which, just as $\mu$, is an integral of the motion; at the same time the value of $\tilde{\mathcal{F}}$ for trapped particles $\tilde{\mathcal{F}}(r, \psi, \varepsilon, \mu_*(r))$ remains practically unchanged when r is changed by a value of the order of the plasma radius R (more accurately, the order of magnitude of the change of $\tilde{\mathcal{F}}$ should not exceed, in this case, the order of magnitude of $\alpha^{1/2}\tilde{\mathcal{F}}$). A change of r at constant $\tilde{\mathcal{F}}$ would not take a point in phase space out of the trapped-particle region, and we would therefore have $r(\partial\delta f/\partial r)_{\tilde{\mathcal{F}}} \sim \delta f$. Changing from the variables r, $\psi$, $\varepsilon$, $\mu$ and r, $\psi$, $\varepsilon$, $\tilde{\mathcal{F}}$ (using the fact that $\tilde{\mathcal{F}}$ is constant along the phase trajectory $u_r\partial\tilde{\mathcal{F}}/\partial r + u_\psi\partial\tilde{\mathcal{F}}/\partial\psi = 0$), we can neglect the terms $u_r(\partial\delta f/\partial r)_{\tilde{\mathcal{F}}}$ compared with $u_\psi(\partial\delta f/\partial\psi)_{\tilde{\mathcal{F}}}$, and the left-hand side of the kinetic equation (4.59) takes the form

$$u_r\frac{\partial f_M}{\partial r} + u_\psi\left(\frac{\partial \delta f}{\partial \psi}\right)_{\tilde{\mathcal{F}}}.$$

We define the variable $\tilde{\mathcal{J}}$ that satisfies the foregoing requirements in the following manner:

$$\tilde{\mathcal{J}} = I(r, \psi, \varepsilon, \mu) - I(r_\mu, 0, \varepsilon, \mu), \tag{4.60}$$

where the radius $r_\mu$ is obtained from the condition $\mu_*(r_\mu) = \mu$. In the vicinity of the line $\mu = \mu_*(r)$ the value of $r_\mu$ differs little from r and can be obtained from the expansion

$$\mu - \mu_*(r) = (r_\mu - r)\frac{\partial \mu_*}{\partial r}, \tag{4.61}$$

where, as follows from the definition of $(\partial I_0/\partial r)(r, \varepsilon, \mu_*) = 0$,

$$\frac{\partial \mu_*}{\partial r} = -\frac{\partial^2 I_0/\partial r^2}{\partial^2 I_0/\partial\mu\partial r}. \tag{4.62}$$

Now expanding I in (4.60), accurate to terms of second order, in powers of the small difference $(r_\mu - r)$ and using (4.61) and (4.62), we easily arrive at the relation

$$\tilde{\mathcal{J}}(r, \psi, \varepsilon, \mu) = \frac{1}{2}\left(\frac{\partial^2 I_0}{\partial r^2}\right)^{-1}\left(\frac{\partial^2 I_0}{\partial\mu\partial r}\right)^2 [\mu - \mu_*(r)]^2 + \delta I(r, \psi) - \delta I(r, 0) \tag{4.63}$$

(for brevity we do not indicate the values of the arguments $\varepsilon$ and $\mu$ in $\delta I$).

Since $\tilde{\mathcal{J}}$ differs from I by a term independent of r and $\psi$ [see (4.60)], the velocity $u_\psi$ is expressed in terms of $\tilde{\mathcal{J}}$, in accord with (4.37"), in which $I_0$ must be replaced by $\tilde{\mathcal{J}}_0$. When differentiating here $\tilde{\mathcal{J}}_0$ with respect to r, it suffices to retain only the derivative of $[\mu - \mu(r)]^2$; the discarded terms will be small in the parameter $\alpha^{1/2}$:

$$u_\psi = -\frac{mc}{ert_\parallel B_0}\frac{\partial^2 I_0}{\partial\mu\partial r}[\mu - \mu_*(r)] = -2\omega\sigma\sqrt{\mathcal{J} - y}. \tag{4.64}$$

The following notation was introduced in the last expression:

$$\mathcal{J} \equiv \frac{2}{r^2}\left(\frac{\partial^2 I_0}{\partial r^2}\right)^{-1}\tilde{\mathcal{J}}, \tag{4.65}$$

$$y \equiv \frac{2}{r^2}\left(\frac{\partial^2 I_0}{\partial r^2}\right)^{-1}[\delta I(r, \psi) - \delta I(r, 0)], \tag{4.66}$$

$$\omega \equiv \frac{mc}{2et_\parallel B_0}\frac{\partial^2 I_0}{\partial r^2}, \tag{4.67}$$

$$\theta \equiv \frac{1}{r}\mu_*(r)\frac{\partial^2 I_0}{\partial\mu\partial r}\left(\frac{\partial^2 I_0}{\partial r^2}\right)^{-1}, \tag{4.68}$$

$$\sigma \equiv \operatorname{sign}\{\theta[\mu - \mu_*(r)]\}.$$

Finally, changing in the right-hand side of the kinetic equation (4.59) to the variable $\tilde{\mathcal{Y}}$ (in this case it is likewise sufficient to differentiate only $[\mu - \mu_*(r)]^2$ in the derivatives $\partial\tilde{\mathcal{Y}}/\partial\mu$), we get

$$\left(\frac{\partial\delta f}{\partial\psi}\right)_{\mathcal{Y}} = -\frac{2\sigma\nu\theta^2}{\omega}\frac{\partial}{\partial\mathcal{Y}}\sqrt{\mathcal{Y}-y}\,\frac{\partial\delta f}{\partial\mathcal{Y}} + \frac{\sigma r}{2\sqrt{\mathcal{Y}-y}}\frac{\partial y}{\partial\psi}\frac{\partial f_M}{\partial r}. \tag{4.69}$$

Since the function y (4.66) is periodic in $\psi$, with a period equal to $\pi/2$, it suffices to consider the variation of $\psi$ in the interval $[-\pi/4, \pi/4]$.

An equation of form (4.69) is encountered in various problems of plasma physics [34-37], and the method of its solution is well known. We shall not dwell, therefore, on the details of the calculation.

We represent $\delta f$ as a series in the small parameter $\nu$:

$$\delta f = \delta f^{(0)} + \delta f^{(1)} + \ldots.$$

We need only the zeroth term $\delta f^{(0)}$ of the expansion. This function differs in form for trapped and untrapped particles. The separatrix that divides these two regions is defined by the condition

$$\mathcal{Y} = \mathcal{Y}_c \equiv \max_{\psi} y(\psi).$$

Denoting the values of $\delta f^{(0)}$ at $\mathcal{Y}<\mathcal{Y}_c$ by $\delta f_t$ and at $\mathcal{Y}>\mathcal{Y}_c$ by $\delta f_u$, and proceeding in the same manner as in [35, 36], we easily obtain

$$\delta f_t = -\sigma r\sqrt{\mathcal{Y}-y}\,\frac{\partial f_M}{\partial r}, \tag{4.70}$$

$$\delta f_u = -\sigma r\left(\sqrt{\mathcal{Y}-y} - \frac{\pi}{4}\int_{\mathcal{Y}_c}^{\mathcal{Y}}\frac{d\mathcal{Y}'}{E(\mathcal{Y}')}\right)\frac{\partial f_M}{\partial r},$$

where

$$E(\mathcal{Y}) = \int_{-\pi/4}^{-\pi/4} d\psi \sqrt{\mathcal{Y} - y(\psi)}. \tag{4.71}$$

The function $\delta f^{(0)}$ is continuous on the separatrix, whereas its derivative has a discontinuity there. As $\mathcal{Y} \to \infty$ [i.e., $|\mu - \mu_*(r)| \gg \Delta\mu_0$, but $|\mu - \mu_*(r)| \ll \mu_*(r)$], $\delta f_u$ tends to a finite limit independent of $\psi$:

$$\lim_{\mathcal{Y}\to\infty} \delta f_u(\mathcal{Y}, \psi) = \sigma \frac{\pi r}{4} \frac{\partial f_M}{\partial r} \left[ \int_{\mathcal{Y}_c}^{\infty} d\mathcal{Y} \left( \frac{1}{E(\mathcal{Y})} - \frac{2}{\pi \mathcal{Y}^{1/2}} \right) - \frac{4}{\pi} \mathcal{Y}_c^{1/2} \right]. \tag{4.72}$$

Note that $\delta f^{(0)}$ is of the same order as $\alpha^{1/2} f_M$.

We can now calculate the flux q due to the particles from that region of phase space for which we solved above the kinetic equation. Using (4.26), we can write the expression for q in the form

$$q = \frac{2\pi c r^2}{em} \int d\varepsilon d\mu d\psi \frac{\partial^2 I_0}{\partial r^2} \delta f \left( \frac{\partial y}{\partial \psi} \right)_{r, \varepsilon, \mu}, \tag{4.73}$$

where the integration with respect to $\mu$ is carried out in the vicinity of $|\mu - \mu_*(r)| \ll \mu_*$. With the aid of relations (4.63) and (4.65), we replace the integration variable $\mu$ by $\mathcal{Y}$, and then the derivative $(\partial y/\partial\psi)_{r,\varepsilon,\mu}$ can be set, at the required accuracy, equal to $(\partial y/\partial\psi)_{r,\varepsilon,\mathcal{Y}}$:

$$q = \frac{\pi c \mu_*}{em} \int d\varepsilon d\psi \frac{\partial^2 I_0/\partial r^2}{|\theta|} \sum_{\sigma} \int_{y}^{\infty} d\mathcal{Y} \delta f \frac{(\partial y/\partial\psi)_{r,\varepsilon,\mathcal{Y}}}{\sqrt{\mathcal{Y} - y}}.$$

Integrating now by parts and using Eq. (4.69), we obtain

$$q = -\frac{8\pi^2 c \mu_* r^2}{em} \int d\varepsilon \frac{\nu |\theta|}{\omega} \frac{\partial^2 I_0}{\partial r^2}$$
$$\times \sum_{\sigma} \sigma \left\langle \int_{y}^{\infty} d\mathcal{Y} \sqrt{\mathcal{Y} - y} \frac{\partial}{\partial \mathcal{Y}} \left( \sqrt{\mathcal{Y} - y} \frac{\partial \delta f}{\partial \mathcal{Y}} \right) \right\rangle.$$

The second integral in this equation can be calculated in elementary fashion and is equal to $-\frac{1}{2} \lim_{\mathcal{Y}\to\infty} \delta f$. Now using relation (4.72) and substituting the equations for $\omega$, $\theta$, $\partial f_M/\partial r$ in accordance with (4.67), (4.68), and (4.48), we write the final expression for q in the form

$$q = \frac{2\sqrt{2}\pi^3\mu_*^2 rB_0 L_0}{m^{3/2}} \left\{ \left( \frac{1}{n}\frac{\partial n}{\partial r} + \frac{e}{T}\frac{\partial \varphi_0}{\partial r} \right) \int\limits_{\mu_* B_0 + e\varphi_0}^{\infty} d\varepsilon \nu\gamma \left| \frac{\partial^2 I_0/\partial\mu\partial r}{\partial^2 I_0/\partial r^2} \right| \right.$$

$$\times \frac{f_M}{\sqrt{\varepsilon - \mu_* B_0 - e\varphi_0}} + \frac{1}{T}\frac{\partial T}{\partial r}$$

$$\left. \times \int\limits_{\mu_* B_0 + e\varphi_0}^{\infty} d\varepsilon \nu\gamma \left| \frac{\partial^2 I_0/\partial\mu\partial r}{\partial^2 I_0/\partial r^2} \right| \left( \frac{\varepsilon - e\varphi_0}{T} - \frac{3}{2} \right) \frac{f_M}{\sqrt{\varepsilon - \mu_* B_0 - e\varphi_0}} \right\}, \quad (4.74)$$

where $\gamma = \gamma(r, \varepsilon)$ denotes the content of the square brackets in (4.72).

To complete the investigation of the banana regime it remains for us to determine the contribution of particles with $|\mu - \mu_*(r)| \gg \Delta\mu_0$ to the diffusion fluxes. To determine the correction $\delta f$ in this phase-space region is here a much more complicated problem, since the approximation (4.57) no longer holds here, and to solve the kinetic equation we must use the nonsimplified collision integral. Fortunately, as we shall presently show, this region makes no contribution to the diffusion fluxes, so that Eq. (4.74) is the final expression for the diffusion rate in the banana regime.

In the left-hand side of the kinetic equation (4.36) for particles with $|\mu - \mu_*(r)| \gg \Delta\mu_0$ we can neglect $u_r \partial\delta f/\partial r$ compared with the two remaining terms, and we obtain, as a result,

$$u_r \frac{\partial f_M}{\partial r} + u_\psi \frac{\partial \delta f}{\partial \psi} = \mathrm{St}\,\delta f. \quad (4.75)$$

We note that since the sought function $\delta f$ should be made continuous with the function $\delta f^{(0)}$ obtained above in the region $\Delta\mu_0 \ll |\mu - \mu_*| \ll \mu_*$, we have in order of magnitude $\partial f \sim \alpha^{1/2} f_M$. The second term in the left-hand side of (4.75) is therefore larger than the first by $\alpha^{-1/2}$ times. We nevertheless retain the term $u_r \partial f_M/\partial r$ so as to be able to determine $\delta f$ accurate to terms $\sim\alpha f_M$. This accuracy will be needed to find the density perturbation $\delta n$ (see below).

Equation (4.75) can be solved in the following manner. We define the function g:

$$g \equiv -\int d\psi \frac{u_r}{u_\psi} \frac{\partial f_M}{\partial r} = \frac{\delta I}{\partial I_0/\partial r} \cdot \frac{\partial f_M}{\partial r} \tag{4.76}$$

and the difference $h = \delta f - g$. We then get from (4.75)

$$u_\psi \frac{\partial h}{\partial \psi} = \mathrm{St}\,(h + g).$$

We seek the solution of this equation in the form of a series in the low collision frequency $\nu$:

$$h = h^{(0)} + h^{(1)} + \ldots .$$

In the zeroth approximation we have $\partial h^{(0)}/\partial\psi = 0$; therefore, the function $h^{(0)}$ does not depend on $\psi$. In the next-order approximation,

$$u_\psi \frac{\partial h^{(1)}}{\partial \psi} = \mathrm{St}\,(h^{(0)} + g).$$

The condition for the last equation to have a solution is obtained by averaging both parts over the azimuth

$$0 = \langle \mathrm{St}\,(h^{(0)} + g)\rangle.$$

Recognizing that $\langle g \rangle = 0$ and hence $\langle \mathrm{St}\, g \rangle = 0$, and that $h^{(0)}$ does not depend on $\psi$, we arrive at the following equation for $h^{(0)}$:

$$\mathrm{St}\, h^{(0)} = 0. \tag{4.77}$$

To determine the function $h^{(0)}$ uniquely we must also take into account the constraints imposed on it by relations (4.38). It is easily seen that the localized part of the distribution function (4.70) and the increment g do not contribute to the integrals (4.38). In fact, the integrals of $\delta f^{(0)}$ are taken over a small vicinity of the value $\mu = \mu_*$, and when integrating with respect to $\mu$ we can take the weighting factors preceding $\delta f^{(0)}$ outside the integral sign. As a result, relations (4.38) for $\delta f^{(0)}$ are satisfied identically, since $\delta f^{(0)}$ is an odd function of the variable $\mu - \mu_*$, and

$$\int d\mu \delta f^{(0)} = 0. \tag{4.78}$$

The function g also transforms (4.38) into identities by virtue of the fact that $\langle g \rangle = 0$. Thus, the constraints (4.38) reduce to the following conditions on $h^{(0)}$:

$$\int h^{(0)} t_{\parallel} \, d\varepsilon d\mu = \int h^{(0)} t_{\parallel} (\varepsilon - e\varphi_0) \, d\varepsilon d\mu = 0,$$

which, together with (4.77) and the condition that $h^{(0)}$ be joined to $\delta f^{(0)}$ at $\Delta\mu_0 \ll |\mu - \mu_*| \ll \mu$, make it possible to determine $h^{(0)}$ uniquely.

We express now the contribution made to the flux q by the nonlocalized part of the distribution function in terms of the collision integral of $\delta f$. To this end, we substitute in (4.26) the expression for $\partial \delta f / \partial \psi$ from (4.75) and note that the term containing $\partial f_M / \partial r$ vanishes on averaging over $\psi$. As a result, we obtain

$$q = -\frac{8\pi^2 cr}{em} \left\langle \int \frac{d\varepsilon d\mu}{u_\psi} \, \delta f \, \mathrm{St} \, \delta f \right\rangle.$$

Together with (4.77), this means that the nonlocalized part of the distribution function $h^{(0)}$ makes no contribution to the fluxes. As for the increment g, it causes a particle flux that is smaller by a factor $\alpha^{-3/2}$ than (4.74) [see the estimate (4.41)].

To conclude this section, we indicate that in the banana regime, just as in the rotational one, there is a perturbation of the ion density and an ensuing perturbation $\delta\varphi$ of the potential. As already noted, by virtue of (4.77), the function $\delta f^{(0)}$ makes no contribution to $\delta n$; therefore, the density perturbation that depends on the azimuth is connected with the correction g. The latter, on the other hand, coincides with the function $\delta f^0$ of the rotational regime. Thus, the perturbation $\delta\varphi$ of the potential is calculated in the banana regime in exactly the same manner as in Section 4.4.†

4.6. "Plateau" Regime. In this and in the succeeding sections we neglect, for simplicity, the perturbation $\delta\varphi$ of the potential when solving the kinetic equation (this is valid, e.g., at $T_e = 0$).

---

†The singularity appearing in this case in the integral (4.53) and due to vanishing of $\partial I_\alpha / \partial r$ must be taken in the sense of the principal value.

Simple estimates show that in the plateau regime one can neglect in the kinetic equation the derivative $\partial\delta f/\partial r$ compared with the remaining terms:

$$u_r \frac{\partial f_M}{\partial r} + u_\psi \frac{\partial \delta f}{\partial \psi} = \mathrm{St}\,\delta f. \tag{4.79}$$

As noted in Section 4.3, in the plateau regime the transport is determined by particles with small values of the difference $\mu - \mu_*(r)$, for which the second term of the left-hand side of (4.79) is of the order of magnitude of the collision term. For such particles we can simplify the expression for $u_\psi$:

$$u_\psi = -\frac{1}{r\omega_H t_\parallel}\frac{\partial I_0}{\partial r} \simeq -\frac{\Delta\mu}{r\omega_H t_\parallel}\frac{\partial^2 I_0}{\partial\mu\partial r}, \quad \Delta\mu \equiv \mu - \mu_*(r),$$

where $\omega_H = eB_0/mc$, the derivative $\partial^2 I_0/\partial\mu\partial r$ is calculated at the point $\mu = \mu_*(r)$, and the collision term, just as in the banana regime, can be taken in the form (4.57). Thus, the final expression for the kinetic equation takes the form

$$\frac{1}{r\omega_H t_\parallel}\frac{\partial \delta I}{\partial\psi}\frac{\partial f_M}{\partial r} - \frac{\Delta\mu}{r\omega_H t_\parallel}\frac{\partial^2 I_0}{\partial\mu\partial r}\frac{\partial\delta f}{\partial\psi} = \nu\mu_*^2\frac{\partial\delta f}{\partial\Delta\mu^2}. \tag{4.80}$$

Taking (3.3) into account, we expand $\delta I$ in a Fourier series:

$$\delta I = \sum_{k=1}^{\infty} I^{(k)}(r, \varepsilon, \mu)\cos 4k\psi. \tag{4.81}$$

We seek the solution of (4.80) in the form

$$\delta f = \sum_{k=1}^{\infty}(g_k \cos 4k\psi + h_k \sin 4k\psi). \tag{4.82}$$

Averaging of (4.82) over $\psi$ yields zero, so that such a solution satisfies relations (4.38). Substituting (4.81) and (4.82) in (4.80), we obtain

$$\left.\begin{aligned} &\frac{4k\Delta\mu}{r\omega_H t_\parallel}\frac{\partial^2 I_0}{\partial\mu\partial r} g_k - \nu\mu_*^2\frac{\partial^2 h_k}{\partial\Delta\mu^2} = \frac{4kI^{(k)}}{r\omega_H t_\parallel}\frac{\partial f_M}{\partial r}, \\ &\frac{4k\Delta\mu}{r\omega_H t_\parallel}\frac{\partial^2 I_0}{\partial\mu\partial r} h_k + \nu\mu_*^2\frac{\partial^2 g_k}{\partial\Delta\mu^2} = 0. \end{aligned}\right\} \tag{4.83}$$

The values of $I^{(k)}$ are calculated in these equations at $\mu = \mu_*(r)$.

Introduction of the dimensionless variables

$$\begin{pmatrix} \tilde{g}_k \\ \tilde{h}_k \end{pmatrix} = \left(\frac{\partial^2 I_0}{\partial\mu\partial r}\right)^{2/3} \left(\frac{r\omega_H \nu\mu_*^2 t_\parallel}{4k}\right) \left(\frac{\partial f_M}{\partial r}\right)^{-1} \begin{pmatrix} g_k \\ h_k \end{pmatrix},$$

$$\zeta = \left(\frac{4k}{r\omega_H \nu\mu_*^2 t_\parallel} \frac{\partial^2 I_0}{\partial\mu\partial r}\right)^{1/3} \Delta\mu$$

transforms (4.83) into the following system of differential equations:

$$\frac{\partial^2 \tilde{h}_k}{\partial\zeta^2} - \zeta\tilde{g}_k = -I^{(k)}, \qquad \frac{\partial^2 \tilde{g}_k}{\partial\zeta^2} + \zeta\tilde{h}_k = 0. \tag{4.84}$$

The boundary conditions for the system (4.84), which determine uniquely the sought solution, are that $\tilde{g}_k$, $\tilde{h}_k \to 0$ as $\zeta \to \pm\infty$. To solve (4.84) it is convenient, after fixing k, to introduce the complex variable $v = (-i\tilde{h}_k + \tilde{g}_k)/I^{(k)}$. The system (4.84) is transformed thereby into the equation

$$-i\frac{d^2 v}{d\zeta^2} + \zeta v = 1. \tag{4.85}$$

It is easy to obtain an explicit expression for v by using the Laplace method; this expression is given, e.g., in [38]. We need only the following properties of the function v (see [38]):

$$\int_{-\infty}^{\infty} \operatorname{Im} v\, d\zeta = -\pi, \qquad \int_{-\infty}^{\infty} \operatorname{Re} v\, d\zeta = 0.$$

From this we find that

$$\int_{-\infty}^{\infty} \tilde{g}_k d\zeta = 0, \qquad \int_{-\infty}^{\infty} \tilde{h}_k d\zeta = \pi I^{(k)}. \tag{4.86}$$

We note also that, as follows from the system (4.84), at $|\zeta| \gg 1$ [in dimensional variables at $\Delta\mu \gg \mu_*(\nu\tau_D)^{1/3}$] we have

$$\tilde{g}_k = \frac{1}{\zeta} I^{(k)}.$$

As is already verified, this means that in the region $\mu_*(\nu\tau_D)^{1/3} \ll |\Delta\mu| \ll \mu_*$ the function $\delta f$ is matched to the correction $\delta f^{(0)}$ for the Maxwellian distribution function from the rotational regime (see Section 4.4), which describes in the plateau regime particles with $\Delta\mu \gg \mu_*(\nu\tau_D)^{1/3}$.

In the calculation of the flux (4.26) it must be borne in mind that the function $\delta f$ differs from zero only in a narrow vicinity of the value $\mu = \mu_*(r)$. In the integration with respect to $\mu$ the weighting factors in (4.26) can therefore be regarded as constant. Equations (4.86) enable us then to calculate the ensuing integrals. The result takes the form

$$q = -\frac{8\pi^3 B_0}{m^3 \omega_H L_0}\left\{\left(\frac{1}{n}\frac{\partial n}{\partial r} + \frac{e}{T}\frac{\partial \varphi_0}{\partial r}\right)\sum_{k=1}^{\infty} k \int_{\mu_* B_0 + e\varphi_0}^{\infty} d\varepsilon f_M (I^{(k)})^2 \left|\frac{\partial^2 I_0}{\partial \mu \partial r}\right|^{-1} + \frac{1}{T}\frac{\partial T}{\partial r}\sum_{k=1}^{\infty} k \int_{\mu_* B_0 + e\varphi_0}^{\infty} d\varepsilon f_M \left(\frac{\varepsilon - e\varphi_0}{T} - \frac{3}{2}\right)(I^{(k)})^2 \left|\frac{\partial^2 I_0}{\partial \mu \partial r}\right|^{-1}\right\}. \quad (4.87)$$

4.7. Strong-Collision Regime. At $\nu\tau_D \gg 1$, the solution of the kinetic equation (4.79) can be sought, as in the preceding section, in the form (4.82). Substituting (4.82) and (4.81) in (4.79), we obtain

$$-\frac{4k}{r\omega_H t_\parallel} I^{(k)} \frac{\partial f_M}{\partial r} - 4ku_\psi g_k = \mathrm{St}\, h_k, \quad (4.88)$$

$$4ku_\psi h_k = \mathrm{St}\, g_k. \quad (4.89)$$

We can look for a solution of the system (4.88), (4.89) in the form of a series in the small parameter $(\nu\tau_D)^{-1}$. Since it follows from the simple estimates of Section 4.3 that the particle flux (4.26),

$$q = -\frac{8\pi^2 B_0}{m^2 \omega_H}\sum_{k=1}^{\infty} k \int d\varepsilon d\mu I^{(k)} h_k, \quad (4.90)$$

is proportional to $\nu^{-1}$, we must assume that the series expansion of the functions $h_k$ begins with first-order terms

$$\left.\begin{aligned} h_k &= h_k^{(1)} + h_k^{(2)} + \ldots, \\ g_k &= g_k^{(0)} + g_k^{(1)} + \ldots \end{aligned}\right\} \quad (4.91)$$

[the superscript indicates the power of the parameter $(\nu\tau_D)^{-1}$].

Substituting now the expansions (4.91) in (4.89), we find that the function $g_k^{(0)}$ is determined from the equation

$$\mathrm{St}\, g_k^{(0)} = 0 \quad (4.92)$$

and, consequently (see, e.g., [39]),

$$g_k^{(0)} = \frac{\delta n}{n} f_M + \frac{v^2}{v_T^2} \frac{\delta T}{T} f_M , \qquad (4.93)$$

where we have introduced the still undetermined constants $\delta n$ and $\delta T$, as well as the thermal velocity $v_T = \sqrt{2T/m}$. The solution (4.93) describes the azimuthal density and temperature perturbations that appear in the course of diffusion. The quantities $\delta n$ and $\delta T$ are determined from a system of linear equations that are obtained by integrating (4.88) with weights 1 and $v^2$.

$$\left. \begin{aligned} -\frac{1}{r\omega_H} \int \frac{I^{(k)}}{t_\parallel} \frac{\partial f_M}{\partial r} d^3v &= \frac{\delta n}{n} \int u_\psi f_M d^3v + \frac{\delta T}{T v_T^2} \int u_\psi v^2 f_M d^3v, \\ -\frac{1}{r\omega_H} \int \frac{v^2 I^{(k)}}{t_\parallel} \frac{\partial f_M}{\partial r} d^3v &= \frac{\delta n}{n} \int u_\psi v^2 f_M d^3v + \frac{\delta T}{T v_T^2} \int u_\psi v^4 f_M d^3v. \end{aligned} \right\} \qquad (4.94)$$

It is easily seen that, in order of magnitude,

$$\frac{\delta n}{n} \sim \frac{\delta T}{T} \sim \frac{\delta r}{r}.$$

With the aid of $g_k^{(0)}$ obtained in this manner, the function $h_k^{(1)}$ is determined from Eq. (4.88):

$$-\frac{4k}{r\omega_H t_\parallel} I^{(k)} \frac{\partial f_M}{\partial r} - 4k u_\psi g_k^{(0)} = \mathrm{St}\, h_k^{(1)}. \qquad (4.95)$$

For a unique determination of $h_k^{(1)}$ it is necessary to supplement (4.95) with the conditions that arise upon integration of (4.89) with weights 1 and $v^2$:

$$\int u_\psi h_k^{(1)} d^3v = \int u_\psi v^2 h_k^{(1)} d^3v = 0.$$

Finally, once $h_k^{(1)}$ is obtained, calculation of the flux q by means of Eq. (4.90) solves our problem.

Calculations by the scheme described above require specification of the dependences of the functions $I^{(k)}$ and $u_\psi$ on $\varepsilon$ and $\mu$, and are apparently impossible without resorting to numerical methods. The problem becomes somewhat simpler if the Bhatnager–Gross–Krook model collisional term [40] is used. An example of such a calculation can be found in [41].

## 5. Resonant Stochastic Transport in Open Traps

5.1. Qualitative Treatment. Before we calculate the resonant transfer coefficients, we classify qualitatively the possible regimes and estimate the diffusion coefficient. Bearing it in mind that the resonant regime is realized for ions in an ambipolar trap, we refer in this section, to be specific, to the ion component.

Assume satisfaction of the inequality

$$\left| a \frac{\partial \Delta\psi}{\partial r} \right| \ll 1; \tag{5.1}$$

then, as shown in Section 3.5, the radial random walks $\Delta x_r$ (3.46) of the trapped particles are large compared with the radial drift displacement, of the order of $a$, of the nonresonant particles. It will be shown below that the main contribution to the transport is made in this case precisely by the resonant particles. The situation when the inverse of inequality (5.1) holds is the subject of Section 5.5.

In the plane of the variables $\varepsilon$, $\mu$ resonant particles on a given radius r correspond to the vicinities of certain curves, also called hereafter resonant. They are obtained from the equation

$$\Delta\psi(r, \varepsilon, \mu) = (2k+1)\frac{\pi}{2}. \tag{5.2}$$

The widths of the vicinities $\Delta\mu_r$ along the $\mu$ axis and $\Delta\varepsilon_r$ along the $\varepsilon$ axis are determined by the condition that a change $\Delta\mu_r$ in the magnetic moment of the particle or $\Delta\varepsilon_r$ in its energy causes a displacement in terms of $\Delta\psi$ by an amount equal to the width $|a\partial\Delta\psi/\partial r|^{1/2}$ of the resonance:

$$\left.\begin{aligned} \frac{\Delta\mu_r}{\mu} &\sim \frac{1}{\mu}\left|\frac{\partial\Delta\psi}{\partial\mu}\right|^{-1}\left|a\frac{\partial\Delta\psi}{\partial r}\right|^{1/2} \sim \sqrt{\frac{|a|}{R\Delta\psi}}, \\ \frac{\Delta\varepsilon_r}{\varepsilon} &\sim \frac{1}{\varepsilon}\left|\frac{\partial\Delta\psi}{\partial\varepsilon}\right|^{-1}\left|a\frac{\partial\Delta\psi}{\partial r}\right|^{1/2} \sim \sqrt{\frac{|a|}{R\Delta\psi}}. \end{aligned}\right\} \tag{5.3}$$

For rough estimates we have introduced in (5.3) a certain characteristic value of $\Delta\psi$ and assumed that the function $\Delta\psi(r, \varepsilon, \mu)$ has a smooth dependence on $\varepsilon$ and $\mu$ in the significant region of its variation: $|\partial\Delta\psi/\partial\mu| \sim \Delta\psi/\mu$ and $|\partial\Delta\psi/\partial\varepsilon| \sim \Delta\psi/\varepsilon$. With the aid of (5.3) we find that the effective collision frequency for the resonant particles, of the order of magnitude of the reciprocal time of stay inside the resonant strip, is of the form

$$\nu_{ef} \sim \nu_i \left( \frac{\mu}{\Delta\mu_r} \right)^2,$$

where $\nu_i$ is the reciprocal time of ion scattering through an angle of order unity.

Estimates of the ion diffusion coefficient depend on the ratio of $\nu_{ef}$ to the period $\tau_r$ of the resonant-particle motion. The case $\nu_{ef} \lesssim \tau_r^{-1}$ is the analog of the banana regime of the neoclassical theory. The contribution made to the diffusion coefficient by one resonance curve is then equal to

$$(\Delta x_r)^2 \nu_{ef} \frac{\Delta\mu_r}{\mu}$$

(the last factor allows for the small fraction of the trapped particles). This expression must be multiplied by the number $N_{res}$ of the resonance curves, which accounts for the possibility of various values of k in Eq. (5.2). In order of magnitude we have

$$N_{res} \sim \Delta\psi/\pi.$$

With the aid of (5.3) we now obtain an estimate of the diffusion coefficient

$$\mathscr{D} \sim \nu_i R^2 \left| a \frac{\Delta\psi}{R} \right|^{1/2}. \tag{5.4}$$

The inverse limiting case $\tau_r^{-1} \lesssim \nu_{ef}$ is an analog of the plateau regime of the neoclassical transport theory. In this regime the diffusion is determined by particles with values of $\Delta\mu$ ($\Delta\mu \gg \Delta\mu_r$), for which the time $\tau$ of encircling the trajectory becomes comparable with reciprocal effective collision frequency $\tilde{\nu}_{ef}^{-1} \sim \nu_i^{-1} (\Delta\mu/\mu)^2$. The time $\tau$ and the size of the radial random walk $\Delta x$ of particles with $\Delta\mu \gg \Delta\mu_r$ (but with $\Delta\mu \ll \mu$) are estimated as before with the aid of (3.44). It must only be taken into account that for particles whose magnetic moment differs by $\Delta\mu$ from the resonant value $\mu$ on a given radius r (at fixed $\varepsilon$) the resonant value $R_k$ is shifted from r by a distance $x \sim \Delta\mu (\partial\Delta\psi/\partial\mu)(\partial\Delta\psi/\partial r)^{-1}$. Therefore,

$$\tau \sim t_{\parallel} \frac{\mu}{\Delta\mu\Delta\psi}, \quad \Delta x \sim \frac{a\mu}{\Delta\mu\Delta\psi}. \tag{5.5}$$

From the condition $\tau \sim \tilde{\nu}_{ef}^{-1}$ we find that

$$\Delta\mu \sim \mu \left| \frac{\nu_i t_\parallel}{\Delta\psi} \right|^{1/2}. \tag{5.6}$$

Estimating now the diffusion coefficient to be $\mathcal{D} \sim N_{res}(\Delta x)^2 \times \tilde{\nu}_{ef}\Delta\mu/\mu$, we get

$$\mathcal{D} \sim a^2/t_\parallel. \tag{5.7}$$

Further increase of the collision frequency upsets the conditions for applicability of the plateau regime. The reason is that the width of the region (5.6) in which the particles that effect the transport are located becomes comparable between the distance $\Delta\mu$ between the resonance curves, viz.,

$$\Delta\mu \sim \frac{\pi}{\partial\Delta\psi/\partial\mu}.$$

This occurs when

$$\nu_i \gtrsim t_\parallel^{-1}\left(\mu\frac{\partial\Delta\psi}{\partial\mu}\right)^{-2} \sim \frac{1}{t_\parallel(\Delta\psi)^2}. \tag{5.8}$$

The situation at higher collision frequencies is discussed in Section 5.5.

5.2. Calculation of the Diffusion Flux. We note first that, in contrast to neoclassical regimes, in the resonant regime the ion distribution function in the equatorial plane depends not only on the variables $r$, $\psi$, $\varepsilon$, and $\mu$ but also on the sign of the $z$ component of the particle velocity. We shall distinguish between particles with different sign of $v_z$ of the discrete variable $\sigma = \text{sign}\, v_z$. For example, where necessary, the dependence of $f$ on $\sigma$ will be designated by a subscript: $f_\sigma(r, \psi, z, \varepsilon, \mu)$.

The kinetic equation for finding the distribution function of the leading centers in the equatorial plane is obtained from the stationary drift kinetic equation with a collision term:

$$v_\parallel \frac{\partial f}{\partial s} + \mathbf{v}_D \nabla f = \operatorname{St} f, \tag{5.9}$$

where $v_\parallel$ and $v_D$ are the longitudinal and drift velocities of the particle. We divide Eq. (5.9) by $v_\parallel$ and integrate it along a phase-trajectory segment whose initial and final points lie in the equatorial plane. It is necessary to choose the segment negotiated by the particle in a time $4t_{\|}$ for resonant particles, and in a time $2mt_{\|}$ for nonresonant ones (see Section 3.5), and then the initial and final points are close to each other. Since the ion mean free path is much longer than the trap length L,* the difference between the values of the distribution function on the ends of the segment is small and can be represented as $\Delta \mathbf{r} \nabla f$, where $\Delta \mathbf{r}$ is the vector drawn from the initial to the final point:

$$\Delta \mathbf{r} \nabla f = \int_{\text{init}}^{\text{fin}} \frac{ds}{v_{\|}} \operatorname{St} f.$$

Dividing the last equation by the time of motion $4t_{\|}$ (or $2mt_{\|}$ for nonresonant particles), we arrive at the equation

$$\dot{\psi} \frac{\partial f}{\partial \psi} + \dot{r} \frac{\partial f}{\partial r} = \overline{\operatorname{St} f}, \qquad (5.10)$$

where $\overline{\operatorname{St} f}$ is the collision term averaged along the phase trajectory. In the left-hand side of (5.10) we must set the value of z in f equal to zero, which is its value in the equatorial plane.

Expressions for $\dot{r}$ and $\dot{\psi}$, corresponding to particles with $\sigma = +$, were obtained in Section 3.5 and are given by Eqs. (3.44) and (3.51) for resonant and nonresonant particles, respectively. As for the function $f_-$, it can be expressed in terms of $f_+$ if it is allowed that a particle with $\sigma = -$ and with coordinates r, $\psi$ had crossed the equatorial plane during the preceding step in a positive direction at the point $r - a \sin(2\psi - \Delta\psi)$, $\psi - \Delta\psi$ (see Section 3.5), and using the approximate constancy of f along a phase-trajectory segment limited to two successive transits through the equatorial plane:

$$f_-(r, \psi, z=0, \varepsilon, \mu) = f_+(r - a \sin(2\psi - \Delta\psi), \ \psi - \Delta\psi, \ z=0, \ \varepsilon, \ \mu). \qquad (5.11)$$

---

*Obviously, it should also be large compared with mL. In other words, $\delta\psi_0$ of Section 3.5 must be chosen such as to have $2\pi L/\delta\psi_0 \ll \lambda$, and then $mL \ll \lambda$. Since $L \ll \lambda$, this can always be done without violating the condition $\delta\psi_0 \ll 2\pi$.

Since the radial random walks are maximal for resonant particles, the maximum distortion of the Maxwellian distribution function in the presence of radial density and temperature gradients takes place along ($\varepsilon$, $\mu$)-plane lines that are defined by the resonance condition (5.2). When solving the kinetic equation along a resonance curve with a given value of k, it is convenient to transform from $\varepsilon$ and $\mu$ to new variables, one of which remains constant along this curve, e.g., $R_k$ and $\mu$. In terms of these variables the correction $\delta f$ to the Maxwellian distribution function changes by a factor of 2 when $R_k$ changes by a small amount, and depends little on $\mu$. This allows us to retain in the general expression for St f only terms that contain two differentiations with respect to $R_k$ (cf. the corresponding approximation in Section 4.5). We can then put $R_k = r$ in the coefficient of the second derivative with respect to $R_k$. The result takes the form (see Appendix 5)

$$\mathrm{St} f \simeq \nu_t(r, \mu) r^2 \frac{\partial^2 \delta f}{\partial R_k^2}, \tag{5.12}$$

where

$$\nu_t(r, \mu) = \frac{\nu_0(r)}{r^2}\left[\mu^2\left(\frac{\partial R_k}{\partial \mu}\right)^2 G_1 + \mu\varepsilon \frac{\partial R_k}{\partial \mu}\frac{\partial R_k}{\partial \varepsilon} G_2 + \varepsilon^2\left(\frac{\partial R_k}{\partial \varepsilon}\right)^2 G_3\right],$$

$$G_2 = 2h/\chi^{5/2}, \quad G_3 = h/\chi^{5/2},$$

while the functions $\nu_0(r)$, $G_1$, and h, and the parameters $\zeta$ and $\chi$ are defined as in Section 4.5.

We now note that we cannot simply remove the averaging symbol from the right-hand side of kinetic equation (5.10), as was done in Section 4.5. The reason is that the value of the distribution function and the phase-space point r, $\psi$, z, $\varepsilon$, $\mu$ is equal, by virtue of the constancy of f along the phase trajectory, to the value of $f_+$ in the equatorial plane at the point r, $\psi - \Delta\psi(r, \varepsilon, \mu)z/L$, $\varepsilon$, $\mu$. Therefore, in view of the integral character of the collision term, in the calculation of $\overline{\mathrm{St} f}$ the values of $f_+$ corresponding to different azimuths in the equatorial plane become "entangled." However, the collision term (5.12) in the solution of the kinetic equation in a small vicinity of the resonance curves contains only the value of the distribution function of those resonant particles for which one can neglect the dependence of $\Delta\psi$ on $\varepsilon$ and $\mu$, and we can put $\Delta\psi = (2k + 1)\pi/2$. In this case the averaged collision term is equal simply to St f from the distribution function taken at the corresponding point of the equatorial plane. The coordinate z thus drops out completely from Eq. (5.10).

To find now the correction $\delta f$ to the Maxwellian distribution function, we must first consider the resonant plateau regime, when $\tau_r^{-1} \ll \nu_{ef}$. To solve (5.10) in the vicinity of the resonance curve (5.1), it is convenient to transform in the collision term (5.2) from the variable $R_k$ to the variable $x = r - R_k(r, \varepsilon, \mu)$, so that $\delta f$ becomes a function of $r, \psi, x(r, \varepsilon, \mu), \mu$. In that case, $\delta f_+$ satisfies the equation

$$\frac{x}{t_\parallel}\frac{\partial \Delta\psi}{\partial r}\frac{\partial \delta f_+}{\partial \psi} + (-1)^k \frac{a}{t_\parallel}\cos 2\psi \frac{\partial f_M}{\partial r} = \nu r^2 \frac{\partial^2 \delta f_+}{\partial x^2}. \quad (5.13)$$

Finding its solution in the form

$$\delta f_+ = w\cos 2\psi + u \sin 2\psi \quad (5.14)$$

and proceeding as in Section 4.6, we arrive at

$$u - iw = v(-1)^{k+1}\frac{a}{2r\dfrac{\partial \Delta\psi}{\partial r}}\left(\frac{\nu_i t_\parallel}{2r\dfrac{\partial \Delta\psi}{\partial r}}\right)^{-1/3}\frac{\partial f_M}{\partial r}, \quad (5.15)$$

where the function $v(\zeta)$ is a solution of Eq.(4.85), and the variable $\zeta$ is connected with x by the relation

$$x = r\left(\frac{\nu_i t_\parallel}{2r\dfrac{\partial \Delta\psi}{\partial r}}\right)^{1/3}\zeta. \quad (5.16)$$

Calculating now the particle flux q from the equation

$$q = 2\pi L_0 r\left\langle \int \delta f \dot{x} d^3v\right\rangle = \frac{4\pi^2 r L_0 B_0}{m_i^2}\sum_\sigma\left\langle \int \delta f_\sigma \dot{x}\frac{d\varepsilon d\mu}{|v_\parallel|}\right\rangle \quad (5.17)$$

[the factor $2\pi r L_0$ in (5.17) is due to the fact that q is the particle flux from the entire length of the setup], we get

$$q = \frac{2\pi^3 r B_0}{m_i^2}\sum_k \int d\varepsilon d\mu a^2 \frac{\partial f_M}{\partial r}\,\delta\left(\Delta\psi - (2k+1)\frac{\pi}{2}\right). \quad (5.18)$$

In the resonant banana regime, when $\tau_r^{-1} \gg \nu_{ef}$, the kinetic equation (5.10), notation apart, coincides with the equation solved for $\delta f$ in Section 4.5. Therefore, omitting intermediate calculations, we present directly the final results:

$$q = -\frac{2^{5/2}\pi^3\gamma r^3 B_0}{m_i^2} \sum_k \int d\varepsilon d\mu v t_{\parallel} \delta\left(\Delta\psi - (2k+1)\frac{\pi}{2}\right) \left| a \frac{\partial \Delta\psi}{\partial r} \right|^{1/2} \frac{\partial f_M}{\partial r}. \quad (5.19)$$

The numerical coefficient $\gamma$ is equal to

$$\gamma = \frac{4}{\pi} - 2\int_0^1 \frac{dt}{t^2}\left[\frac{1}{E(t)} - \frac{2}{\pi}\right] = 0.88,$$

where E(t) is an elliptic integral of the second kind.

The calculation method described above was implemented in [15]. Expressions (5.18) and (5.19) were obtained in [42] by another method (without solving the kinetic equation explicitly).

Just as in the neoclassical regimes, the deviation of the distribution function from Maxwellian in resonant transport leads to a density perturbation $\delta n(r, \psi, z)$ and accordingly to a potential perturbation $\delta\varphi(r, \psi, z)$ on the magnetic surface r = const. The radial electric drift produces on the long part of the magnetic bottle, under the influence of the field $\delta E_\psi = -r^{-1}\partial\delta\varphi/\partial\psi$, an additional radial displacement $\Delta r_\varphi$ whose order of magnitude, generally speaking, is equal to the displacement $a$ in the mirrors. The procedure for calculating $\Delta r_\varphi$ will be illustrated below using as the example the resonant banana regime, under the assumption that the electron density in the entire volume of the magnetic bottle (and not only along the force line), obeys the Boltzmann distribution:

$$n_e = \text{const } e^{\frac{e\varphi}{T_e}}. \quad (5.20)$$

This assumption, while not fundamental, permits a noticeable simplification of the calculation. In addition, calculation of $\Delta r_\varphi$ with the aid of (5.20) is of particular interest for a tandem trap, where the electron Boltzmann distribution can set in as a result of their intermixing through the outer magnetic bottles (see [11]).

As follows from the results of Section 4.5, the localized part of the distribution function does not contribute to $\delta n$, which is due entirely to the nonlocalized part of $\delta f$ (4.76). Substituting for $u_r$ and $u_\psi$ in the integral of (4.76) the expressions (3.51), we find that in the resonance regime the $\psi$-dependent nonlocalized part of the distribution function $g_+$ takes the form

$$g_+ = \frac{1}{2}\,\frac{\partial f_M}{\partial r}\,\frac{a}{\cos\Delta\psi}\sin 2\psi. \tag{5.21}$$

The function $g_-$ is obtained by applying (5.11) to the function $f = f_M + g$:

$$g_-(r,\ \psi) = g_+(r,\ \psi - \Delta\psi) + f_M(r - a\sin(2\psi - \Delta\psi)) - f_M(r) = -\frac{1}{2}\,\frac{\partial f_M}{\partial r}\,\frac{a}{\cos\Delta\psi}\sin 2\psi. \tag{5.22}$$

Moving along the phase trajectories, we can determine g in the entire volume of the magnetic bottle from the distribution functions (5.21), (5.22) obtained in the equatorial plane:

$$g_+(r,\ \psi,\ z) = g_+\left(r,\ \psi - \Delta\psi\frac{z}{L},\ 0\right),$$

$$g_-(r,\ \psi,\ z) = g_-\left(r,\ \psi + \Delta\psi\frac{z}{L},\ 0\right),$$

and then calculate $\delta n_i$ by integrating. The result is

$$\delta n_i = -A(r,\ z)\cos 2\psi,$$

where

$$A(r,\ z) = \frac{2\pi B_0}{m_i^2}\int\frac{d\varepsilon d\mu}{|v_\parallel|}\,\frac{a}{\cos\Delta\psi}\sin\left(\frac{2z}{L}\Delta\psi\right)\frac{\partial f_M}{\partial r}.$$

With the aid of (5.20), we get from the quasineutrality condition

$$\delta\varphi = \frac{Te}{en}\,\delta n.$$

Since the radial component of the drift velocity is given by

$$v_r = c\delta E_\psi / B_0 = -c \frac{\partial \delta\varphi}{\partial \psi} \Big/ rB_0,$$

we have for the additional radial displacement $\Delta r_\varphi$:

$$\Delta r_\varphi = -\frac{c}{rB_0} \int dt \frac{\partial \delta\varphi}{\partial \psi}.$$

Calculating the last integral along the phase-trajectory segment which was considered in the derivation of the transformation (3.38), (3.39) (we designate the coordinates of this segment as $r_0$ and $\psi_0$), we get, after simple calculations,

$$\Delta r_\varphi = \cos 2(\psi_0 + \Delta\psi) \sin \Delta\psi \frac{4\pi c L_0 T_e}{enr_0 \, | v_\parallel | \, m_i^2} \int \frac{d\varepsilon d\mu}{| v_\parallel |} a \frac{\partial f_M}{\partial r} \left(1 - \frac{1}{\Delta\psi}\right) \tan \Delta\psi. \tag{5.23}$$

This quantity should be added to the right-hand side of (3.38), so that the transformation (3.39) leads to replacement of $a$ by the renormalized value $\tilde{a}$:

$$a \to \tilde{a} = a - \frac{2\pi c T_e L}{enrm_i^2 \, | v_\parallel |} \int \frac{d\varepsilon d\mu}{| v_\parallel |} \tilde{a} \tan \Delta\psi \left(1 - \frac{1}{\Delta\psi}\right) \frac{\partial f_M}{\partial r}. \tag{5.24}$$

The quantity $\tilde{a}$ should be substituted for $a$ in all the equations pertaining to the resonant banana regime and, in particular, in the flux (5.19) and in expression (5.23) for $\Delta r_\varphi$ [the last circumstance was already taken into account in (5.24)]. Solving this integral equation with the degenerate kernel, we obtain

$$\tilde{a} = a - \frac{1}{| v_\parallel |} \left( \int \frac{d\varepsilon d\mu}{| v_\parallel |} a \left(1 - \frac{1}{\Delta\psi}\right) \tan \Delta\psi \frac{\partial f_M}{\partial r} \right) \times \left( \frac{enrm_i^2}{2\pi c L T_e} + \int \frac{d\varepsilon d\mu}{v_\parallel^2} \left(1 - \frac{1}{\Delta\psi}\right) \tan \Delta\psi \frac{\partial f_M}{\partial r} \right)^{-1}.$$

We point out, in conclusion, that an example of the calculations of the transport coefficient in the resonance regime can be found for actual installations in [43] (neglecting the renormalization of $a$).

5.3. Influence of Narrow-Peak Transport on the Reflection Function. In the treatment of resonance diffusion in Sections 5.1 and 5.2 above, we have assumed throughout that the reflection function $a(r, \varepsilon, \mu)$ is a smooth function of its arguments. This may not be so in some cases, and the reflection function in individual regions of the $\varepsilon$, $\mu$ plane can change significantly following small changes of its arguments. By way of illustration, we present the calculated form of the function $a$ for the magnetic field of the AMBAL trap [14]. It is convenient in this case to express $a$ not in terms of $\varepsilon$ and $\mu$ but in terms of the velocity v of the particle prior to entry into the trap and the angle $\theta$ between the direction of the velocity and the magnetic field (the pitch angle) in the region of the uniform field. Then

$$a(r, v, \theta) = rvA(\theta);$$

the form of the function $A(\theta)$ is determined by configuration of the magnetic field inside the trap. A plot of $A^2(\theta)$ for the AMBAL is shown in Fig. 13. It can be seen from this figure that the plot of $A(\theta)$ has a strongly pronounced peak at $\cos\theta = 0.75$ with a width $\Delta\theta \simeq 0.15$ that is small compared with $\pi/2$. This peak appears because the mirrors of the AMBAL have rather long transition sections with almost constant value of B, corresponding to the approximate value 2.3 of the local mirror ratio. Particles with pitch angle close to arccos 0.75 stay a long time in this section and move away from the initial magnetic surface by drifting in the nonaxisymmetric magnetic field. Another singularity of the $A(\theta)$ dependence (not shown in Fig. 13) exists in the vicinity of the loss-cone angle $\theta_c = \arcsin k^{-1/2}$, where k is the mirror ratio. Its cause is the same as before: as $\theta$ tends to $\theta_c$, the time spent by the particle near the field maximum increases and leads to a corresponding increase of the displacement $\Delta r$.

As noted first in [14], the presence of sharp peaks in the reflection function can lead to an increase of the transport velocity in the neoclassical regime, at $\Delta\psi \ll 1$ (see Section 3.1). The effect is appreciable if the time $\tau_{\Delta\theta} \sim \nu^{-1}\Delta\theta^2$ of scattering of the ions through an angle $\Delta\theta$ is shorter than their time $t_\parallel \sim L/v$ of passage through the trap. A particle with a value $\Delta\theta$ inside the peak, after a reflection from the mirror and a radial displacement $a$ (by $a$ is meant the characteristic value of the reflection function inside the peak), is scattered on moving toward the opposite mirror and leaves the peak region, so that the cancellation of the displacements from the field-reversed mirrors, which is typical of the neoclassical regime, does not take place. After a

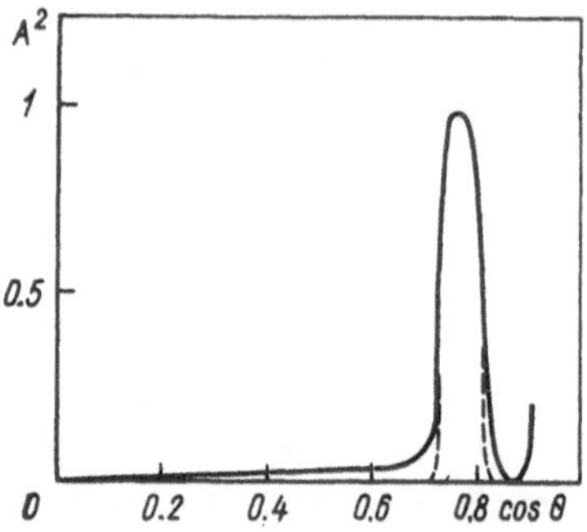


Fig. 13. The function $A^2(\theta)$ for the central magnetic bottle of the AMBAL trap (in relative units).

time $\tau_{\Delta\theta}$, a fraction $\tau_{\Delta\theta}/t_{\parallel}$ of all the particles contained in the trap is reflected from the mirror, and only $\sim\Delta\theta$ of them are located inside the peak. The total diffusion coefficient can therefore be estimated at*

$$\frac{a^2}{\tau_{\Delta\theta}}\frac{\tau_{\Delta\theta}}{t_{\parallel}}\Delta\theta \sim \nu a^2 \frac{\tau_{\Delta\theta}}{t_{\parallel}\Delta\theta}.$$

At small $\Delta\theta$ the inequality $\tau_{\Delta\theta} \ll t_{\parallel}$ is compatible with the inequality $\tau_{\Delta\theta}/t_{\parallel}\Delta\theta \gg 1$, and it is this which enables the transport coefficient to exceed the estimate $\nu a^2$.

Under real conditions, however, the inequality $\tau_{\Delta\theta} \ll t_{\parallel}$ is valid only for very long traps, in which $\Delta\psi$ is already much larger than unity. This raises the question of how the peaks of the reflection function affect the transport coefficients in the resonance regime. The appropriate estimates were made in [44], and more accurate calculations based on the solution of the kinetic equation are given in [45].

We confine ourselves below to an estimate of the effect in a resonance regime as applied to the following model of the reflection function. We assume that the dependence of $a$ on $\theta$ is of the form shown in Fig. 13 by the dashed line, and denote by $a$ and $\Delta\theta$, respectively, the characteristic

*Additional subtle effects can arise in the case not considered here, when the peak of the function $(\theta)$ has abrupt edges (of width less than $\Delta\theta$); see [14].

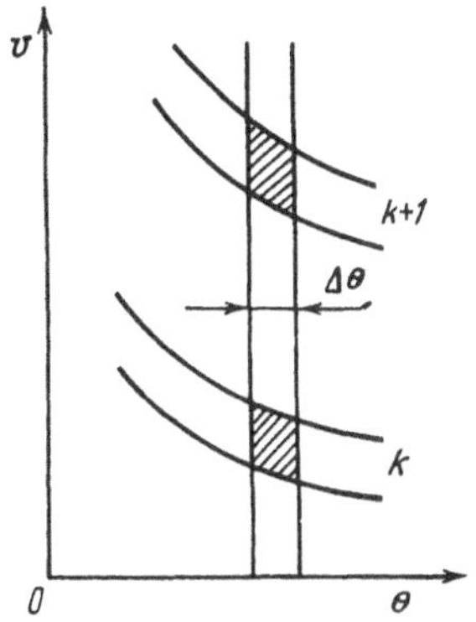


Fig. 14. Plane of the variables v and θ. The regions where the resonance bands cross the band of the peak of the function *a* are shaded.

value of *a* and the width of the peak. The presence of the peak produces on the (v, θ) plane a narrow band in which *a* differs from zero and which intersects the resonance curves (5.2) (Fig. 14).

We consider first the case when the relative width of the resonance band (5.3) is larger than Δθ:

$$\frac{\Delta\mu_r}{\mu},\quad \frac{\Delta\varepsilon_r}{\varepsilon} \gtrsim \Delta\theta. \tag{5.25}$$

When estimating the contribution made by the peak particles to the diffusion coefficient, account must be taken of those particles which satisfy simultaneously the resonance condition (5.2), i.e., lie in the shaded region of Fig. 14. The role of the effective collision frequency $\nu_{ef}$ is played for them by the quantity $\nu/(\Delta\theta)^2$, which equals the reciprocal of the time that the particle stays on the intersection of the two bands. Using the estimate (3.46) of the radial displacement of the resonant particle, we find, just as in Section 5.1, that in the banana regime, when $\nu_{ef} < \tau_r^{-1}$,

$$\mathscr{D} \sim N_{res}\frac{\nu}{(\Delta\theta)^2}(\Delta x_r)^2\frac{\Delta\mu_r}{\mu}\Delta\theta \sim \nu R^2\left(\frac{a}{R}\right)^{3/2}\frac{1}{\Delta\theta\sqrt{\Delta\psi}},$$

$$\nu t_{\parallel} \lesssim \left(\frac{a\Delta\psi}{R}\right)^{1/2}(\Delta\theta)^2. \tag{5.26}$$

In the plateau regime the peak particles contributing to $\mathcal{D}$ will be those located within a certain width $\Delta\mu$ of the resonance curve. The value of $\Delta\mu$ is obtained from the condition $\nu_{ef} \sim \tau^{-1}$, where $\tau$ should be obtained from expression (5.5):

$$\frac{\Delta\mu}{\mu} \sim \frac{\nu t_{\parallel}}{\Delta\psi(\Delta\theta)^2}. \qquad (5.27)$$

With the aid of the estimate (5.5) we obtain for the displacement $\Delta x$ of the particles in question

$$\mathcal{D} \sim N_{res}\frac{\nu}{(\Delta\theta)^2}(\Delta x)^2\frac{\Delta\mu}{\mu}\Delta\theta \sim \frac{a^2}{t_{\parallel}}\Delta\theta, \qquad \nu t_{\parallel} \gtrsim \left(\frac{a\Delta\psi}{R}\right)^{1/2}(\Delta\theta)^2. \qquad (5.28)$$

Finally, at even higher collision frequencies, when $\nu t_{\parallel} \gtrsim (\Delta\theta)^2$, the width (5.27) becomes comparable with the distance $\sim N_{res}^{-1} \sim 1/\Delta\psi$ between the resonance curves, the resonant particles are no longer unique, and equal contributions to $\mathcal{D}$ are made by all the particles from the band $\Delta\theta$. Recognizing that the characteristic value of the radial displacement for them is $\sim a$ and that, as a result of scattering they leave the band $\Delta\theta$ after one pass through the magnetic bottle, we get

$$\mathcal{D} \sim \frac{1}{t_{\parallel}}a^2\Delta\theta, \qquad \nu t_{\parallel} \gtrsim (\Delta\theta)^2,$$

which coincides with the estimate (5.28). Equations (5.26) and (5.28) describe thus the contribution of the peak particles to the diffusion coefficient in the entire range of the parameter $\nu t_{\parallel}$ if the inequality (5.25) is satisfied.

If, however, the width of the peak exceeds the relative width of the resonance bands,

$$\Delta\theta \gtrsim \frac{\Delta\mu_r}{\mu}, \quad \frac{\Delta\varepsilon_r}{\varepsilon},$$

the diffusion coefficients can be calculated as before from Eqs. (5.18) and (5.19) in which the required reflection function is substituted.

To gain an idea of the role of the transport connected with the narrow peak $\Delta\theta$, we compare the diffusion velocity determined by Eqs. (5.26) and (5.28) with the case of resonance diffusion for a smooth $a(\theta)$ dependence (Fig. 15). It can be seen from this figure that the presence of the peak

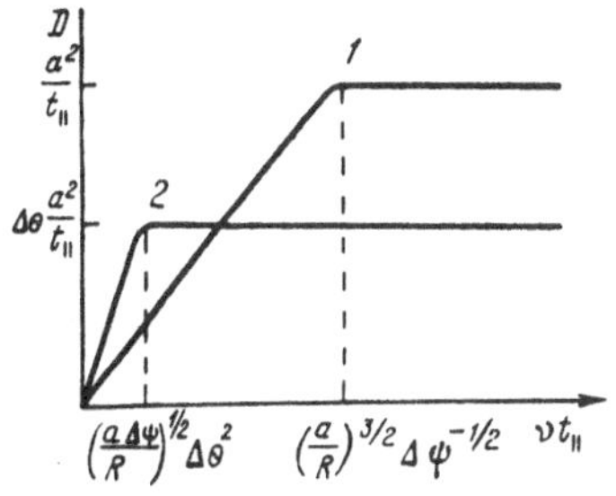

Fig. 15. Diffusion coefficient vs. collision frequency for a smooth dependence of $a$ on $\theta$ (1), and in the case of a peak of width $\Delta\theta$ (2).

is manifested by an increase of the transport into the region of lower collision frequencies.

5.4. Estimate of Transport Coefficients at $\lambda < L$. It was assumed throughout that the ion and electron mean free path in the trap exceeds substantially the length L of the installation. In the first experiments on tandem-mirror traps [46, 47], however, the value of $\lambda$ was either comparable with or somewhat larger than L. It is therefore of interest to determine, at least by way of estimate, the transport coefficients in the region $\lambda \lesssim L$. Following [48], we shall estimate the diffusion coefficient in this regime. We assume that the mean free path $\lambda$, while small compared with L, still exceeds the trap length $\ell$. This allows us to consider, as before, reflection from the mirror as an instantaneous (in terms of the collision time $\nu^{-1}$) "collision" with the end of the trap, an act that leads to a radial displacement of the particle in accordance with (3.37).

A distinctive property of the $\lambda \lesssim L$ regime is that now the particle wanders randomly along the installation and can land several times in succession in the same mirror, undergoing each time a radial displacement (outward or inward). The number of particle "collisions" with the (right- and left-hand) magnetic mirrors per unit time can be easily estimated by dividing the particle flux $nv_T$ entering the mirror by the number nL of particles per magnetic force tube of unit cross section (this yields the reciprocal time per pass through the trap $v/L \sim t_{\parallel}^{-1}$). Imagine now that the particle is located at one of the trap mirrors. The time required for it to reach the opposite end point of the installation is

$$t_{dif} \sim \frac{L^2}{\lambda^2} \nu^{-1}. \tag{5.29}$$

During that time, the number of collisions with the magnetic mirrors is $t_{dif}/t_{\parallel} \sim L/\lambda$. We arrive thus at the conclusion that a particle located at one of the mirrors will return to it $L/\lambda$ times before it reaches the opposite mirror. If the azimuthal-drift frequency $\omega_D = \Delta\psi/t_{\parallel}$ is so short that during the time (5.29) the particle will drift in azimuth through an angle that is small compared with $\lambda/2$,

$$\omega_D \frac{L^2}{\lambda^2} \nu^{-1} \ll 1, \tag{5.30}$$

all these reflections from the mirror will take place on one azimuth. As a result, the total particle displacement $\Delta r$ will be

$$\Delta r \sim \bar{a}L/\lambda,$$

where $\bar{a}$ is the reflection function averaged over the Maxwellian distribution (the averaging is necessary because, at each entry into the mirror, the particle has a different energy and a different magnetic moment). Assuming $\bar{a} \sim a$, and estimating the diffusion coefficient at $(\Delta r)^2/t_{dif}$, we get

$$\mathcal{D} \sim a^2\nu. \tag{5.31}$$

Thus, in the parameter region considered, the diffusion coefficient increases linearly with the collision frequency.

Matters are different when the inequality (5.30) is violated. The successive reflections from the mirror during the time $t_{dif}$ are no longer in phase, and the radial displacements are partially canceled out.

We introduce a length L such that during the time of diffusion over a length $L_1$ the ion drifts in azimuth through an angle $\pi/2$:

$$L_1 \sim \lambda \sqrt{\frac{\nu}{\omega_D}}.$$

Obviously, by virtue of satisfaction of an inequality inverse to (5.30), $L_1$ is shorter than L, but, at the same time, we assume that $L_1 \gtrsim \lambda$ (i.e., $\nu \gtrsim \omega_D$). Reasoning as above, we conclude that a particle situated at a mirror will enter it $\sqrt{\nu/\omega_D}$ times before it departs to a distance $L_1$ away [it must be recognized that the frequency of the collisions with the mirror is now $(L_1/v_T)^{-1}$ and not $(L/v_T)^{-1}$]. All these entries into the mirror are in phase (in view of our choice of $L_1$) and lead to a total displacement

$$\Delta r \sim \bar{a}\sqrt{\nu/\omega_D}. \tag{5.32}$$

At the same time, if the particle is at a distance $L_1$, its next entry into a mirror (the nearest or the farthest) leads to a radial displacement that is not correlated with the preceding one, since they are separated by a time interval longer than $\omega_D^{-1}$. Consequently, the elementary step in the radial random walk is the quantity (5.32), and the average time $t_1$ between the steps is the time between two successive stays in cells of length $L_1$ adjacent to the mirrors. Estimating $t_1$ as the time $L_1^2/\lambda^2\nu \sim \omega_D^{-1}$ of stay in the cell divided by the fraction $L_1/L$ of the time that the cell stays on a section of length $L_1$, we find that

$$t_1 \sim \left(\omega_D \frac{\lambda}{L}\sqrt{\frac{\nu}{\omega_D}}\right)^{-1}.$$

We obtain next $\mathcal{D}$ from the equation

$$\mathcal{D} \sim \frac{(\Delta r)^2}{t_1} \sim \frac{a^2}{t_\parallel}\sqrt{\frac{\nu}{\omega_D}}.$$

Finally, at $\omega_D \gtrsim \nu$ each successive entry into the mirror leads to a radial displacement that is not correlated with the preceding one. In this case,

$$\mathcal{D} \sim a^2/t_\parallel.$$

5.5. Stochastic Diffusion. In this section we consider a regime in which an inequality inverse to (5.1) holds. To understand the character of the particle motion in this case, let us find the relation of the distance between neighboring resonance radii $R_k$ and $R_{k+1}$, on the one hand, and the width $\Delta x_r$ of the "banana." Since on going from $R_k$ to $R_{k+1}$ the value of $\Delta\psi(r, \varepsilon, \mu)$ changes by $\pi$, we have

$$\Delta R_k = R_{k+1} - R_k \sim \frac{\pi}{|\partial\Delta\psi/\partial r|}.$$

Comparing this expression with (3.46), we verify that at

$$\left| a \frac{\partial \Delta \psi}{\partial r} \right| \gg 1 \tag{5.33}$$

$\Delta x_r$ exceeds the distance $\Delta R_k$. In other words, in this case the resonances overlap, so that the resonant structure of the trajectories is upset. At the same time, this means that the motion instability called "stochastic" sets in [49], as a result of which the displacement of a particle in the radial direction acquires a random character.

The criterion (5.33) can also be interpreted from another standpoint. The radial displacement of a particle when reflected from a mirror changes the radially dependent $\Delta\psi$ by an amount $\sim a|\partial\Delta\psi/\partial r|$. The phase $\psi$ of the next entry of the particle into the opposite mirror is changed by the same amount. This is equivalent to a $\psi$-phase relaxation equal, according to (5.33), to $\gtrsim 1$. As a result, the azimuths $\psi$ of the successive entries into the mirror become uncorrelated, and it is this which leads to the random radial displacements on passing through the magnetic mirror. Clearly, under these conditions the radial mixing of the particles is described by the Fokker–Planck equation, which has in cylindrical geometry the form

$$t_{\parallel} \frac{\partial f}{\partial t} = -\frac{1}{r} \frac{\partial}{\partial r} rf \langle \Delta r \rangle + \frac{1}{2r} \frac{\partial^2}{\partial r^2} rf \langle \Delta r^2 \rangle. \tag{5.34}$$

We emphasize that the radial transfer takes place in this regime without any collisions, and is due simply to the stochastic motion of the particles in the trap [8]. This regime was therefore named stochastic diffusion in [8].

The Fokker–Planck equation contains the azimuth-averaged value $\Delta r$. If we use Eq. (3.9), which is based on the drift approximation, we find that $\langle \Delta r \rangle = 0$. This means that the usual drift equations obtained in first order in the parameter $r_H$ are not accurate enough to determine $\langle \Delta r \rangle$; account must be taken of drifts proportional to $r_H{}^2$. It will be shown below, however, that the answer is expressed in terms of the longitudinal adiabatic invariant I introduced above and calculated in first-order drift theory.

We start from the statement that the equations describing the drift, averaged over the period, of a particle trapped in a magnetic bottle can be expressed in Hamiltonian form, and furthermore not only in first but also in higher orders of drift theory [50]. The canonical variables can, in this case, be chosen to be the curvilinear coordinates $\alpha(\mathbf{r})$ and $\beta(\mathbf{r})$ of the guiding center of the particle [51], which are defined by the condition

$$\mathbf{B} = [\nabla\alpha\nabla\beta]. \tag{5.35}$$

Their rate of change, averaged over the period of the longitudinal oscillations in the trap, is equal to

$$\dot{\alpha} = \frac{\partial\mathcal{H}}{\partial\beta}, \quad \dot{\beta} = -\frac{\partial\mathcal{H}}{\partial\alpha}, \tag{5.36}$$

where

$$\mathcal{H} = \mathcal{H}(\alpha, \beta, I, \mu) \tag{5.37}$$

is the Hamiltonian; I and $\mu$ are the longitudinal adiabatic invariant and the magnetic moment, both calculated with the required degree of accuracy. In second-order drift theory, expressions for I and $\mu$ were obtained in [12], but we shall not need their explicit form. The period of the longitudinal oscillations of the particle T is expressed in terms of $\mathcal{H}$ in the usual fashion:

$$T = \left(\frac{\partial\mathcal{H}}{\partial I}\right)^{-1}.$$

Note that the variables $\alpha$ and $\beta$ have the following property that follows from (5.35): the magnetic flux $\Phi$ through a certain surface S is given by

$$\Phi = \int_S d\alpha d\beta. \tag{5.38}$$

The converse is also true: if relation (5.38) is valid for any surface S, then (5.35) is valid. In particular, in a uniform field of strength $B_0$ we can choose the coordinates $\alpha$ and $\beta$ to be

$$\alpha = B_0\frac{r^2}{2}, \quad \beta = \psi, \tag{5.39}$$

where r and $\psi$ are the cylindrical coordinates of the force line.

Let us calculate the change of the coordinate $\alpha$ in one period T of the longitudinal oscillations. Accurate to terms $\sim T^2$ we have

$$\Delta\alpha = \Delta\alpha^{(1)} + \Delta\alpha^{(2)} = \dot{\alpha}T + \frac{1}{2}\ddot{\alpha}T^2. \tag{5.40}$$

The first term in the right-hand side of this relation equals

$$\Delta\alpha^{(1)} = \left(\frac{\partial\mathcal{H}}{\partial I}\right)^{-1}\frac{\partial\mathcal{H}}{\partial\beta} = \frac{\partial I}{\partial\beta}, \tag{5.41}$$

where the function $I = I(\alpha, \beta, \mathcal{H}, \mu)$ is implicitly defined by relation (5.37). The second term can be transformed by using the fact that

$$\ddot{\alpha} = \frac{d}{\partial t}\frac{\partial\mathcal{H}}{\partial\beta} = \dot{\beta}\frac{\partial^2\mathcal{H}}{\partial\beta^2} + \dot{\alpha}\frac{\partial^2\mathcal{H}}{\partial\alpha\partial\beta} = -\frac{\partial\mathcal{H}}{\partial\alpha}\frac{\partial^2\mathcal{H}}{\partial\beta^2} + \frac{\partial\mathcal{H}}{\partial\beta}\frac{\partial^2\mathcal{H}}{\partial\alpha\partial\beta}$$
$$= \frac{\partial\mathcal{H}}{\partial\beta}\frac{\partial}{\partial\beta}\left(\frac{\partial\mathcal{H}/\partial\alpha}{\partial\mathcal{H}/\partial\beta}\right). \tag{5.42}$$

Then

$$\Delta\alpha^{(2)} = \frac{1}{2}\left(\frac{\partial\mathcal{H}}{\partial I}\right)^{-2}\left(\frac{\partial\mathcal{H}}{\partial\beta}\right)^2\frac{\partial}{\partial\beta}\left(\frac{\partial\mathcal{H}/\partial\alpha}{\partial\mathcal{H}/\partial\beta}\right)$$
$$= \frac{1}{2}\left(\frac{\partial I}{\partial\beta}\right)^2\frac{\partial}{\partial\beta}\left(\frac{\partial I/\partial\alpha}{\partial I/\partial\beta}\right). \tag{5.43}$$

Bearing in mind that the coordinate $\beta$ is identified in our case with the azimuth $\psi$, we define

$$\langle\Delta\alpha\rangle = \oint d\beta\Delta\alpha. \tag{5.44}$$

The first term of (5.41) makes no contribution to the averaged value $\langle\Delta\alpha\rangle$, while the second, (5.43), yields, after integrating by parts,

$$\langle\Delta\alpha\rangle = \frac{1}{2}\frac{\partial}{\partial\alpha}\oint d\beta\left(\frac{\partial I}{\partial\beta}\right)^2 = \frac{1}{2}\frac{\partial}{\partial\alpha}\langle(\Delta\alpha^{(1)})^2\rangle.$$

Now, replacing $\alpha$ by $B_0r^2/2$, we obtain $\langle\Delta r\rangle$:

$$\langle\Delta r\rangle = \frac{1}{2r}\frac{\partial}{\partial r}r\langle\Delta r^2\rangle. \tag{5.45}$$

To calculate the right-hand side of this relation, it suffices already to use the approximation (3.9), i.e., first-order drift theory. We note that, as expected, $\langle\Delta r\rangle$ is a quantity of second order in the Larmor radius $r_H$, so that the first and second terms in the right-hand side of the Fokker–Planck equation are of the same order of magnitude.

Substituting (5.45) in (5.34), we write the kinetic equation for f in the form

$$\frac{\partial f}{\partial t} = \frac{1}{2rt_\parallel} \frac{\partial}{\partial r} r \langle \Delta r^2 \rangle \frac{\partial f}{\partial r}. \tag{5.46}$$

Integrating now both sides over $d^3v$ and averaging over the azimuth, we find that

$$\frac{\partial n}{\partial t} = -\frac{1}{r} \frac{\partial}{\partial r} r \left( \frac{q}{2\pi r L_0} \right), \tag{5.47}$$

where

$$q = 2\pi r L_0 \int d^3v \frac{\langle (\Delta r)^2 \rangle}{2t_\parallel} \frac{\partial f}{\partial r} = \frac{2\pi^2 r B_0}{m^2} \int d\varepsilon d\mu a^2 \frac{\partial f_M}{\partial r}. \tag{5.48}$$

In the last equation we used (3.37) for $\Delta r$ and substituted for f its zeroth approximation – the Maxwellian distribution function.

An estimate of the diffusion coefficient in the stochastic regime is given in [8]. Its exact value is calculated in [52], where it is also noted that it coincides with the diffusion coefficient in the resonant plateau regime at $\Delta\psi \gg 1$. To verify this, it suffices to replace the sum of the $\delta$ functions that enter in (5.18) by their mean value (this is possible if the number of resonances is large and the distances between them on the phase plane are small compared with the characteristic scale of change of the integrand):

$$\sum_k \delta\left(\Delta\psi - (2k+1)\frac{\pi}{2}\right) \to \frac{1}{\pi}.$$

Following the arguments advanced above concerning the phase relaxation, which led to stochasticity of the motion, we can now deduce the result of an increase of the collision frequency in the plateau regime (i.e., at $|a\,\partial\Delta\psi/\partial r| \ll 1$), when the inequality (5.8) begins to be satisfied. It is easily seen that, in this case, the change of the magnetic moment $\mu$ (as well as of the energy $\varepsilon$), due to the Coulomb collisions, viz., $\Delta\mu \sim \mu(\nu t_{\|})^{1/2}$, is such that the corresponding change of $\Delta\psi$ (equal to $\Delta\mu\partial\Delta\psi/\partial\mu$) exceeds unity. In other words, a phase relaxation takes place, connected with the change of $\Delta\psi$ in the course of the collisions. Clearly, regardless of the mechanism that causes the phase relaxation, the particle motion becomes stochastic and the radial transport is determined by Eqs. (5.47) and (5.48).

## APPENDICES

### Appendix 1

We calculate the longitudinal current that flows in section 2 of a long magnetic bottle (see Fig. 7). Since the coordinate z is measured from the equatorial plane, plane 2 corresponds to $z = z_2 \equiv -L_0/2$ and plane 1 to $z = z_1 \equiv -(\ell + L_0/2)$. We rewrite (2.32) in the form

$$\frac{\partial}{\partial s}\frac{j_{\|}}{B} = 2\frac{c}{B^4}[\mathbf{B}\nabla p]\,\nabla B$$

and integrate it along a force line from plane 1 to plane 2. Recognizing that the longitudinal current in section 1 is zero, we get

$$j_2 = 2cB_0\int_{z_1}^{z_2} dz\,\frac{ds}{dz}B^{-4}[\mathbf{B},\ \nabla p]\,\nabla B. \qquad \text{(A1.1)}$$

The last integral can be calculated with sufficient accuracy by setting the derivative ds/dz equal to unity and replacing B whenever it enters under the differentiation sign by $\mathscr{B}$. Then

$$j_2 = 2cB_0\int_{z_1}^{z_2}\frac{dz}{\mathscr{B}^4}\left\{-\frac{\mathscr{B}}{r}\frac{\partial B}{\partial r}\frac{\partial p}{\partial\psi}+\frac{\mathscr{B}}{r}\frac{\partial B}{\partial\psi}\frac{\partial p}{\partial r}+\frac{\partial B}{\partial z}\left(\frac{B_r}{r}\frac{\partial p}{\partial\psi}-B_\psi\frac{\partial p}{\partial r}\right)\right\} = -2cB_0\int_{z_1}^{z_2}\frac{dz}{\mathscr{B}^4}\left\{\frac{\mathscr{B}}{r}\frac{\partial(B,\ p)}{\partial(r,\ \psi)}+\frac{\mathscr{B}'}{r}\frac{\partial(\chi,\ p)}{\partial(r,\ \psi)}\right\}$$

$$= -2cB_0\int_{z_1}^{z_2}\frac{dz}{\mathscr{B}^3}\frac{1}{r}\frac{\partial(r_0,\ \psi_0)}{\partial(r,\ \psi)}\left\{\frac{\partial(B,\ p)}{\partial(r_0,\ \psi_0)}+\frac{\mathscr{B}'}{\mathscr{B}}\frac{\partial(\chi,\ p)}{\partial(r_0,\ \psi_0)}\right\}, \qquad \text{(A1.2)}$$

where $\chi$ is the magnetic potential (2.4). In terms of the variables $r_0$, $\psi_0$, and z, expression (2.4) takes the form

$$\chi = r_0^2 \frac{B_0}{\mathcal{B}} \left\{ e^{-\Phi} \left( \frac{1}{4} \mathcal{B}' + b \right) \cos^2 \psi_0 + e^{\Phi} \left( \frac{1}{4} \mathcal{B}' - b \right) \sin^2 \psi_0 \right\}. \quad \text{(A1.3)}$$

Using the relation

$$\frac{\partial (r_0, \psi_0)}{\partial (r, \psi)} = \frac{\mathcal{B}}{B_0} \frac{r}{r_0},$$

which obviously follows from the conservation of the magnetic flux inside a force tube, we rewrite (A1.2) in the form

$$j_2 = -2c \int_{z_1}^{z_2} \frac{dz}{\mathcal{B}^2} \frac{1}{r_0} \frac{\partial (\chi \mathcal{B}' \mathcal{B}^{-1} + B, p)}{\partial (r_0, \psi_0)}. \quad \text{(A1.4)}$$

Substituting in this integral the expressions (2.10) and (A1.3) for B and $\chi$, we ultimately get

$$j_2 = a r_0 \frac{\partial p}{\partial r_0} \sin 2\psi_0 + (a \cos 2\psi_0 + d) \frac{\partial p}{\partial \psi_0}, \quad \text{(A1.5)}$$

where

$$\left. \begin{matrix} a \\ d \end{matrix} \right\} = \pm c B_0 \int_{z_1}^{z_2} \frac{dz}{\mathcal{B}^4} (e^{\Phi} G_2 \mp e^{-\Phi} G_1),$$

$$G_{\substack{1 \\ 2}} = 4b^2 - \frac{1}{2} \mathcal{B}\mathcal{B}'' + \frac{3}{4} (\mathcal{B}')^2 \pm (4\mathcal{B}' b - 2\mathcal{B} b'). \quad \text{(A1.6)}$$

Since, by virtue of the symmetry of the trap, $\mathcal{B}(z)$ and $b(z)$ are even functions of the coordinate z, while $\Phi(z)$ is odd, it is easily seen that the coefficient d for the opposite mirror reverses sign, while the coefficient $a$ remains the same as before.

We prove now the following property of the expression for d: If the force lines intersect the plane 1 at right angles, then d is proportional to the second integral of (2.19). The condition that the force lines must be perpendicular to the plane $z = z_1$ means that

$$b(z_1) = \mathcal{B}'(z_1) = 0. \quad \text{(A1.7)}$$

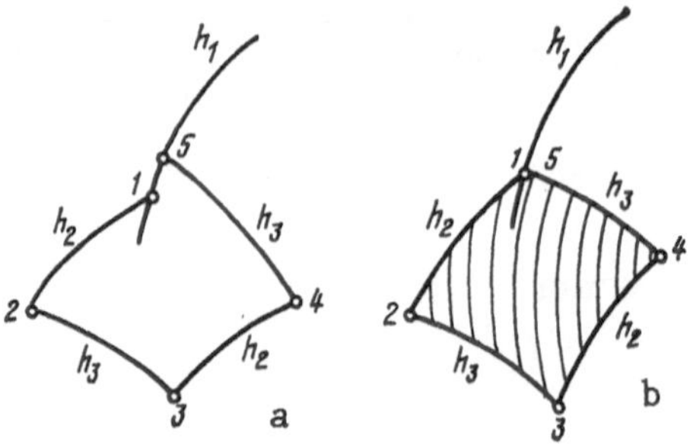


Fig. 16. For use in the analysis of the embedment condition.

Expressing d in the form

$$d = -2cB_0 \int\limits_{z_1}^{z_2} \frac{dz}{\mathcal{B}^4}\left[\left(4b^2 - \frac{1}{2}\mathcal{B}\mathcal{B}'' + \frac{3}{4}(\mathcal{B}')^2\right)\cosh\Phi - (4\mathcal{B}'b - 2\mathcal{B}b')\sinh\Phi\right]$$

$$= 2cB_0 \int\limits_{z_1}^{z_2} \frac{dz}{\mathcal{B}^3}\left[\frac{1}{2}\mathcal{B}''\cosh\Phi - 2b'\sinh\Phi\right]$$

$$- 2cB_0 \int\limits_{z_1}^{z_2} \frac{dz}{\mathcal{B}^4}\left[\left(4b^2 + \frac{3}{4}(\mathcal{B}')^2\right)\cosh\Phi - 4\mathcal{B}'b\sinh\Phi\right],$$

we can readily show, integrating by parts, that the last two integrals are proportional to each other if (A1.7) is valid [it must be recognized here that in the $z = z_2$ plane the magnetic field is already uniform, i.e., $b(z_2) = \mathcal{B}'(z_2) = 0$]. As a result we get

$$d = cB_0 \int\limits_{z_1}^{z_2} \frac{dz}{\mathcal{B}^3}\left(\frac{1}{2}\mathcal{B}''\cosh\Phi - 2b'\sinh\Phi\right).$$

Thus, if d = 0 the function U turns into an identical constant independent of $r_0$ and $\psi_0$.

## Appendix 2

The proof that the condition (3.57) is sufficient for the embedment of the magnetic field is based on an analysis of its geometric meaning. To this end we construct three families of orthogonal lines, designated $h_1$, $h_2$, and $h_3$, such that vectors tangent to them at a given point are directed, respectively, along $\varkappa$ and $\mathbf{B}$, and the binormal $\mathbf{b}$ to the force line (the family $h_2$ coincides with the force lines of the magnetic field). We use two pairs of arcs of the lines $h_2$ and $h_3$ to make up a differentially small contour whose initial and final points lie on the line $h_1$ (Fig. 16). We then have two possibilities: either the end point of the contour 5 does not coincide with the initial point 1 (Fig. 16a), or the two coincide (Fig. 16b), and then the point 5 can be identified with 1. Calculating the circulation of the vector $\varkappa$ along the closed contour 1–2–3–4–5–1, we find that, in the second case, it is equal to zero (since $\varkappa$ is everywhere perpendicular to the contour line), and in the first case a nonzero contribution to the circulation is made by the segment 5–1. Recognizing that the circulation along a small contour is proportional to the projection of the curl in the direction normal to the contour, which in our case is the direction of $\varkappa$, we conclude that the situation illustrated in Fig. 16a,b obtains, depending on whether $\varkappa \operatorname{rot} \varkappa$ is zero or not. In a magnetic field having the embedment property, the lines $h_2$ and $h_3$ (the tangents to which are parallel to $\mathbf{v}_{\parallel}$ and $\mathbf{v}_D$) lie on the drift surface; therefore, the contour made up of their arcs must be closed, as shown in Fig. 16b. If $\varkappa \operatorname{rot} \varkappa \neq 0$ in the vicinity of the point 1, as seen from Fig. 16a, there is no surface on which the lines $h_2$ and $h_3$ would lie.

The above reasoning is simultaneously the proof of the sufficiency of embedment condition (3.57): if this condition is met everywhere in the region occupied by the magnetic field, the magnetic surfaces can be constructed by "stretching" them on the network of lines $h_2$ and $h_3$.

## Appendix 3

It is convenient to derive the equations for $\partial N/\partial t$ and $\partial W/\partial t$ by starting from the kinetic equation written in divergent form (4.2) and from the expressions for N and W in terms of the function g:

$$N = 2\pi \left\langle \int d\varepsilon d\mu g \right\rangle, \quad W = 2\pi \left\langle \int d\varepsilon d\mu g\,(\varepsilon - \overline{e\varphi}) \right\rangle.$$

Calculating the derivative $\partial N/\partial t$, we get

$$\frac{\partial N}{\partial t} = 2\pi \left\langle \int d\varepsilon d\mu \frac{\partial g}{\partial t} \right\rangle = -2\pi \left\langle \int \frac{\partial}{\partial \xi^1} \dot{\xi}^1 g d\varepsilon d\mu \right\rangle +$$

$$+ 2\pi\left\langle\int\left(\frac{\partial g}{\partial t}\right)_{\mathrm{St}} d\varepsilon d\mu\right\rangle. \qquad \text{(A3.1)}$$

The last term can be transformed with the aid of (4.8):

$$\int d\varepsilon d\mu\left(\frac{\partial g}{\partial t}\right)_{\mathrm{St}} = \frac{4\pi}{m^2} B \sqrt{\frac{g}{g_{33}}} \int \frac{ds}{|v_{\parallel}|} \mathrm{St}\, f d\varepsilon d\mu$$

$$= \sqrt{\frac{g}{g_{33}}} \int ds \int d^3v\, \mathrm{St}\, f = 0, \qquad \text{(A3.2)}$$

where we have used the fact that the collision integral does not alter the number of particles. With allowance for (A3.2), substituting in (A3.1) the expressions (3.1) and (4.6) for $\dot{\xi}^1$ and g, we arrive at the first equation of (4.15).

We now calculate $\partial W/\partial t$, using Eq. (4.2) and integrating by parts:

$$\frac{\partial W}{\partial t} = 2\pi\left\langle\int d\varepsilon d\mu\,(\varepsilon - e\bar{\varphi})\frac{\partial g}{\partial t} - eg\frac{\partial\bar{\varphi}}{\partial t}\right\rangle$$

$$= 2\pi\left\langle -\int d\varepsilon d\mu\,\frac{\partial}{\partial \xi^1}(\varepsilon - e\bar{\varphi})\, g\dot{\xi}^1\right\rangle - 2\pi\left\langle e\int d\varepsilon d\mu\left(\dot{\xi}^1\frac{\partial\bar{\varphi}}{\partial\xi^1} + \dot{\xi}^2\frac{\partial\bar{\varphi}}{\partial\xi^2}\right)g\right\rangle$$

$$+ 2\pi\left\langle\int d\varepsilon d\mu\left[\left(1 - e\frac{\partial\bar{\varphi}}{\partial\varepsilon}\right)\overline{\frac{\partial\varphi}{\partial t}} - e\frac{\partial\bar{\varphi}}{\partial t}\right]g\right\rangle$$

$$+ 2\pi\left\langle\int d\varepsilon d\mu\,(\varepsilon - e\bar{\varphi})\left(\frac{\partial g}{\partial t}\right)_{\mathrm{St}}\right\rangle. \qquad \text{(A3.3)}$$

The first two terms of this expression can be transformed into (4.16) and (4.17) by using (3.1) and (4.6), respectively. The third term, which is proportional to the rate of change of the potential, can be interpreted as the change of the internal energy of the plasma as a result of its adiabatic heating (or cooling) when the potential distribution along the force line is changed:

$$\dot{W}_{\mathrm{ad}} \equiv 2\pi\left\langle\int d\varepsilon d\mu g\left[\left(1 - e\frac{\partial\bar{\varphi}}{\partial\varepsilon}\right)\overline{\frac{\partial\varphi}{\partial t}} - e\frac{\partial\bar{\varphi}}{\partial t}\right]\right\rangle. \qquad \text{(A3.4)}$$

This term vanishes in the case of a stationary process.

Finally, it is convenient to separate in the last term the part which we designate by Γ and which describes the heat exchange with the other plasma component:

$$\Gamma \equiv \frac{1}{d\xi^1}\int d\Sigma ds \int (\varepsilon - e\varphi)\,\mathrm{St}\, f d^3v.$$

In view of (4.1), we can rewrite the expression for Γ in the form

$$\Gamma = \frac{8\pi^2}{m^2}\left\langle B\sqrt{\frac{g}{g_{33}}}\int d\varepsilon d\mu \frac{ds}{|v_\parallel|}(\varepsilon - e\varphi)\,\mathrm{St}\,f\right\rangle. \qquad \text{(A3.5)}$$

Denoting the remaining part by $\dot{W}_{St}$, we obtain

$$\dot{W}_{\mathrm{St}} = 2\pi\left\langle \int d\varepsilon d\mu\,(\varepsilon - e\bar{\varphi})\left(\frac{\partial g}{\partial t}\right)_{\mathrm{St}}\right\rangle - \Gamma$$

$$= \frac{8\pi^2}{m^2}\left\langle B\sqrt{\frac{g}{g_{33}}}\int d\varepsilon d\mu \frac{ds}{|v_\parallel|}(\varepsilon - e\varphi)\,(\overline{\mathrm{St}\,f} - \mathrm{St}\,f)\right\rangle. \qquad \text{(A3.6)}$$

It follows from the last expression that the term $\dot{W}_{St}$ is due to the difference between the averaged collision term and St f.

## Appendix 4

We rewrite (4.54) in the form (recalling that in our case $\delta\varphi = \delta\varphi$)

$$\delta n = \frac{4\pi B_0}{emEL_0}\left(\int d\varepsilon d\mu \delta I_0 \frac{\partial f_M}{\partial r} - \frac{e}{m}\delta\varphi \int d\varepsilon d\mu t_\parallel \frac{\partial f_M}{\partial r}\right), \qquad \text{(A4.1)}$$

where $\delta I_0$ stands for $\delta I$ calculated without allowance for $\delta\varphi$. It is convenient to represent the first integral of (A4.1) in the form

$$\int d\varepsilon d\mu \delta I_0 \frac{\partial f_M}{\partial r} = R - \langle R\rangle,$$

where

$$R = \int d\varepsilon d\mu I \frac{\partial f_M}{\partial r}.$$

Using the expression for I:

$$I = \int ds \sqrt{\frac{2}{m}(\varepsilon - e\varphi_0 - \mu B)}$$

and changing the sequence of the integration

$$R=\sqrt{\frac{2}{m}}\int ds\int_{e\varphi_0}^{\infty} d\varepsilon\,\frac{\partial f_M}{\partial r}\int_0^{(\varepsilon-e\varphi_0)/B} d\mu\,\sqrt{\varepsilon-e\varphi_0-\mu B},$$

we readily find that

$$R=\left(\frac{1}{n}\frac{\partial n}{\partial r}+\frac{e}{T}\frac{\partial\varphi_0}{\partial r}+\frac{1}{T}\frac{\partial T}{\partial r}\right)\frac{mTn}{4\pi}U, \qquad \text{(A4.2)}$$

where $U = \int ds/B$.

The second integral in (A4.1) can be calculated by using expression (3.33) for $t_{\|}$:

$$\int d\varepsilon d\mu t_{\|}\frac{\partial f_M}{\partial r}=\frac{nm^2L}{2\pi B_0}\left(\frac{1}{n}\frac{\partial n}{\partial r}+\frac{e}{T}\frac{\partial\varphi_0}{\partial r}-\frac{3}{4}\frac{1}{T}\frac{\partial T}{\partial r}\right). \qquad \text{(A4.3)}$$

Combining (A4.1), (A4.2), and (A4.3), we obtain (4.55).

## Appendix 5

We start from the Landau collision integral written in the form

$$\mathrm{St}\,f=L\left[\frac{\partial}{\partial v_\alpha}\left(f\frac{\partial\varphi_1}{\partial v_\alpha}\right)-\frac{\partial}{\partial v_\alpha}\left(\frac{\partial f}{\partial v_\beta}\frac{\partial^2\varphi_2}{\partial v_\alpha\partial v_\beta}\right)\right], \qquad \text{(A5.1)}$$

where

$$\varphi_1(\mathbf{v})=-\frac{1}{4\pi}\int d^3v'\frac{f(\mathbf{v}')}{|\mathbf{v}-\mathbf{v}'|},$$

$$\varphi_2(\mathbf{v})=-\frac{1}{8\pi}\int d^3v' f(\mathbf{v}')\,|\mathbf{v}-\mathbf{v}'|,$$

$$L=\left(\frac{4\pi e^2}{m}\right)^2\Lambda,$$

and $\Lambda$ is the Coulomb logarithm. We simplify (A5.1) by using the fact that the distribution function f differs little from Maxwellian:

$$f=f_M+\delta f(\varepsilon,\mu),\quad \delta f\ll f_M,$$

where a feature of $\delta f$ is that it changes abruptly when it is displaced in narrow bands that overlie certain lines existing on the $\varepsilon$, $\mu$ plane. This means that the largest terms remaining in the collision term after linearization are those containing the second derivatives of $\delta f$:

$$\mathrm{St}\, f = -L \frac{\partial^2 \varphi_{2M}}{\partial v_\alpha \partial v_\beta} \frac{\partial^2 \delta f}{\partial v_\alpha \partial v_\beta}, \tag{A5.2}$$

where $\varphi_{2M}$ denotes the potential $\varphi_2$ calculated from the Maxwell function:

$$\varphi_{2M} = -\frac{nT}{4\pi v m}\left[(x+1)\, h'(x) + \left(x + \frac{1}{2}\right) h(x)\right], \tag{A5.3}$$

where

$$h(x) = \frac{2}{\sqrt{\pi}} \int_0^x e^{-t} \sqrt{t}\, dt,$$

$$x = \frac{mv^2}{2T}.$$

Using the explicit expression (A5.3) for $\varphi_{2M}$ we can rewrite the collision term (A5.2) in the following form:

$$\mathrm{St}\, f = \frac{2\pi n e^4 \Lambda}{m^2} \frac{\partial^2 \delta f}{\partial v_\alpha \partial v_\beta}\left[\left(h' + h - \frac{h}{2x}\right)\left(\frac{\delta_{\alpha\beta}}{v} - \frac{v_\alpha v_\beta}{v^3}\right) + \frac{v_\alpha v_\beta}{v^3} \frac{h}{x}\right]. \tag{A5.4}$$

We recognize now that $\delta f$ is a function of the energy $\varepsilon$ and of the magnetic moment $\mu$, and express the derivatives with respect to the velocity components in (A5.4) in terms of derivatives with respect to $\varepsilon$ and $\mu$:

$$\mathrm{St}\, f = \nu_0 \left[\mu^2 \frac{\partial^2 \delta f}{\partial \mu^2} G_1 + \mu\varepsilon \frac{\partial^2 \delta f}{\partial \varepsilon \partial \mu} G_2 + \varepsilon^2 \frac{\partial^2 \delta f}{\partial \varepsilon^2} G_3\right], \tag{A5.5}$$

where

$$G_1(\zeta, x) = \frac{1}{\zeta x^{5/2}}\left[h\left(1 - \frac{1}{2x}\right) + h'\right] + \frac{h}{x^{5/2}},$$

$$G_2(x) = 2G_3(x) = \frac{2h}{x^{5/2}},$$

$$\zeta = \frac{\mu B}{\varepsilon - e\varphi - \mu B},$$

$$\nu_0 = \frac{2\sqrt{2}\, \pi n e^4 \Lambda}{m^{1/2} T^{3/2}}.$$

The expression obtained is valid, of course, only in a small vicinity of the aforementioned lines in the $\varepsilon$, $\mu$ plane. In the neoclassical regime these are the lines $\mu$ = const, so that the derivatives with respect to $\varepsilon$ are small compared with the derivatives with respect to $\mu$. It suffices, therefore, to retain in (A5.5) only the first term. In the resonance regime, the function $\delta f$ depends on $R_k(\varepsilon, \mu)$ and $\mu$, and

$$R_k \frac{\partial \delta f}{\partial R_k} \gg \mu \frac{\partial \delta f}{\partial \mu}.$$

We can assume, on the basis of the last expression, that

$$\frac{\partial^2 \delta f}{\partial \varepsilon^2} = \left(\frac{\partial R_k}{\partial \varepsilon}\right)^2 \frac{\partial^2 \delta f}{\partial R_k^2}, \quad \frac{\partial^2 \delta f}{\partial \mu^2} = \left(\frac{\partial R_k}{\partial \mu}\right)^2 \frac{\partial^2 \delta f}{\partial R_k^2}, \quad \frac{\partial^2 \delta f}{\partial \varepsilon \partial \mu} = \frac{\partial R_k}{\partial \varepsilon} \frac{\partial R_k}{\partial \mu} \frac{\partial^2 \delta f}{\partial R_k^2}.$$

As a result, we obtain the collision term (5.12).

## REFERENCES


1. Yu. V. Gott, M. S. Ioffe, and V. G. Tel'kovskii, Nucl. Fusion Suppl., Part 3, 1045 (1962).
2. F. H. Coensgen et al., in: Plasma Physics and Controlled Nuclear Fusion Research, 1976, IAEA, Vienna (1977), Vol. 2, p. 135
3. G. I. Dimov, Preprint No. 77-46, Inst. Nucl. Phys., Siberian Branch, Acad. Sci. USSR, Novosibirsk (1977).
4. F. H. Coensgen, TMX Major Project Proposal, LLL-Prop-148, Lawrence Livermore Laboratory (1977).
5. S. Miyoshi et al., in: Plasma Physics and Controlled Nuclear Fusion Research, 1978, IAEA, Vienna (1979), Vol. 2, p. 437.
6. J. Kesner et al., Preprint PFC/JA81-11, Massachusetts Institute of Technology (1981).
7. J. C. Riordan, A. J. Lichtenberg, and M. A. Lieberman, Nucl. Fusion, 19, 21 (1979).
8. D. D. Ryutov and G. V. Stupakov, Pis'ma Zh. Eksp. Teor. Fiz., 26, 182 (1977).
9. A. A. Galeev and R. Z. Sagdeev, in: Reviews of Plasma Physics, Vol. 7, M. A. Leontovich (ed.), Consultants Bureau, New York (1979).
10. B. B. Kadomtsev, in: Plasma Physics and the Problem of Controlled Thermonuclear Reactions [in Russian], Izd. Akad. Nauk SSSR, Moscow (1958), Vol. 3, p. 285.
11. D. D. Ryutov and G. V. Stupakov, Fiz. Plazmy, 4, 501 (1978).
12. T. G. Northrop, C. S. Liu, and M. D. Kruskal, Phys. Fluids, 9, 1503 (1966).

13. R. J. Hastie, J. B. Taylor, and F. A. Haas, Ann. Phys. (N.Y.), 41, 302 (1967).
14. P. B. Lysyanskii and M. A. Tiunov, Fiz. Plazmy, 8, 963 (1982).
15. D. D. Ryutov and G. V. Stupakov, Dokl. Akad. Nauk SSSR, 240, 1086 (1978).
16. H. P. Furth and M. N. Rosenbluth, Phys. Fluids, 7, 764 (1964).
17. M. N. Rosenbluth and C. L. Longmire, Ann. Phys. (N.Y.), 1, 120 (1957).
18. T. F. Volkov, in: Reviews of Plasma Physics, Vol. 4, M. A. Leontovich (ed.), Consultants Bureau, New York (1966).
19. D. A. Panov, Preprint IAE-3535/6, Moscow (1981).
20. V. D. Shafranov, in: Reviews of Plasma Physics, Vol. 2, M. A. Leontovich (ed.), Consultants Bureau, New York (1966).
21. L. D. Pearlstein, T. B. Kaiser, and W. A. Newcomb, Phys. Fluids, 24, 1326 (1981); W. A. Newcomb, J. Plasma Phys., 26, 529 (1981).
22. G. V. Stupakov, Fiz. Plazmy, 5, 871 (1979).
23. A. I. Morozov and L. S. Solov'ev, in: Reviews of Plasma Physics, Vol. 2, M. A. Leontovich (ed.), Consultants Bureau, New York (1966).
24. P. B. Lysyanskii and B. M. Fomel', Preprint Inst. Nucl. Phys., Siberian Branch, Acad. Sci. USSR, No. 79-58, Novosibirsk (1979).
25. J. A. Byers, Nucl. Fusion, 22, 49 (1982).
26. D. A. Panov, Pis'ma Zh. Eksp. Teor. Fiz., 35, 70 (1982).
27. G. V. Stupakov, in: Fusion Energy, 1981 Selected Lectures Presented at the Spring College on Fusion Energy, Trieste, May 26 to June 19, IAEA, Vienna (1982), p. 255.
28. G. V. Stupakov, Fiz. Plazmy, 5, 958 (1979).
29. P. J. Catto and R. D. Hazeltine, Phys. Rev. Lett., 46, 1002 (1981).
30. T. G. Northrop and E. Teller, Phys. Rev., 117, 215 (1960).
31. D. A. Spong, E. G. Harris, and C. L. Hedric, Nucl. Fusion, 19, 665 (1979).
32. R. D. Hazeltine et al., Nucl. Fusion, 19, 1597 (1979).
33. R. D. Hazeltine and P. J. Catto, Phys. Fluids, 24, 290 (1981).
34. A. A. Galeev and R. Z. Sagdeev, Zh. Eksp. Teor. Fiz., 53, 348 (1967).

35. B.N. Breizman, V. V. Mirnov, and D. D. Ryutov, Zh. Eksp. Teor. Fiz., 58, 1770 (1970).
36. V. V. Mirnov and D. D. Ryutov, Nucl. Fusion, 12, 627 (1972).
37. V. E. Zakharov and V. I. Karpman, Zh. Eksp. Teor. Fiz., 43, 490 (1962).
38. F. L. Hinton and M. N. Rosenbluth, Phys. Fluids, 16, 836 (1973).
39. S. I. Braginskii, in: Reviews of Plasma Physics, Vol. 1, M. A. Leontovich (ed.), Consultants Bureau, New York (1965).
40. P. L. Bhatnagar, E. P. Gross, and M. Krook, Phys. Rev., 94, 511 (1954).
41. D. D. Ryutov and G. V. Stupakov, Preprint Inst. Nucl. Phys., Siberian Branch, Acad. Sci. USSR, No. 78-30, Novosibirsk (1978).
42. R. H. Cohen and G. Rowland, Phys. Fluids, 24, 2295 (1981).
43. R. H. Cohen, Nucl. Fusion, 19, 1579 (1979).
44. D. D. Ryutov and G. V. Stupakov, in: Plasma Physics and Controlled Nuclear Fusion Research 1980, IAEA, Vienna (1981), Vol. 1, p. 119.
45. L. S. Pekker, Preprint Inst. Nucl. Phys., Siberian Branch, Acad. Sci. USSR, No. 81-131, Novosibirsk (1981).
46. F. H. Coensgen et al., Phys. Rev. Lett., 44, 1132 (1980).
47. K. Yatsu et al., Phys. Rev. Lett., 43, 627 (1979).
48. R. H. Cohen, W. M. Nevins, and G. V. Stupakov, Nucl. Fusion, 22, 611 (1982).
49. G. M. Zaslavskii and B. V. Chirikov, Usp. Fiz. Nauk, 105, 3 (1971).
50. M. Kruskal, Adiabatic Invariants [Russian translation], Inostr. Lit., Moscow (1962).
51. G. V. Stupakov, Pis'ma Zh. Eksp. Teor. Fiz., 36, 318 (1982).
52. B. V. Chirikov, Fiz. Plazmy, 5, 880 (1979).

# CLASSICAL LONGITUDINAL PLASMA LOSSES FROM OPEN ADIABATIC TRAPS

V. P. Pastukhov

## Introduction

Traps are called magnetic if the magnetic force lines leave the main volume of the plasma and go off to the chamber walls or to some special receiving units. From the standpoint of containment of high-temperature plasma these traps have a large number of attractive properties. These include, first, a high permissible value of $\beta$ (ratio of the plasma and magnetic-field pressures), greater simplicity and ease of construction compared with closed system, and stability to rough hydrodynamic disturbances (in "minimum-B" systems). At the same time, open traps have also substantial shortcomings, principal among which is the considerable loss of plasma along the force lines.

All open traps can be divided into two large classes that differ greatly in the character of the longitudinal loss. The first includes traps in which the Larmor radius of the particles is much shorter than the force-line curvature radius, as a result of which the quantity $v_\perp^2/B$ (where $v_\perp$ is the transverse velocity of the particles and B is the absolute value of the magnetic field) is an adiabatic invariant. Traps of this class are therefore frequently called adiabatic. By virtue of conservation of the energy and of the adiabatic invariant, the particles, whose velocity vector in the main volume of the trap makes a rather large angle with the magnetic-field direction, are reflected from regions where the magnetic field is stronger as they move along a force line. These regions are frequently called magnetic mirrors. In the absence of collisions with other particles or of some electric oscillations that alter the angle between the velocity vector and the magnetic field, a particle trapped between mirrors could be confined almost infinitely. Thus, the longitudinal plasma loss in an adiabatic trap is determined mainly by the configuration of the region of adiabatic containment of the particles in velocity space, as well as by the rate of isotropization of the distribution function.

The second class consists of open traps in which $v_{\perp}^2/B$ is invariant for most particles, so that the particles are rapidly scattered away from the inhomogeneities of the magnetic field. Typical representatives of this class are traps with cusp geometry of the magnetic field, in which the magnetic field in the bulk of the plasma is close to zero, and the longitudinal loss constitutes outflow of the plasma through narrow strong-magnetic-field "gaps." The longitudinal loss is then determined by the width of the "gaps," which is determined in turn by the diffusion of the particles across the magnetic field.

The present survey is devoted to longitudinal losses of plasma from adiabatic traps. The physical processes that determine the longitudinal loss in magnetic-cusp traps differ substantially from the corresponding processes in adiabatic traps and are not considered here. The principal features of longitudinal confinement of plasma in magnetic-cusp traps can be found in [42-46].

As already mentioned, longitudinal plasma loss in an adiabatic trap is due to particle scattering that changes the angle between the particle velocity vector and the magnetic field direction. The loss due to Coulomb scattering is usually called classical. Other possible losses are anomalous ones that result from development of kinetic instabilities due to disequilibrium of the distribution function of any of the plasma components. We consider here only classical losses. It must be emphasized here that the classical longitudinal loss is of interest not simply as one of the particular loss mechanisms, but is of more fundamental significance. First, the classical mechanism, in contrast to the anomalous ones, is always present and determines thereby the minimum possible level of the plasma loss. Analysis of any type of adiabatic trap therefore begins with an estimate of the classical longitudinal loss. Second, Coulomb collisions tend to establish a maximum equilibrium of the particle distribution function, so that the class of possible kinetic instabilities is limited. Consideration of the classical loss is therefore an important prerequisite for the subsequent analysis of the anomalous-loss mechanisms. Third, results based on the Coulomb-scattering mechanism must, in principle, not be subject to the significant uncertainties that characterize, as a rule, the anomalous mechanisms. These results are therefore quite fundamental in character.

The class of adiabatic traps has a long history and includes systems such as simple mirror traps. The ancestor of this class is the simple mirror trap proposed in the early 1950s by Budker [1] and independently by York and Post [2]. The simple mirror trap is also a structural element in all other adiabatic traps. The main laws governing the classical longitudinal loss of plasma in adiabatic traps are therefore considered in the present paper first with the simple mirror trap as the example, and are next generalized to include more complicated systems.

The mathematical problem of classical longitudinal losses has been fully formalized and reduces to solution of a system of coupled Boltzmann equations for different plasma components, with a collision term in the Fokker–Planck or in the equivalent Landau form. The real solution of the problem, however, requires a number of physical and mathematical simplifications, as well as a reasonable combination of analytic and numerical methods. The construction of approximate models, description of methods for their solution, and a discussion of the accuracy of the result constitute the main content of the present review, with principal attention paid to analytic methods.

## 1. Basic Principles of Plasma Confinement in Simple Mirror Traps

The magnetic-field configuration of a simple mirror trap is shown schematically in Fig. 1. The magnetic field in the trap is not uniform along the force lines and increases from $B_0$ at the center to $B_m$ in the mirror. The quantity $R = B_m/B_0$ is called the mirror ratio. If the particle Larmor radius is much less than the curvature radius of the force line, the particle motion can be described in the drift approximation [3]. By virtue of the conservation of the energy and of the adiabatic invariant $J_\perp = mv_\perp^2/2B$, the particle should lose longitudinal velocity as it moves into the

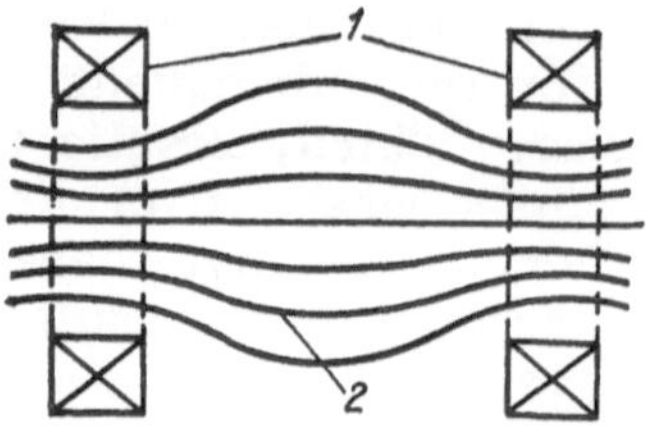


Fig. 1. Magnetic configuration of simple mirror trap: 1) current-carrying coils; 2) force lines.

region of the stronger magnetic field. It is easily seen that the particle cannot leave the trap along a force line if its velocity components at the center satisfy the relation

$$v_{\perp 0}^2(R-1) \geqslant v_{\parallel 0}^2, \tag{1.1}$$

where $v_\perp$ and $v_\parallel$ are, respectively, the transverse and longitudinal components of the velocity relative to the magnetic field. In the absence of collisions, such particles will oscillate along the force lines between the reflection points for an infinite time. It is assumed here, of course, that the particles will not drift transversely toward the walls of the traps; in other words, it is assumed that the drift surfaces of the confined particles are closed. The velocity-space region in which (1.1) is not satisfied is called the loss cone. Particles with velocities in this region will leave the trap within one pass.

When speaking of plasma confinement in simple mirror traps, one usually has in mind conditions such that the time of particle scattering into the loss cone exceeds considerably the transit time along the trap. These are indeed the most typical conditions for hot-plasma physics and for thermonuclear applications. In the other limiting case of fast scattering into the loss cone, the adiabatic invariance of $J_\perp$ does not play the main role, and the longitudinal loss has the character of gasdynamic outflow of the isotropic plasma along the force line. This situation can be realized in sufficiently long systems with dense plasma, in the so-called dynamic trap [4]. We shall be interested hereafter mainly in the case of rare scattering.

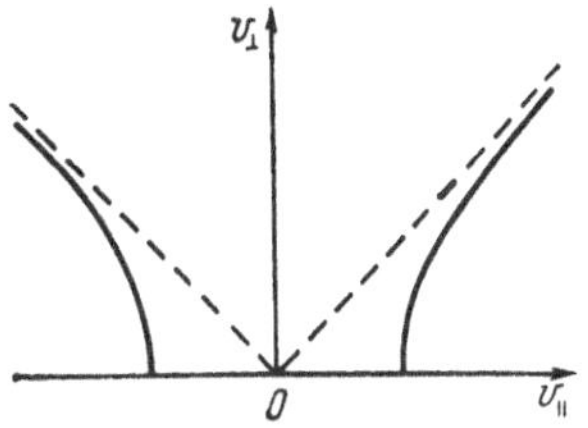


Fig. 2. Boundary of the electron-confinement region.

The classical loss is determined by Coulomb scattering into the loss cone. The ion scattering, however, is slower than that of electrons by an approximate factor $\sqrt{m_i/m_e}$ (where $m_i$ and $m_e$ are, respectively, the ion and electron masses). To equalize the ion and electron losses and preserve by the same token the plasma quasineutrality, the plasma takes on a positive potential relative to the end faces through which the force lines leave the main volume of the plasma. For a positive plasma potential, the region of adiabatic confinement of the electrons in velocity space differs from (1.1) and takes the form

$$v_{\parallel 0}^2 - (R-1)\, v_{\perp 0}^2 \leqslant 2e\Phi_0/m_e, \tag{1.2}$$

where e is the electron charge and $\Phi_0$ is the potential difference between the plasma center and the wall. The analysis that follows will show that usually $e\Phi_0$ is several times larger than the average electron energy, so that the electron-loss region in velocity space, which takes the form of a two-cavity hyperboloid (Fig. 2), lands in the zone of higher velocities. The electron distribution function formed as a result of the Coulomb collision must therefore not differ greatly in the region $v^2 < 2e\Phi_0/m_e$ from Maxwellian. The characteristic time of electron longitudinal loss will exceed the time of their Coulomb scattering into the loss cones by an approximate factor $\exp\{e\Phi_0/T_e\}$, where $T_e$ is the electron temperature; it is this which ensures equalization of the electron and ion losses.

While improving the electron confinement, the positive plasma potential must, on the other hand, facilitate the escape of the ions from the trap. Indeed, taking into account the conservation of the energy and of $J_\perp$, and assuming that the potential in the mirror is equal to the wall potential, we

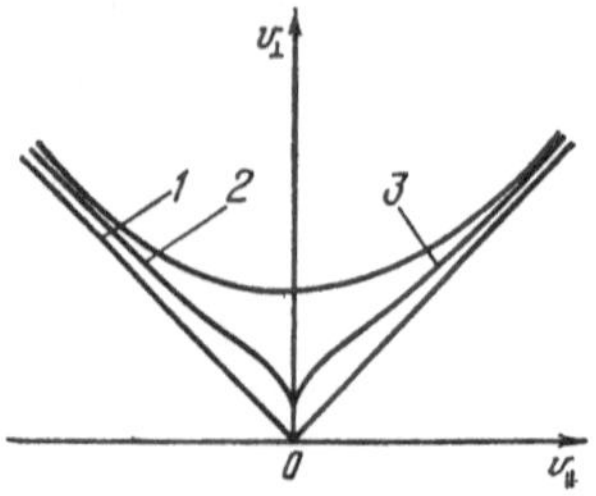


Fig. 3. Boundary of the ion-confinement region: 1) cone; 2) hyperboloid; 3) self-consistent boundary.

arrive at the following expression for the ion-velocity space region:

$$(R-1)v_{\perp 0}^2 - v_{\parallel 0}^2 \geqslant 2Z_i e\Phi_0/m_i, \qquad (1.3)$$

where $eZ_i$ is the ion charge. According to (1.3), at positive plasma potential the loss region in ion-velocity space expands and takes the form of a single-cavity hyperboloid (Fig. 3). The ions with energy lower than $Z_i e\Phi_0/(R-1)$ should then escape from the trap regardless of the direction of their velocity. To prevent the positive plasma from affecting catastrophically the ion confinement, their average energy must exceed $Z_i e\Phi_0/(R-1)$ considerably. Under the conditions of a simple mirror trap, it is necessary for this purpose to feed energy mainly to the ion component of the plasma (for example, by injecting a beam of high-energy neutral atoms). The equilibrium electron temperature, determined by the balance between the electron heating by the ions and the longitudinal losses (see Section 2), turns out then to be lower by approximately an order of magnitude than the average ion energy. This restricts the potential and the minimum energy of the ions to an acceptable level.

The derivation of Eq. (1.3) for the ion confinement region is not quite legitimate. The point is that the conservation of the energy and of $J_\perp$ yields the following expression for the dependence of $v_\parallel{}^2$ on the longitudinal coordinate s:

$$v_\parallel^2(s) = v_{\parallel 0}^2 - \frac{2}{m_i}\left[Z_i e(\Phi(s)-\Phi_0) + \frac{m_i v_{\perp 0}^2}{2}\left(\frac{B(s)}{B_0}-1\right)\right]. \qquad (1.4)$$

The boundary of region (1.3) corresponds to vanishing of $v_{\parallel}^2$ in the mirror. This condition would indeed determine the boundary of the confinement region if the expression in the square brackets, which plays the role of the effective potential energy $U_{ef}(s)$, were to reach its maximum in the mirror. For electrons, the corresponding $U_{ef}(s)$ increases monotonically from the center toward the mirror, so that (1.2) determines the confinement region correctly. In the case of ions, as first shown in [5], $U_{ef}(s)$ varies in general nonmonotonically and can have intermediate maxima and minima. The presence of additional maxima of $U_{ef}(s)$ extends the confinement region compared with (1.3), so that the self-consistent boundary of this region occupies an intermediate position between the cone (1.1) and the hyperboloid (1.3) (Fig. 3). The second consequence of the nonmonotonic $U_{ef}(s)$ is the appearance of ions trapped near the minima of $U_{ef}(s)$. These ions form a special group that is absent from the central section of the trap.

The explicit form of $U_{ef}(s)$ depends on $\Phi(s)$, which can be found from the quasineutrality condition and from the approximate Boltzmann distribution of the electron density in the longitudinal direction:

$$e(\Phi(s) - \Phi_0) = T_e \ln(n_i(s)/n_{i0}). \tag{1.5}$$

The ion density $n_i(s)$ is determined by the form of their distribution function, which depends in turn on the form of the boundary of the confinement region in velocity space, and hence on $U_{ef}(s)$. The determination of the exact self-consistent boundary of the ion-confinement region is thus a rather complicated problem. As a rule, therefore, the confinement regions considered are of the form (1.1) or (1.3), which can be called, respectively, the "optimistic" and "pessimistic" regions.

We proceed now to the problem of Coulomb scattering in the loss region. The evolution of the distribution function is described by the Boltzmann equation

$$\frac{\partial f}{\partial t} + \mathbf{v}\frac{\partial f}{\partial \mathbf{r}} + \dot{\mathbf{v}}\frac{\partial f}{\partial \mathbf{v}} = \mathrm{St}[f] + S, \tag{1.6}$$

where St [f] is the collision term in the Landau or Fokker–Planck form; $S(\mathbf{r}, \mathbf{v})$ is the density of the particle sources in phase space; and $\dot{\mathbf{v}} \equiv d\mathbf{v}/dt$. In the zeroth approximation in the ratio of the particle time of flight to the collision time, the distribution function should satisfy the Vlasov stationary equation:

$$\mathbf{v}\frac{\partial f^{(0)}}{\partial \mathbf{r}}+\dot{\mathbf{v}}\frac{\partial f^{(0)}}{\partial \mathbf{v}}=0, \qquad (1.7)$$

a general solution of which is an arbitrary function of the integrals of motion. We consider first, for simplicity, an axisymmetric trap with three integrals of motion. The first two, the energy $\varepsilon$ and the approximate integral $J_\perp$ have already been mentioned. The third is the generalized moment, which is conveniently represented in the form of the magnetic flux bounded by the drift surface of the particle's Larmor center:

$$\Psi = 2\pi\int_0^r B_z r dr. \qquad (1.8)$$

The first-order correction $f^{(1)}$ to the distribution function is now no longer a function of the integrals of motion only, but it can be represented in the form $f^{(1)} = f^{(1)}(\varepsilon, J_\perp, \Psi, s)$. In addition, in first order in the ratio of the time and collision times, account must be taken of the slow dependence of $f^{(0)}$ on the time. Equations (1.6) then take the form

$$\frac{\partial f^{(0)}}{\partial t}+v_\parallel\frac{\partial f^{(1)}}{\partial s}=\mathrm{St}\,[f^{(0)}]+S(\varepsilon,\ J_\perp,\ \Psi,\ s). \qquad (1.9)$$

To eliminate $f^{(1)}$ from (1.9), we integrate the latter along the force line over the period of the particle oscillations between the mirrors:

$$\oint\frac{\partial f^{(0)}}{\partial t}\,\frac{ds}{v_\parallel(s)}=\oint\{\mathrm{St}\,[f^{(0)}]+S\}\frac{ds}{v_\parallel(s)}. \qquad (1.10)$$

Here $v_\parallel(s) = [2(\varepsilon - J_\perp B)/m]^{1/2}$, and the integration is over a closed contour between the reflection points, where $v_\parallel(s) = 0$.

The operator $\mathrm{St}\,[F^{(0)}]$ and the operation of averaging along a force line do not affect the variable $\Psi$, so that Eq. (1.10) can be considered independently for different values of $\Psi$. (The influx of particles with a given value of $\Psi$ onto a surface by transverse transport can be taken into account by a suitable choice of the density of states S.) We shall hereafter omit the variable $\Psi$ and the zeroth-approximation superscript of the distribution function.

Being a constant on the trajectory of the leading center of the particle, f can be expressed in terms of the distribution function $f_0$ in the central section; $f_0$ depends only on $v_{\perp 0}$, $v_{\parallel 0}$, and t: $f = f_0(v_{\perp 0}, v_{\parallel 0}, t)$. Taking next $\partial f_0/\partial t$ outside the integral in the left-hand side of (1.10), we arrive at the following equation for $f_0$:

$$\frac{\partial f_0(v_{\perp 0}, v_{\parallel 0}, t)}{\partial t} = \oint \{\mathrm{St}[f_0] + S\} \frac{ds}{v_\parallel} \Big/ \oint \frac{ds}{v_\parallel}. \tag{1.11}$$

Note that on going from (1.10) to (1.11), we have in essence excluded from consideration the particles trapped in the local minima of $U_{ef}$ at $s \neq 0$, since $f_0$ is not defined for them. We can write for the distribution function of these particles a separate equation similar to (1.11), expressing f in terms of the distribution function at the minimum of $U_{ef}(s)$

The specific form of the right-hand side of (1.11) depends on the profile of the magnetic field. This side has the simplest form when $B(s) = B_0$ practically over the entire length of the trap, and B(s) changes from $B_0$ to $B_m$ only on short sections near the mirrors. Then, neglecting in (1.11) the contribution made to the integral by the regions where $B(s) \neq B_0$, we arrive at the following equation for $f_0(v_{\perp 0}, v_{\parallel 0}, t)$:

$$\left.\begin{aligned} &\frac{\partial f_0}{\partial t} = \mathrm{St}_0[f_0] + \overline{S}, \\ &\overline{S} = \frac{1}{2L} \int_{-L}^{L} S(v_{\perp 0}, v_{\parallel 0}, s)\, ds, \end{aligned}\right\} \tag{1.12}$$

where $\mathrm{St}_0[f_0]$ is the collision term in the central cross section, and the mirrors are located at the points $s = \pm L$. The approximation in which (1.12) was obtained is usually called the "square-well" approximation. It should be noted that the region where the results obtainable with the aid of (1.12) are valid is not confined to a particular profile of the magnetic field, but has a more general character. First, these results remain qualitatively in force also in cases when the square-well approximation is not valid. Second, as will be shown below, in a number of important yet rather general cases Eq. (1.11) can be reduced effectively to a form similar to (1.12). Equation (1.12) will therefore be the center of our attention from now on.

Let us dwell briefly on the properties of nonaxisymmetric traps. In nonsymmetric traps the generalized moment is no longer an integral of the motion. Nonetheless, in systems that are not too long, where the period of the particle oscillations between the mirrors is much less than the time required by the Larmor center to circuit around the trap axis, the role of $\Psi$ is played in part by the adiabatic invariant $J_{\parallel} = \oint v_{\parallel}(s)ds$. In contrast to the case of axial symmetry, however, $f_0$ is now no longer constant on the drift surface and is defined by the condition $J_{\parallel}$ = const. If the period of circuiting around the magnetic axis is short compared with the collision time, averaging is possible over the circuiting period, analogous to the averaging over the longitudinal-oscillation period. The variations of the longitudinal profile of the magnetic field on circuiting along the drift surface are usually small, so that a condition analogous to the square-well approximation can be used in the average along the circuit. The equation for the distribution function in a nonaxisymmetric trap reduces then to Eq. (1.11) [or (1.12)] obtained for a symmetric trap. In other words, axial asymmetry should not influence noticeably the integral value of the longitudinal loss. Nonetheless, if the subtler effects are of interest, allowance for the asymmetry can turn out to be very important. It is thus shown in [6] that even small variations of the mirror ratio on the drift surface lead to a strong spatial inhomogeneity of the flux of the particles leaving the trap along the magnetic field.

The basic meaning of the averaging carried out above is that the problem of the evolution of the distribution function in six-dimensional phase space in the rare-collision approximation could be reduced to the problem of evolution of the function $f_0$ in the space of the two variables $v_{\perp 0}$ and $v_{\parallel 0}$. Since small-angle scattering dominates in Coulomb collisions, the evolution of $f_0$ has the character of diffusive spreading. In the rare-collision approximation, $f_0$ differs from zero only inside the confinement region, so that particle and energy losses are determined by the diffusion flux through the boundary of this region.

A general expression for the operator St [f] in the case of Coulomb collisions was first obtained by Landau [7]. For the exposition that follows, however, it is more convenient to use an equivalent form, obtained in [8, 9], of the operator St [f]:

$$\mathrm{St}\,[f_\alpha] = 2\pi \sum_\beta \Lambda^{\alpha/\beta} \left( \frac{Z_\alpha Z_\beta e^2}{m_\alpha} \right)^2 \frac{\partial}{\partial v_j} \left( \frac{\partial^2 g_\beta}{\partial v_j \partial v_k} \frac{\partial f_\alpha}{\partial v_k} - 2 \frac{m_\alpha}{m_\beta} f_\alpha \frac{\partial h_\beta}{\partial v_j} \right). \tag{1.13}$$

The indices α and β here label the particle species, and $\Lambda^{\alpha/\beta}$ is a dimensionless coefficient called the Coulomb logarithm. The functions

$$g_\beta = \int |\mathbf{v} - \mathbf{v}'|\, f_\beta(\mathbf{v}')\, d\mathbf{v}', \quad h_\beta = \int \frac{f_\beta(\mathbf{v}')}{|\mathbf{v} - \mathbf{v}'|} d\mathbf{v}', \tag{1.14}$$

which are connected by the relations

$$\Delta_v g \equiv 2h, \quad \Delta_v h \equiv -4\pi f, \tag{1.15}$$

where $\Delta_v$ is the Laplacian in velocity space, are usually called the Rosenbluth–Trubnikov potentials. In analogy with the theory of Brownian motion, Eq. (1.11) with St [f] in the form (1.13) is frequently called the Fokker–Planck equation. A detailed derivation of (1.13) is given in [10].

Many integral conservation laws hold in systems with Coulomb scattering. In the present case we are interested in the conservation of the number of particles and of the energy. The particle-number conservation law states that the collision term has the form of the divergence of the particle flux in velocity space, so that a change in the total number of particles of any plasma component is determined by the flux of particles through the confinement-region boundary. To clarify the laws connected with the energy redistribution, we obtain the change of the total energy of the particles of species α as a result of collisions with particles of species β. We integrate to this end $\mathrm{St}_\beta[f_\alpha]$ over the velocity space with weight $v^2$. Simple identity transformations yield

$$\int v^2 \mathrm{St}_\beta\,[f_\alpha]\, d\mathbf{v} \equiv 2\pi\Lambda^{\alpha/\beta} \left( \frac{Z_\alpha Z_\beta e^2}{m_\alpha} \right)^2 \left\{ \int \frac{\partial}{\partial v_k} \left[ v^2 \left( \frac{\partial^2 g_\beta}{\partial v_k \partial v_j} \frac{\partial f_\alpha}{\partial v_j} - \frac{m_\alpha}{m_\beta} f_\alpha \frac{\partial}{\partial v_k} \Delta_v g_\beta \right) - 2 v_j \frac{\partial^2 g_\beta}{\partial v_j \partial v_k} f_\alpha \right] d\mathbf{v} + \right.$$

$$+\frac{4}{m_\beta}\int \partial \mathbf{v} d\mathbf{v}' \frac{f_\alpha(\mathbf{v}) f_\beta(\mathbf{v}')\left(m_\beta v'^2 - m_\alpha v^2\right)}{|\mathbf{v}-\mathbf{v}'|^3}\Bigg\}. \qquad (1.16)$$

It is easily seen that when particles of like species collide, the last term of (1.16) vanishes. In this case, the total change of energy is determined by the particle flux through the confinement-region boundary. In collisions between unlike particles, the last term of (1.16) describes the energy transfer from component $\beta$ to component $\alpha$, and in this case the energy given up by the component $\beta$ and acquired by the component $\alpha$ is conserved.

A few words concerning the Coulomb logarithm. $\Lambda^{\alpha/\beta}$ is equal to the natural logarithm of the ratio of the Debye radius to the impact parameter, at which the particles are scattered through an angle on the order of $\pi/2$. This impact parameter is, in turn, equal to the larger of the quantities $Z_\alpha Z_\beta e^2/m_{\alpha\beta}|\mathbf{v}_\alpha - \mathbf{v}_\beta|^2$ and $\hbar/m_{\alpha\beta}|\mathbf{v}_\alpha - \mathbf{v}_\beta|$, where $m_{\alpha\beta}$ is the reduced mass. Strictly speaking, $\Lambda^{\alpha/\beta}$ is a function of the relative velocity of the particles, but the large value $\Lambda^{\alpha/\beta} = 10\text{-}20$ allows us to regard it as approximately constant and to replace the relative velocity by its mean value. This has, in fact, already been taken into account in (1.13), where $\Lambda^{\alpha/\beta}$ was taken outside the interval in the expressions for g and h. It is nonetheless not always obvious what must be substituted for the average value of the relative velocity. This pertains primarily to the scattering of electrons by ions where, as will be shown below, a distinction must be made between the values of $\Lambda$ that enter in the angular scattering and in the energy exchange. In addition, the condition $\Lambda^{\alpha/\beta} = \Lambda^{\beta/\alpha}$ must be satisfied in all the relations that describe energy exchange between different species of particles. Questions pertaining to Coulomb logarithm are dealt with in greater detail in [10, 11].

## 2. Electron Confinement in Simple Mirror Traps

Using the relations derived in the preceding section, we consider electron confinement in simple mirror traps. The approach described here to the solution of Eq. (1.11) was first developed in [12] and was subsequently generalized to include the case of a multicomponent plasma and an arbitrary profile B(s) in [13], where it was shown, in particular, that the results on electron confinement can be applied with slight modifications to ion confinement in the central section of a tandem-mirror trap.

We consider first, for simplicity, the case when there is only one species of ions with $Z_i = 1$ and the square-well approximation is valid. We omit here the zero index that labels quantities pertaining to the central section of the trap.

In view of the high positive plasma potential, the distribution function of the main group of electrons should be close to Maxwellian. Therefore, taking into account the integral dependence of g on f, we can use for $g_e$, with good accuracy, an expression corresponding to a Maxwellian distribution function:

$$g_e = 4\pi n\left(\frac{m_e}{2\pi T_e}\right)^{3/2}\left\{\int_0^v vv'^2\left(1+\frac{v'^2}{3v^2}\right)\exp\left(-\frac{m_e v'^2}{2T_e}\right)dv'\right.$$

$$\left.+\int_v^\infty v'^3\left(1+\frac{v^2}{3v'^2}\right)\exp\left(-\frac{m_e v'^2}{2T_e}\right)dv'\right\}. \quad (2.1)$$

The average electron velocity exceeds considerably the average ion velocity, and in the zeroth approximation in the parameter $m_e/m_i$ we have $g_i \approx nv$. In this approximation the electrons colliding with the ions change the velocity direction, but conserve their energy. The energy exchange between the ions and electrons is accounted for by the next terms of the expansion. As will be shown below, the ions exchange energy most effectively here with low-energy electrons, and for scattering into the loss paraboloid an important role is played by the form of the distribution function at $v \geq \sqrt{2e\Phi/m_e}$.

Thus, scattering into the loss hyperboloid and energy exchange with ions can be treated quite independently. The connection between these processes is via the redistribution of the energy over the electron spectrum as a result of electron–electron collisions, and reduces to maintenance of a near-Maxwellian distribution function.

We consider first scattering into the loss hyperboloid. In view of the assumptions made, it is easy to reduce Eq. (1.12) with St[f] in the form (1.13) to the following dimensionless form:

$$\frac{\tau_e v_{Te}^3}{n}\,\frac{\partial\left(n v_{Te}^3 F\right)}{\partial t} = \hat{C}(F) \equiv 4\pi F^2 + \frac{1}{2}\,\frac{\partial^2 G}{\partial x^2}\,\frac{\partial^2 F}{\partial x^2} + \frac{1}{x^2}\,\frac{\partial G}{\partial x}\,\frac{\partial F}{\partial x} +$$

$$+\frac{1}{2x^3}\left(\frac{\partial G}{\partial x}+1\right)\frac{\partial}{\partial \mu}(1-\mu^2)\frac{\partial F}{\partial \mu}, \qquad (2.2)$$

$$x = v/v_{Te}, \quad \mu = v_{\parallel}/v, \quad G = g_e/nv_{Te},$$

$$v_{Te} = \sqrt{2T_e/m_e}, \quad F = v_{Te}^3 f_e/n, \quad \tau_e = \sqrt{m_e}\, T_e^{3/2}/\sqrt{2}\, \pi n e^4 \Lambda_e.$$

It is recognized in (2.2) that the electron–ion Coulomb logarithm in the angular-scattering problem should be approximately equal to the electron–electron Coulomb logarithm $\Lambda_e$, since in both cases the reduced mass corresponds to $m_e$, and the relative velocity to $v_{Te}$. Moreover, the source term has been left out; this is justified when the electrons enter (e.g., by ionization) predominantly the low-energy part of the distribution function. In this case, the influence of the electron source, just as exchange with the ions, can be taken into account additively in the particle-balance and energy-balance equations.

The characteristic time of variation of the plasma parameters, particularly n, $\Phi$, and $T_e$, should be, as noted in Section 1, the time of ion collision, which greatly exceeds the value $\tau_e$ for the electron–electron collisions. Thus, the left-hand side of (2.2) contains a small parameter corresponding to the ratio of these times and equal, in order of magnitude, to $\exp\{-e\Phi/T_e\}$. In the zeroth approximation in this parameter the function F should satisfy the equation

$$\hat{C}(F) = 0 \qquad (2.3)$$

with the boundary condition that F vanish on the boundary of the hyperboloid (1.2). Knowing the solution (2.3), we easily obtain expressions for $dn/dt$ and $dT_e/dt$ by calculating the particle and energy fluxes, corresponding to this quasistationary solution, through the surface of the hyperboloid.

Since the solution of Eq. (2.3) differs noticeably from a Maxwell function only at $v \gtrsim \sqrt{2e\Phi/m_e}$, we can restrict the consideration of the function F to the region $x \gg 1$. In view of the smallness of $\exp\{-x^2\}$, Eq. (2.3) assumes the substantially simpler form

$$\frac{1}{x^2}\frac{\partial}{\partial x}\left(F+\frac{1}{2x}\frac{\partial F}{\partial x}\right)+\frac{1}{x^3}\frac{\partial}{\partial \mu}(1-\mu^2)\frac{\partial F}{\partial \mu} = 0. \qquad (2.4)$$

This equation, together with the boundary conditions

$$\left.\begin{aligned} &F\,|_{x\to 0} \to \frac{e^{-x^2}}{\pi^{3/2}} + O\left(e^{-x_\Phi^2}\right), \quad F\,\Big|_{\mu^2=\mu_b^2;\ x=x_\Phi} = 0, \\ &\mu_b^2 = \frac{R-1}{R} + \frac{x_\Phi^2}{Rx^2}, \quad x_\Phi^2 = e\Phi/T_e \end{aligned}\right\} \tag{2.5}$$

constitutes the complete mathematical formulation of the problem. To be able to effect in the future various modifications of the results, we shall consider, in lieu of (2.4), an equation of more general form:

$$\frac{1}{x^2}\frac{\partial}{\partial x}\left(F + \frac{1}{2x}\frac{\partial F}{\partial x}\right) + \frac{\alpha}{x^3}\frac{\partial}{\partial \mu}(1-\mu^2)\frac{\partial F}{\partial \mu} = 0, \tag{2.6}$$

where $\alpha$ is a dimensionless parameter of order unity.

The solution of (2.6) is made complicated by the fact that the boundary conditions (2.5) prevent separation of the variables. The most general method of solving differential equations in similar cases is to construct and solve some auxiliary integral equation. To use the method, we continue the function F into the loss-hyperboloid region $\{x > x_\Phi;\ \mu_b^2 < \mu^2 \leq 1\}$, such that F is continuous together with its first derivative on the surface of the hyperboloid. In this region, F need not satisfy Eq. (2.6), but we can stipulate that, in the wider range in which it is defined, which constitutes all of velocity space, it satisfy the equation

$$\hat{C}(F) + Q(x, \mu) = 0, \tag{2.7}$$

where $Q(x, \mu)$ is a certain source density that differs from zero in the region $\{x > x_\Phi;\ \mu_b^2 < \mu^2 \leq 1\}$. Knowing the basic solution (the Green's function) of Eq. (2.7), we can express F in terms of Q. Then, taking the boundary conditions into account, we obtain a first-order Fredholm equation for $Q(x, \mu)$.

Solution of such an integral equation by analytic methods is, generally speaking, no simpler a task than the solution of the initial equation. In our case, however, we can use the following approximate procedure. We assume that the source density $Q(x, \mu)$ is specified such that the function F vanishes on a certain surface $\Sigma$ that is, generally speaking, different from the surface of the hyperboloid. Since F decreases exponentially with increasing x, the main contribution to the loss of electrons from the trap is made by the region directly adjacent to the vertex of the hyperboloid. If the surface $\Sigma$ differs little from that of the hyperboloid near its vertex, the particle and energy fluxes through this surface should also differ little from the fluxes through the hyperboloid surface. We consider in this connection a source (more accurately, drain) distribution of the form

$$Q(x,\ \mu) = -qe^{-x^2}\delta(1-\mu^2)\,\eta(x-a), \tag{2.8}$$

where $\eta(x - a)$ is the Heaviside step function. We determine the constants q and $a$ from the conditions that the surface $\Sigma$ corresponding to a source of the form (2.8) coincide with the loss-hyperboloid surface at the point $\{x = x_\Phi,\ \mu^2 = 1\}$ and have the same curvature radius in the vicinity of this point:

$$F(x_\Phi;\ \mu^2=1)=0,\quad \left(\frac{\partial F}{\partial x}-\frac{1}{Rx}\frac{\partial F}{\partial \mu}\right)_{\mu^2=1,\ x=x_\Phi}=0. \tag{2.9}$$

We note that the factor $\exp\{-x^2\}$ in Q allows us to preserve on the surface $\Sigma$ the topological properties of the hyperboloid.

To obtain a solution for (2.7) with a source distribution (2.8), we must find the fundamental solution of (2.7), i.e., a function $\tilde{F}$ that satisfies the following equation:

$$\widehat{C}(\widetilde{F})-\frac{\delta(x-\xi)\,\delta(1-\mu^2)}{\pi x^2}=0. \tag{2.10}$$

A change of variable $r = \exp\{x^2\}$ transforms (2.10) into

$$\frac{1}{r^2}\frac{\partial}{\partial r}r^2\frac{\partial\widetilde{F}}{\partial r}+\frac{\alpha}{r^2\ln r^2}\frac{\partial}{\partial\mu}(1-\mu^2)\frac{\partial\widetilde{F}}{\partial\mu}-\frac{\delta(r-r_0)\,\delta(1-\mu^2)}{\pi r_0}=0, \tag{2.11}$$

where $r_0 = \exp\{\xi^2\}$. Equation (2.11) is very similar to the equation of the Coulomb potential of a point source (with the Maxwellian distribution corresponding to a source located at the origin). In our case, however, the Coulomb potential cannot be used, since the operator acting on $\tilde{F}$ is not a Laplacian. Nonetheless, we make use of this helpful analogy, and use for this purpose the change of variables $\rho = r\sqrt{1-\mu^2}$ and $\zeta = r\mu\sqrt{\alpha}/\sqrt{\ln r^2}$ to transform (2.11) into

$$\frac{1}{\rho}\frac{\partial}{\partial\rho}\rho\frac{\partial}{\partial\rho}(\zeta\widetilde{F})+\frac{\partial^2}{\partial\zeta^2}(\zeta\widetilde{F})+O[(1-\mu^2)x^2/\alpha]$$

$$-\frac{r_0^2}{2\pi\rho}[\delta(\zeta-r_0)-\delta(\zeta+r_0)]\delta(\rho)=0. \quad (2.12)$$

Since greatest interest attaches in our case to a solution of (2.12) in the vicinity of sufficiently small angles $(1-\mu^2) \lesssim [T_e/R(e\Phi+T_e)]$, we can neglect at large R the terms $O[(1-\mu^2)x^2/\alpha]$, and obtain then the following approximate expression for $\widetilde{F}$:

$$\widetilde{F}\approx-\frac{r_0^2}{4\pi\zeta}\left\{[\rho^2+(\zeta-r_0)^2]^{-\frac{1}{2}}-[\rho^2+(\zeta+r_0)^2]^{-\frac{1}{2}}\right\}. \quad (2.13)$$

Recognizing that as $x\to 0$ the function F should be Maxwellian accurate to terms of order $\exp\{-x^2_\Phi\}$, we obtain

$$F=\frac{e^{-x^2}}{\pi^{3/2}}+\pi\int_a^\infty qe^{-\xi^2}\widetilde{F}\xi^2d\xi\approx\frac{e^{-x^2}}{\pi^{3/2}}-\frac{qa^2xe^{-x^2}}{4\alpha\mu}$$

$$\times\ln\left\{\frac{\dfrac{e^{a^2}}{\sqrt{2a^2}}+\mu\dfrac{e^{x^2}}{\sqrt{2x^2}}+\left[\dfrac{1-\mu^2}{\alpha}e^{2x^2}+\left(\dfrac{e^{a^2}}{\sqrt{2a^2}}+\mu\dfrac{e^{x^2}}{\sqrt{2x^2}}\right)^2\right]^{\frac{1}{2}}}{\dfrac{e^{a^2}}{\sqrt{2a^2}}-\mu\dfrac{e^{x^2}}{\sqrt{2x^2}}+\left[\dfrac{1-\mu^2}{\alpha}e^{2x^2}+\left(\dfrac{e^{a^2}}{\sqrt{2a^2}}-\mu\dfrac{e^{x^2}}{\sqrt{2x^2}}\right)^2\right]^{\frac{1}{2}}}\right\}. \quad (2.14)$$

Substituting (2.14) in Eqs. (2.9) and putting $a = x_\Phi$ in the power-law factors, we obtain, after straightforward but somewhat laborious transformations,

$$\exp\{a^2\}\approx\frac{2\alpha R+1}{2\alpha R}\exp\{x_\Phi^2\},\quad q\approx\frac{4\alpha}{\pi^{3/2}x_\Phi^3\ln(4\alpha R+2)}. \quad (2.15)$$

The source distribution (2.8) and the distribution function (2.14) are thus completely determined.

To obtain an expression for the change of the electron density as a result of their departure from the trap, we integrate Eq. (2.2) over the volume of the confinement region in velocity space. Since the collision term takes the form of the divergence of the particle flux in velocity space, the quantity dn/dt is determined by the particle flux through the hyperboloid surface. It suffices then to consider a flux that corresponds to a quasistationary distribution function and, with allowance for (2.7), is equal to the integral of the density of fictitious sources $Q(x, \mu)$ located inside the loss hyperboloid. As a result we obtain the following expression:

$$\frac{\tau_e}{n_e}\left(\frac{dn_e}{dt}\right)_{\text{dep}} = 4\pi \int_{x_\Phi}^{\infty} x^2 dx \int_{\mu_b}^{1} d\mu Q(x, \mu). \tag{2.16}$$

Each electron intersecting the loss-hyperboloid surface in the interval from v to v + dv carries away from the plasma an energy $m_e v^2/2$. Recognizing that the particle-flux vector can have a preferred direction in the region of the loss hyperboloid, we can write the expression for the energy loss due to electron departure from the trap in a form similar to (2.16):

$$\frac{3\tau_e}{2n_e T_e}\left(\frac{dn_e T_e}{dt}\right)_{\text{dep}} = 4\pi \int_{x_\Phi}^{\infty} x^4 dx \int_{\mu_b}^{1} d\mu Q(x, \mu). \tag{2.17}$$

Substituting (2.8), (2.15) in (2.16), (2.17) and integrating, we arrive at the expressions

$$\left(\frac{dn_e}{dt}\right)_{\text{dep}} \approx -\frac{4\alpha_e n_e}{\sqrt{\pi}\,\tau_e}\,\frac{2\alpha_e R}{2\alpha_e R + 1}\,\frac{\exp\{-e\Phi/T_e\}}{\ln(4\alpha_e R + 2)}\,\frac{T_e}{e\Phi} \times\left[1 + \frac{T_e}{2e\Phi} + O\left(\frac{T_e^2}{e^2\Phi^2}\right)\right], \tag{2.18}$$

$$\left(\frac{dn_e T_e}{\partial t}\right)_{\text{dep}} \approx -\frac{4\alpha_e n_e T_e}{\sqrt{\pi}\,\tau_e}\,\frac{2\alpha_e R}{2\alpha_e R + 1}\,\frac{\exp\{-e\Phi/T_e\}}{\ln(4\alpha_e R + 2)} \times\left[\frac{2}{3} + \frac{T_e}{e\Phi} + O\left(\frac{T_e^2}{e^2\Phi^2}\right)\right]. \tag{2.19}$$

The terms $O(T_e^2/e^2\Phi^2)$ in (2.18) and (2.19) can be omitted, since they are exceeded by the intermediate-calculation errors. We also present an expression for the average energy carried away by an electron from the plasma:

$$\langle\varepsilon\rangle_e = \frac{3}{2}\left(\frac{dn_e T_e}{dt}\right)_{\text{dep}} \Big/ \left(\frac{dn_e}{dt}\right)_{\text{dep}} \approx e\Phi + T_e. \tag{2.20}$$

The first term in (2.20) corresponds to the energy expended by the electron to surmount the potential barrier, while the second corresponds to the above-barrier energy carried away by the electron to the wall.

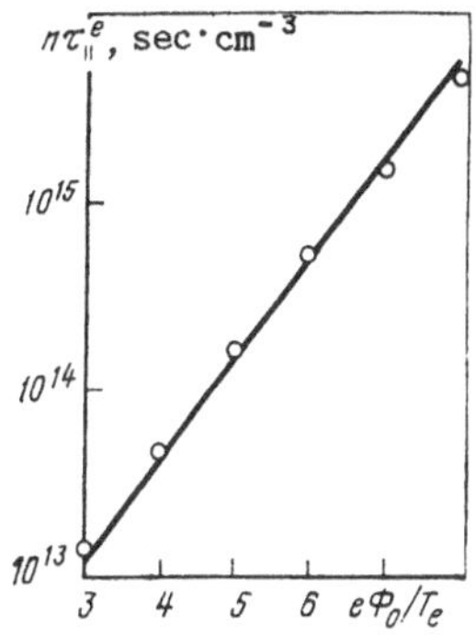


Fig. 4. Dependence of $n\tau_{\parallel}{}^{e}$ on $e\Phi/T_e$ at R = 10 and $T_e$ = 45 keV. The points correspond to the numerical results of [13].

In a plasma with one species of ions having $Z_i = 1$, we have in the square-well approximation $\alpha_e = 1$, which corresponds to equal effectiveness of angular scattering of electrons with $v \gg v_{T_e}$ by thermal electrons and by ions. In a multicomponent plasma with $Z_i \neq 1$, all the changes that must be made in the expressions above can be easily seen to reduce only to a change of the quantity $\alpha_e$, which takes the form

$$\alpha_e = \frac{1}{2}\left(1 + \frac{\Sigma Z_i^2 n_i}{n_e}\right), \tag{2.21}$$

where $n_i$ are the partial densities of the different species of ions, and are connected with the electron density $n_e$ by the relation $n_e = \Sigma Z_i n_i$, where the summation is over all ion species.

We introduce the characteristic time $\tau_{\parallel}{}^{e}$ of the longitudinal electron loss, defined such that $n_e/\tau_{\parallel}{}^{e}$ is equal to the right-hand side of (2.18). According to (2.18), the quantity $\alpha_e n_e \tau_{\parallel}{}^{e}$ should depend only on two combinations of the plasma parameters, $e\Phi/T_e$ and $\alpha_e R$. These relations were checked in [13] by solving a two-dimensional Fokker–Planck equation, and the approximate analytic expression (2.18) turned out to agree well with the results of the numerical calculations. The corresponding analytic and numerical dependences are shown in Figs. 4 and 5.

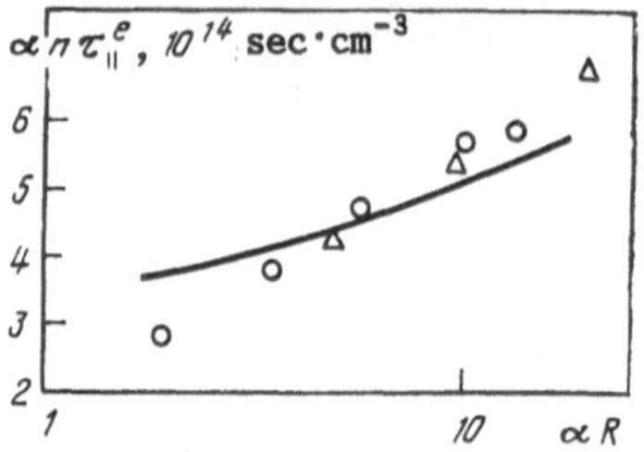


Fig. 5. Dependence of $\alpha n\tau_{\parallel}^{e}$ on $\alpha R$ at $e\Phi/T_e = 6$ and $T_e = 45$ keV. The circles and triangles show, respectively, the values obtained in [13] at fixed $\alpha$, and at fixed R, respectively.

Let us generalize the results to the case of an arbitrary magnetic-field profile. According to the results of Section 1, we should then replace (2.2) by

$$\frac{v_{Te}^3}{n_0}\frac{\partial f_0(v_0,\ \mu_0,\ t)}{\partial t}\oint\frac{ds}{v_\parallel}=\frac{1}{\tau_{e0}}\oint\frac{ds}{v_\parallel}\frac{n}{n_0}\hat{G}(F_0), \tag{2.22}$$

where the zero index pertains to the central section of the plasma. We change in the operator $\hat{C}$ from the variables x and $\mu$, which characterize the particle velocity at the point s, to the variables $x_0$ and $\mu_0$ given by

$$x_0^2=x^2+\frac{e(\Phi_0-\Phi(s))}{T_e},\quad \mu_0^2=1-\frac{x^2B_0}{x_0^2B(s)}(1-\mu^2). \tag{2.23}$$

Just as in the analysis of Eq. (2.6) in the square-well approximation, we are interested in the region $x_0 \gg 1$, $(1-\mu_0^2) \ll 1$, where the $\hat{C}$ can be expressed in the following approximate form:

$$\hat{C}(F_0)\approx\frac{x_0}{x}\frac{1}{x_0^2}\frac{\partial}{\partial x_0}\left(F_0+\frac{1}{2x_0}\frac{\partial F_0}{\partial x_0}\right)$$
$$+\frac{\alpha\mu^2B_0}{xx_0^2\mu_0^2B}\frac{\partial}{\partial\mu_0}\left(1-\mu_0^2\right)\frac{\partial F_0}{\partial\mu_0}+O\left[(1-\mu^2)\frac{B-B_0}{B_0}\right]. \tag{2.24}$$

Retaining the approximations used to solve (2.6), we can leave out the last term of (2.24), and when averaging along a force line we can use the approximation $x \approx x_0 \times \sqrt{\Phi(s)/\Phi_0}$. As a result, the equation for the quasistationary function $F_0(x_0,\ \mu_0)$ is again reduced to the form (2.6), where $\alpha$ is replaced by $\overline{\alpha_e}$, defined by the expression

$$\bar{\alpha}_e \approx \frac{1}{2} \oint \left(1 + \frac{\Sigma Z_i^2 n_i(s)}{n_e(s)}\right) \frac{n_e(s) B_0}{\Phi(s) B(s)} ds \Big/ \oint \frac{n_e(s)}{\Phi(s)} ds. \qquad (2.25)$$

Integrating (2.22) over velocity space in the central section of the trap and using the previously obtained solution of (2.6), we easily verify that for an arbitrary magnetic-field profile the longitudinal electron losses are given by Eqs. (2.18) and (2.19), in which we must replace $n_e$ by $n_{e0}$, $\Phi$ by $\Phi_0$, $\alpha_e$ by $\bar{\alpha}_e$, and $\tau_e$ by $\bar{\tau}_e$, where

$$\bar{\tau}_e^{-1} = \frac{1}{\tau_{e0}} \oint \frac{n_e(s)\Phi_0}{n_{e0}\Phi(s)} ds \Big/ \oint \sqrt{\frac{\Phi_0}{\Phi(s)}} ds. \qquad (2.26)$$

In the bulk of the plasma, where, $n_e/n_{e_0} \gg \exp\{-e\Phi_0/T_e\}$, the longitudinal distribution of the electron density is approximately described by the Boltzmann formula, so that $\Phi(s)$ can be readily expressed in terms of $n_e(s)$. However, the use of the Boltzmann equation to calculate the integrals along the force lines in (2.25) and (2.26) can lead to the appearance of divergences. To avoid them, we must take into account the deviation from the Boltzmann distribution in the low-density region. Recognizing that the low-density regions should make a small contribution to the averaged quantities, we can use for the connection of $\Phi(s)$ with $n_e(s)$ the following model relation that causes $\Phi(s)$ to vanish as $n_e(s) \to 0$:

$$\frac{e\Phi(s)}{T_e} \approx \ln\left[1 + \frac{n_e(s)}{n_{e0}}\left(\exp\left(\frac{e\Phi_0}{T_e}\right) - 1\right)\right]. \qquad (2.27)$$

We now consider the energy exchange between the ions and electrons, which plays an important role in the energy balance of the electron component. In the simplest case of a homogeneous plasma and of a Maxwellian distribution of the electrons and ions in velocity, the energy exchange is described by Spitzer's known equation [14]. Under the conditions of a simple mirror trap, however, where the ion distribution function is essentially non-Maxwellian and the plasma is strongly inhomogeneous along a force line, Spitzer's equation calls for a certain generalization.

The energy transfer from the ions to the electrons is not uniform either along the trap or over the electron spectrum. This can lead, generally speaking, to a certain distortion of the electron distribution function, predominantly at low velocities. This distortion, however, should not be large, since the energy transfer is determined by terms of order $m_e/m_i$ compared with the electron–electron collision term. Allowance for these distortions, as well as for the deviation of the electron distribution function from Maxwellian at $v \geq v_\Phi$, can necessitate only small corrections to the integrated energy transfer. We assume, therefore, from the very outset that the electron distribution function is Maxwellian, and describe the energy transfer in averaged fashion in terms of the change of the electron temperature. To this end, we multiply Eq. (1.6) with $S = 0$ by $m_e v^2/2$ and integrate it over the velocities as well as over the volume of the force tube with unity cross-section at the center, i.e., over $B_0 ds/B$. Substitution of the Maxwellian function in Eq. (16) leads to vanishing of term $(\mathbf{v}\,\partial f/\partial \mathbf{r} + \dot{\mathbf{v}}\,\partial f/\partial \mathbf{v})$ in the left-hand side, and also the electron–electron collision term and of the term with the oblique scattering by ions. The averaged equation, which contains, for simplicity, only one ion species, takes then the form

$$\left(\frac{\partial T_e}{dt}\right)_{ei} \int \frac{n_e(s)}{B(s)}\,ds = \frac{8\pi^2 Z_i^2 e^4}{3m_e} \Lambda^{e/i} \int \frac{ds}{B(s)} \int_0^\infty v^2 dv \frac{\partial}{\partial v} \times v^2 \left(\frac{\partial f_e}{\partial v}\frac{\partial^2 \bar{g}_i}{\partial v^2} - \frac{m_e}{m_i} f_e \frac{\partial}{\partial v}\frac{1}{v^2}\frac{\partial}{\partial v} v^2 \frac{\partial}{\partial v}\bar{g}_i\right), \qquad (2.28)$$

where

$$\bar{g}_i = \int_0^1 g_i(v, \mu, s)\,d\mu = \int_0^v v'^2 v \left(1 + \frac{v'^2}{3v^2}\right) \bar{f}_i dv' + \int_v^\infty v'^3 \left(1 + \frac{v^2}{3v'^2}\right) \bar{f}_i dv', \qquad (2.29)$$

$$\bar{f}_i = 4\pi \int_0^{\mu_{b,i}} f_i d\mu; \quad \mu^2_{b,i}(v, s) = 1 - \frac{B(s)}{B_m}\left(1 + \frac{2eZ_i\Phi(s)}{m_i v^2}\right).$$

The expression given here for $\mu_{b,i}(v, s)$ corresponds to the ion loss hyperboloid (1.3). It should be noted, however, that the explicit form of $\mu_{b,i}$ does not play a major role in all the expressions that follow.

Substituting (2.29) in (2.28) and carrying out an identity transformation that includes integration by parts, we reduce Eq. (2.28) to the form

$$\left(\frac{dT_e}{dt}\right)_{ei}\int\frac{n_e(s)}{B(s)}ds=\frac{2(4\pi e^2Z_i)^2}{3m_e}\Lambda^{e/i}\int\frac{ds}{B(s)}\int_0^\infty dv\bar{f}_i(v)$$

$$\times\left(v\int_0^v f_e(v')v'^2dv'-\frac{m_e}{m_i}v^2\int_v^\infty f_e(v')v'dv'\right). \qquad (2.30)$$

Since the average electron velocity exceeds considerably the average ion velocity, we can calculate the outer integral with respect to v in (2.30) by expanding the expression in the parentheses in powers of v. Recall here that $\Lambda^{e/i}$ depends, strictly speaking, on $|\mathbf{v}-\mathbf{v}'|$. Taking it outside the integrals with respect to v and v' and replacement of the relative velocity of the colliding particles by a certain mean value were only approximations. The first term in the right-hand side of (2.30) describes energy transfer from the ions to slow electrons, and the corresponding relative particle velocity is approximately equal to the ion velocity. Recognizing that the reduced mass in electron–ion collisions is approximately equal to $m_e$, we can write the following expression for the Coulomb logarithm $\Lambda_{ie}^{\varepsilon}$ corresponding to the term with energy transfer from the ions to the slow electrons:

$$\Lambda_{ie}^{\varepsilon}\approx\Lambda_i-\ln(m_i/m_e), \qquad (2.31)$$

where $\Lambda_i$ is the ion–ion Coulomb logarithm. The second term in the right-hand side of (2.30) describes energy transfer to ions from electrons of velocity higher than the average ion velocity. This term corresponds to a relative colliding-particle velocity approximately equal to the average electron velocity, and to a Coulomb logarithm approximately equal to $\Lambda_e$. Taking the foregoing remarks into account and introducing the notation

$$\langle v^k\rangle_i=n_i^{-1}\int_0^\infty v^{k+2}\bar{f}_i dv, \qquad (2.32)$$

we reduce (2.30) to the form

$$\left(\frac{dT_e}{dt}\right)_{ei}\int\frac{n_e}{B}ds=\frac{2(4\pi e^2Z_i)^2}{3m_e}\int\frac{n_i}{B}ds\left[\left(\frac{\langle v^2\rangle_i}{3}f_e\Big|_{v=0}\right.\right.$$

$$\left.\left.+\frac{\langle v^4\rangle_i}{10}\frac{\partial^2 f_e}{\partial v^2}\Big|_{v=0}\right)\Lambda_{ie}^{\varepsilon}-\frac{m_e}{m_i}\Lambda_e\int_0^\infty f_e(v)v\,dv\right]. \qquad (2.33)$$

It is convenient to express the quantities $\langle v^2\rangle_i$ and $\langle v^4\rangle_i$ in terms of the average ion energy $\varepsilon_i$ inasmuch as $\langle v^2\rangle_i = 2\varepsilon_i/m_i$, and $\langle v^4\rangle_i$ is proportional to $(\varepsilon_i/m_i)^2$, with a proportionality coefficient that depends little on the form of $\overline{f_i}$. Taking into account the relative smallness of the term with $\langle v^4\rangle_i$, it can be calculated assuming that $\overline{f_i}$ is Maxwellian. Substituting next the Maxwellian $f_e$ in (2.33), we obtain a final expression for electron heating by the ions:

$$\left(\frac{dT_e}{dt}\right)_{ei} = \frac{16}{9\sqrt{\pi}\,\tau_{e0}}\,\frac{m_e}{m_i}\int ds\,\frac{n_e n_i Z_i^2}{n_{e0}B}$$
$$\times\left\{\varepsilon_i\left(1-\frac{m_e\varepsilon_i}{m_i T_e}\right)\frac{\Lambda_{ie}^{\varepsilon}}{\Lambda_e}-\frac{3}{2}T_e\right\}\Big/\int ds\,\frac{n_e}{B}. \qquad (2.34)$$

In the approximation with a square-well potential and a Maxwellian distribution of the ion velocity, Eq. (2.34) is equivalent to the first two terms of the expansion of the Spitzer equation in powers of $T_i m_e/T_e m_i$. In a multicomponent plasma, the right-hand side of (2.34) must be summed over all ion species.

We estimate now the errors introduced in (2.34) by the assumption that the electron distribution function is strictly Maxwellian. According to (2.30), the ions transfer the bulk of their energy to slow electrons, and this should "superheat" the electrons and make their distribution function flatter than Maxwellian at low velocities. On the other hand, frequent electron–electron collisions tend to establish a Maxwellian distribution and redistribute by the same token the excess energy among the slow and fast electrons. The relative deviation of $f_e$ from Maxwellian at low velocities should be of the order of the ratio $\tau_e$ to the characteristic time of the change of $T_e$. According to (2.34), this ratio has a scale $\varepsilon_i m_e/T_e m_i$ and under typical mirror-trap conditions it amounts to $10^{-2}$. This estimate agrees well with the numerical results of [15]. The error, of the order of $\varepsilon_i m_e/T_e m_i$, and due to substitution of a Maxwellian distribution for $f_e$ in the first term of (2.33), is offset in part by the second term of (2.33), which is negative for a Maxwellian distribution and is likewise of the order of $\varepsilon_i m_e/T_e m_i$.

We now estimate $\varepsilon_i/T_e$ and $e\Phi_0/T_e$ in the simplest case, but one quite typical of simple mirror traps, when the electron energy balance is governed only by the heat from the ions and by the longitudinal losses. Recognizing that the longitudinal losses of the electrons and ions are equal and neglecting the weak dependence on the mirror ratio, we arrive with the aid of (2.18), (2.19), and (2.34) at the following approximate relations:

$$\varepsilon_i \frac{m_e}{m_i} \sim T_e \exp\left\{-\frac{e\Phi_0}{T_e}\right\}; \quad \left(\frac{T_e}{\varepsilon_i}\right)^{3/2} \left(\frac{m_e}{m_i}\right)^{1/2} \sim \frac{T_e}{e\Phi_0} \exp\left\{-\frac{e\Phi_0}{T_e}\right\}. \qquad (2.35)$$

For a hydrogen plasma, Eq. (2.35) yields $(e\Phi_0/T_e) \approx 5.3$, $(\varepsilon_i/T_e) \approx 8.8$, and for a deuterium–tritium plasma these values are, respectively, 6 and 11. Thus, in a simple mirror trap without additional electron heating the average ion and electron energies should differ greatly. Under real experimental conditions, this difference may turn out to be even larger. The point is that the ions that leave the trap along the force line have an energy higher than $e\Phi_0$, so that they can knock out of the wall secondary electrons that are subsequently accelerated to by the wall potential difference and enter the plasma. The influx of secondary electrons into the plasma should be compensated for by an increased electron loss, so as to preserve the quasineutrality of the plasma. Thus, the energy loss considered above is augmented by the electron exchange between the plasma in the wall, wherein a plasma electron that has an energy $(e\Phi_0/T_e)$ in the central section of the trap is replaced by an electron of energy $e\Phi_0$. To suppress the secondary emission and to increase accordingly the electron heat exchange with the wall, it is proposed in [16] to use the so-called "post-mirror expansion nozzle," in which the magnetic field and the electric potential decrease rapidly in the direction from the mirror to the wall.

The main results of this section can be briefly formulated as follows. Relatively simple analytic expressions have been obtained and permit a fairly complete description of the classical longitudinal electron loss in a simple mirror trap, as well as the energy exchange between the ions and electrons. The relative error of the expressions does not exceed several percent, so that these expressions can be used for detailed calculations of the particle- and energy-balance in the trap. If the electrons are not additionally heated, their equilibrium temperature is approximately an order of magnitude lower than the average ion energy.

## 3. Ion Confinement in Simple Mirror Traps

The first analysis of the classical ion loss in simple mirror traps was carried out in the pioneering papers of Budker, York, and Post [1, 2]. The initial ideas concerning the classical ion loss were refined and substantially supplemented in many subsequent papers. It must be noted right away that, in view of the appreciable deviation of the ion distribution function from Maxwellian, and of the linearity of the Fokker–Planck equation, one cannot count on obtaining sufficiently accurate quantitative results by using analytic methods only. An important role in the calculations of the ion losses is therefore occupied by numerical methods.

Before we proceed to analyze the ion losses, we write the concrete form of the Fokker–Planck equation for the ions. We consider, for simplicity, one ion species with $Z_i = 1$ and confine ourselves to the square-well approximation. It is convenient next to introduce the effective ion temperature $T_i = 2\varepsilon_i/3$ and $v_{Ti} = \sqrt{2T_i/m_i}$, and also reduce the ion distribution function $f_i$, the function $g_i$, and the Fokker–Planck equation itself to a dimensionless form, by introducing the notation

$$\left.\begin{aligned} &F(x, \mu, t) = v_{Ti}^3 f_i/n, \quad Q(x, \mu, t) = v_{Ti}^3 \tau_i S_i/n, \\ &G = g_i/n v_{Ti}, \quad x = v/v_{Ti}, \\ &\tau_i = \sqrt{m_i}\, T_i^{3/2}/\sqrt{2}\, \pi n e^4 \Lambda_i. \end{aligned}\right\} \tag{3.1}$$

In the calculation of $g_e$ we can assume a Maxwellian electron distribution function and confine ourselves to the first two terms of the expansion of $g_e$ in the parameter $m_e T_i/m_i T_e$. This corresponds exactly to the approximation used in Section 2 in the analysis of energy exchange between the ions and electrons. As a result, Eq. (1.12) takes the form

$$\tau_i \frac{\partial F}{\partial t} + \frac{\tau_i}{n} F \frac{dn}{dt} - \frac{\tau_i}{v_{Ti}} \frac{dv_{Ti}}{dt} \frac{1}{x^2} \frac{\partial}{\partial x}(x^3 F) = \widehat{C}_i(F) + Q$$
$$+ \frac{4}{3\sqrt{\pi}} \frac{\Lambda_{ie}^{\varepsilon}}{\Lambda_i} \left(\frac{m_e T_i^3}{m_i T_e^3}\right)^{\frac{1}{2}} \frac{1}{x^2} \frac{\partial}{\partial x}\left[x^3 F\left(1 - \frac{3 m_e T_i}{5 m_i T_e} x^2\right)\right]$$
$$+ \frac{2}{3\sqrt{\pi}} \frac{\Lambda_e}{\Lambda_i} \left(\frac{m_e T_i}{m_i T_e}\right)^{\frac{1}{2}} \Delta_x F. \tag{3.2}$$

The quantities $\Lambda_e$, $\Lambda_i$, and $\Lambda_{ie}^{\varepsilon}$ were defined in the preceding section, $\Delta_x$ denotes the Laplacian in x-space, and the expression for the ion–ion collision term $\hat{C}_i(F)$ is of the form

$$\hat{C}_i(F) = \frac{1}{2x^2}\frac{\partial}{\partial x}x^2\Big[\frac{\partial F}{\partial x}\frac{\partial^2 G}{\partial x^2} + (1-\mu^2)\frac{\partial F}{\partial \mu}\frac{1}{x}\frac{\partial^2}{\partial\mu\partial x}\Big(\frac{G}{x}\Big) - F\frac{\partial}{\partial x}\Delta_x G\Big] + \frac{1}{2x^2}\frac{\partial}{\partial\mu}(1-\mu^2)\Big[\frac{\partial F}{\partial x}x\frac{\partial^2}{\partial\mu\partial x}\Big(\frac{G}{x}\Big) + \frac{1}{x}\frac{\partial F}{\partial\mu}\Big(\frac{\partial G}{\partial x} + \frac{\mu}{x}\frac{\partial G}{\partial\mu} + \frac{1}{x}\frac{\partial}{\partial\mu}(1-\mu^2)\frac{\partial G}{\partial\mu}\Big) - F\frac{\partial}{\partial\mu}\Delta_x G\Big]. \quad (3.3)$$

The functions G and $\Delta_x G$ can be written in the following integral form:

$$\left.\begin{aligned} G &= \int_0^\infty \xi^2 d\xi \int_0^1 d\mu' F(\xi,\mu')\, 4qE(p/q),\\ \Delta_x G &= \int_0^\infty \xi^2 d\xi \int_0^1 d\mu' F(\xi,\mu')\,\frac{8}{q}K(p/q),\\ p &= 2(x\xi)^{1/2}\left[(1-\mu^2)(1-\mu'^2)\right]^{1/4},\\ q &= \left[x^2+\xi^2-2x\xi\left(\mu\mu' - \sqrt{1-\mu^2}\sqrt{1-\mu'^2}\right)\right]^{1/2}, \end{aligned}\right\} \quad (3.4)$$

where $E(\zeta)$ and $K(\zeta)$ are complete elliptic integrals.

The function G, and hence also $\Delta_x G$, can also be represented by an expansion in Legendre polynomials

$$G = \sum_{k=0}^{\infty}\frac{4\pi P_{2k}(\mu)}{(1-16k^2)}\left[\int_0^1 P_{2k}^2 d\mu\right]^{-1}\left\{\int_0^x \frac{\xi^{2(k+1)}}{x^{2k-1}}\left(1-\frac{4k-1}{4k+3}\frac{\xi^2}{x^2}\right)d\xi \times \int_0^{\mu_b(\xi)} F(\xi,\mu')P_{2k}(\mu')\,d\mu' + \int_x^\infty x^{2k}\xi^{(3-2k)}\left(1-\frac{4k-1}{4k+3}\frac{x^2}{\xi^2}\right)d\xi \times \int_0^{\mu_b(\xi)} F(\xi,\mu')P_{2k}(\mu')\,d\mu'\right\}. \quad (3.5)$$

The representation (3.5) is frequently more convenient than (3.4). In particular, it leads to expression (2.1) used in the preceding section.

The penultimate term in the right-hand side of (3.2) describes the deceleration of the ions by the slow electrons.

In the terminology of [10], this term corresponds to dynamic friction by the electrons. As shown in Section 2, for classical energy losses the equilibrium electron temperature is approximately an order of magnitude lower than $T_i$, and in this case the term in question turns out to be of the same order as $\hat{C}_i(F)$. In the presence of additional electron-cooling channels the ratio $T_i/T_e$ can exceed $(m_i/m_e)^{1/3}$ noticeably, so that the deceleration of the ions by the electrons assumes a dominant role in Eq. (3.2).

The last term in the right-hand side of (3.2) describes the diffusion of the ions in velocity space, due to scattering of Maxwellian electrons from the ions. The relative value of this term does not exceed 10% of the principal terms of the right-hand side of (3.2), so that the diffusion of the ions by the electrons does not play a principal role and can be regarded as a correction.

Equation (3.2) should be supplemented by the condition that F vanishes on the boundary of the ion-confinement region in velocity space. The question of the shape of the ion-confinement-region boundary was discussed in Section 1, where it was noted, in particular, that the exact boundary has a rather complicated shape that depends on the longitudinal profile of the potential. It is therefore customary to consider a simplified shape of this boundary, in the form of the cone (1.1) or the hyperboloid (1.3). In terms of the dimensionless variables introduced above, the confinement regions (1.1) and (1.3) take the form

$$\mu^2 < \mu_b^2 = (R-1)/R, \quad 0 \leqslant x < \infty, \tag{3.6}$$

$$\mu^2 \leqslant \mu_b^2 = \frac{R-1}{R} - \frac{1}{Rx^2}\frac{e\Phi_0}{T_i}, \quad \frac{e\Phi_0}{T_i(R-1)} \leqslant x < \infty. \tag{3.7}$$

Since the exact boundary of the confinement region is intermediate between (3.6) and (3.7), we can also introduce a certain "effective" confinement region, replacing in (3.7) $\Phi_0$ by $\Phi_{ef} < \Phi_0$. In this case, in a range $2 \lesssim R \lesssim 6$ of moderate mirror ratios $\Phi_{ef}$ can be quite well approximated by $\Phi_{ef} = \Phi_0/2$. This approximation takes into account the fact that the change of the potential in the bulk of the plasma is approximately $\Phi_0/2$, and the remaining change of the potential takes place directly in the mirror and on the force-line segment from the mirror to the wall.

Numerical calculations of classical ion losses are the subject of a large number of papers. The two-dimensional character of Eq. (3.2) in x space, the rather cumbersome forms of the operator $\hat{C}(F)$ and of the function G, and also the need for averaging the right-hand side of (3.2) over the period of the oscillations of the particles between the mirrors in the case of an arbitrary magnetic-field profile, all make the determination of the classical ion loss a rather complicated computational problem. The initial computation models included therefore, as a rule, a number of simplifying assumptions whose influence was analyzed in subsequent papers. Without dwelling in detail on all of them, we shall trace the principal stages of the development of the computation models.

In the first stage, the calculations were performed in the square-well approximation and did not take into account the ion–electron collisions, nor the influence of the ambipolar potential. The main results pertaining to this stage are given in [17]. The second stage is covered in [18], where a method was developed for averaging the Boltzmann equation over the period of the oscillations between the mirrors. This made it possible to consider various magnetic-field configurations. In addition, an approximate method, based on separation of the variables x and $\mu$, was developed in [18] to solve the Fokker–Planck equation for ions. In the third stage, assuming a square well and separation of the variables, account was taken of ion–electron collisions and of the ambipolar potential [19, 20]. Complete two-dimensional (i.e., without separation of the variables x and $\mu$) calculations of the ion losses, with allowance for the factors listed above, were performed in [21].

According to the results of [18], in a wide range of variation of the magnetic-field profile, from a cosine plot to a square well, the ion–ion collision term averaged over the period of the oscillations between the mirrors [see (1.11)] can be replaced with good accuracy by its value in the central section. In other words, for an arbitrary magnetic-field profile, one can use for the ion–ion collision term in a simple mirror trap an expression corresponding to the square-well approximation. This nontrivial fact does not follow directly from the form of $\hat{C}_i(F)$, but is a certain integral property of the solution of the Fokker–Planck equation with a corresponding boundary condition. It can be qualitatively understood as an approximate mutual cancellation of the two factors – the lowering of the collision frequency with increasing distance from the central section, and the simultaneous increase of $\partial f_i/\partial\mu$. On the other hand, according to [18], the effective form of the ion source, referred to the central section [see (1.11)], depends substantially on the magnetic-field profile. This must be taken into account in the calculations, since the plasma in the stationary state is most sensitive to the angular distribution of the reduced source. The last circumstance is quite obvious, since ions with $|\mu| \ll \mu_b$ injected in the trap are lost much more slowly than ions with $|\mu| \sim \mu_b$.

Among the other factors, the ion losses are most strongly influenced by ion–electron collisions and by the change of the confinement-region boundary due to the presence of an ambipolar potential. According to [19-21], in the stationary state at $R \approx 3$ allowance for these factors decreases $n\tau_{\parallel}{}^i$ by approximately six times, where $\tau_{\parallel}{}^i$ is the ion lifetime. The main contribution to the lowering of $n\tau_{\parallel}{}^i$ is made by allowance for the dynamic mutual friction of the ions and electrons, which lowers the equilibrium temperature of the ions at $T_i/T_e \sim (m_i/m_e)^{1/3}$ by an approximate factor of 2. The corresponding change of the confinement-region boundary leads to a relative increase of the loss of the low-energy ions. This, in turn, increases the average ion energy in the trap and compensates in part for the increased loss. Note that in [19-21] they considered the most pessimistic model with the ion-confinement region (3.7). Replacing $\Phi_0$ in (3.7) by $\Phi_{ef} \approx \Phi_0/2$ weakens noticeably the effect of the ambipolar potential on the ion loss. Both the cooling of the ions by the electrons and the effect of the ambipolar potential depend strongly on $T_i/T_e$. The cooling predominates at $T_i/T_e \gtrsim (m_i/m_e)^{1/3}$, and the influence of the ambipolar potential is more substantial at $T_i/T_e < (m_i/m_e)^{1/3}$.

Let us examine the influence of the magnetic-field profile on the ion cooling by electrons. In the square-well approximation this process is described by the penultimate term of Eq. (3.2). Since the cooling of ions by electrons is not connected with oblique scattering, the principal role in the averaging of the ion–electron collision term over the period of the oscillations between the mirror is played by the lowering of the frequency of the ion–electron collisions with increasing distance from the central cross section of the trap. This effect can be integrally taken into account by replacing the averaging of the ion–electron collision term by simple multiplication of this value at the central section by the following coefficient:

$$\int \frac{n^2(s)}{B(s)} ds \Big/ n_0 \int \frac{n(s)}{B(s)} ds. \tag{3.8}$$

It is easily seen that, in accordance with (2.34), this substitution conserves the energy transferred from the ions to the electrons. As a result, the errors introduced into the final results by this substitution do not exceed several percent.

The foregoing reasoning indicates that the square-well approximation in a simple mirror trap is quite adequate both for the ion–ion collision term and, when the factor (3.8) is considered, for the ion–electron collision term. This conclusion, as already noted, is confirmed by the results of numerical calculation. Averaging the source ions over the period of the oscillations between the mirrors does not greatly complicate matters, since the source is a given function of the velocity and of the coordinate s.

The ion–ion collision term is frequently simplified by using the so-called Rosenbluth–Trubnikov isotropic-potentials approximation. It consists of replacing the function $G(x, \mu)$ in (3.3) by its average $\overline{G}(x)$ over the angle. This approximation is justified by the fact that the functions G and $\Delta_x G$ depend much less on $\mu$ than the distribution function F itself. This follows directly from (3.5), according to which $\overline{G}$ is the coefficient at $P_0(\mu) \equiv 1$:

$$\left.\begin{aligned} \overline{G}(x) &\equiv \int_0^1 G(x, \mu)\, d\mu = \int_0^x x\xi^2 \left(1 + \frac{\xi^2}{3x^2}\right) \overline{F}(\xi)\, d\xi \\ &+ \int_x^\infty \xi^3 \left(1 + \frac{x^2}{3\xi^2}\right) \overline{F}(\xi)\, d\xi, \\ \overline{F}(x) &= 4\pi \int_0^{\mu_b(x)} F(x, \mu)\, d\mu. \end{aligned}\right\} \tag{3.9}$$

Replacement of $\overline{G}$ by G causes $\hat{C}_i(F)$ to take the form

$$\hat{C}_i(F) = \frac{1}{2x^2} \frac{\partial}{\partial x} x^2 \left( \frac{\partial F}{\partial x} \frac{\partial^2 \overline{G}}{\partial x^2} - F \frac{\partial}{\partial x} \Delta_x \overline{G} \right) +$$

$$+\frac{\partial \bar{G}}{2x^3\partial x}\,\frac{\partial}{\partial\mu}(1-\mu^2)\frac{\partial F}{\partial\mu}. \qquad (3.10)$$

It must be emphasized that replacement of G by $\bar{G}$ does not violate the particle-number and energy conservation laws.

If the influence of the ambipolar potential on the confinement-region boundary is neglected, $\mu_b$ ceases to depend on x, and the variables x and $\mu$ can be separated in Eq. (3.2) with $\hat{C}_i(F)$ in the form (3.10). In this case, $F(x, \mu)$ can be represented in the form

$$F(x, \mu, t)=\sum_{k=0}^{\infty} X_k(x, t)\, M_k(\mu), \qquad (3.11)$$

where $M_k(\mu)$ are the eigenfunctions of the Legendre equation

$$\frac{\partial}{\partial\mu}(1-\mu^2)\frac{\partial M_k(\mu)}{\partial\mu}+\lambda_k(\lambda_k+1)\, M_k(\mu)=0, \qquad (3.12)$$

and satisfy the boundary conditions

$$(\partial M_k/\partial\mu)_{\mu=0}=0, \quad M_k(\mu_b)=0. \qquad (3.13)$$

Substituting (3.11) in Eq. (3.2), multiplying it by $M_j(\mu)$, and integrating with respect to $\mu$, it is easy to obtain, in view of the orthogonality of the functions $M_k(\mu)$, equations for each of the functions $X_j(x, t)$. According to the analysis of [18], the first six terms of (3.11) ensure, as a rule, the required accuracy of the numerical calculations. Moreover, in a number of cases it suffices to consider only the zeroth term of the expansion (3.11). Thus, in the problem without a source, or with a source having a smooth distribution in $\mu$ and close to $M_0(\mu)$, the function $F(x, \mu, t)$ relaxes within a time shorter than $\tau_i$ to the form $X_0(x, t) \times M_0(\mu)$. The angular dependence of $F(x, \mu, t)$ is close to $M_0(\mu)$ at $x \sim 1$ in the case when the complete two-dimensional solution of (3.2) is considered without replacement of G by $\bar{G}$ (see [18]).

Thus, the functions $M_j(\mu)$ and their corresponding eigenvalues $\lambda_j$ play an important role in the analysis of ion losses. In the general case, $M_j(\mu)$ cannot be expressed in terms of elementary functions and can be represented in series form:

$$\left.\begin{aligned} M_j(\mu) &= 1 + \sum_{l=1}^{\infty} a_l(\lambda_j)\left(\frac{\mu^2}{2}\right)^l, \\ a_l(\lambda_j) &= \prod_{k=0}^{l-1} \frac{(2k-\lambda_j)(2k+1+\lambda_j)}{(k+1)(2k+1)}. \end{aligned}\right\} \tag{3.14}$$

At $\lambda_j = 2k$, where $k = (j + 1), (j + 2), \ldots,$ the series (3.14) has a finite number of terms and, in this case, the function $M_j(\mu)$ differs only by a constant factor from $P_{2k}(\mu)$. At $\lambda_j = 2k + 1$, where $k = j, (j + 1), \ldots,$ the function $M_j(\mu)$ can be expressed, accurate to a constant factor, in terms of a Legendre-equation solution that is linearly independent relative to $P_{2k+1}(\mu)$:

$$M_j(\mu) = AP_{2k+1}(\mu) \int \frac{d\mu}{(1-\mu^2)\, P_{2k+1}^2(\mu)}, \quad \lambda_j = 2k+1. \tag{3.15}$$

Plots of the first four eigenvalues $\lambda_j$ vs. $\mu_b = (1 - R^{-1})^{1/2}$ are shown in Fig. 6. Greatest interest attaches to the lowest eigenvalue $\lambda_0$, which determines, in essence, the ion loss. In the range of typical mirror ratios, $\lambda_0$ assumes twice (at $R = 1.5$ and at $R \approx 3.28$) integer values. To facilitate analytic estimates, it is useful to write for $M_0(\mu)$ the expressions corresponding to these values of $R$:

$$R = 1{,}5, \quad \lambda_0 = 2; \quad M_0(\mu) = 1 - 3\mu^2, \tag{3.16}$$

$$R \approx 3.28, \quad \lambda_0 = 1, \quad M_0(\mu) = 1 - \frac{\mu}{2} \ln \frac{1+\mu}{1-\mu}. \tag{3.17}$$

We also present a simple empirical relation, accurate to 3%, obtained in [22] for the dependence of $\lambda_0$ on $R$ in the interval $1.4 \leq R < \infty$:

$$\lambda_0(\lambda_0 + 1) \approx (\log R)^{-1}. \tag{3.18}$$

A detailed description of the actual difference schemes used to solve the Fokker–Planck equation for ions can be found in the reviews [23, 24], which deal specifically with this question. It is noteworthy that, in addition to these reviews, there have recently been developed difference schemes, called fully conservative, constructed such that the conservation laws contained in the initial equations are exactly satisfied also in their difference analog regardless of the mesh of the difference net. For the Fokker–Planck equation the pertinent conservation laws are those for the

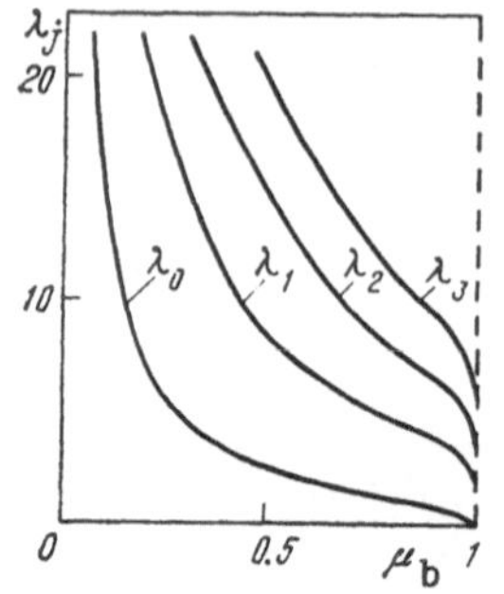


Fig. 6. Dependence of the first four eigenvalues $\lambda_j$ on $\mu_b$.

energy and number of particles, which were discussed in Section 1. The use of nonconservative difference schemes leads to the appearance of fictitious particle and energy sources proportional to the square of the mesh of the difference net. The fictitious sources can introduce considerable systematic errors even for a very fine mesh, particularly if we are interested in the behavior of the system over long time intervals exceeding the collision time. They can lead, in particular, to physically incorrect results, such as the absence of a stationary source and a decrease of the entropy in a closed system [25]. Fully conservative difference schemes are free of these shortcomings and yield sufficiently accurate results even for a relatively coarse net, a most important factor for two-dimensional calculations. Fully conservative difference schemes for the solution of the Fokker–Planck equation are given in [26] for the one-dimensional problem (with separation of the variables) with isotropic G, in [27-29] for a two-dimensional problem with isotropic G, and in [30] for a fully two-dimensional problem.

We proceed now to a direct analysis of Eq. (3.2) and to a derivation of the relations that describe the ion loss. For greater clarity we confine ourselves to approximate analytic models of the distribution function. In real cases, the relations obtained below can be refined by numerical solution of Eq. (3.2).

The first approximate analytic expression for the ion loss was obtained by Budker [1]. A derivation of this expression with a detailed discussion is given also in [11]. Budker considered a stationary state with a specially chosen source function Q, such that the distribution function was of the form $F = \psi(\mu)\exp\{x^2\}$. With such a distribution function and with G replaced by $\overline{G}$ it is easy to verify, with the aid of (3.10), that only one term, containing the derivatives with respect to $\mu$, is different from zero in the operator $\hat{C}_i(F)$. Nor was account taken in [1] of ion–electron collisions and of the influence of the ambipolar potential. With this approach, the problem was reduced to finding the angular part of the distribution function $\psi(\mu)$. Budker considered two limiting cases of angular distribution of the source function: 1) a narrow source proportional to $\delta(\mu)$; 2) a wide source isotropic within the confinement region. Two expressions were accordingly obtained for the ion lifetime:

$$(\tau_{\parallel}^i)_1 = 0.72\tau_i \ln R, \quad (\tau_{\parallel}^i)_2 \approx 0.72\tau_i \ln(R/1.8). \tag{3.19}$$

The results of later numerical calculations have shown that (3.19) accounts correctly for the dependence of $\tau_{\parallel}{}^i$ on R and on the width of the angular distribution of the source. However, the specific dependence of the source function on the variable x exerted a noticeable influence on the numerical coefficient in (3.19) and, more importantly, on the ion temperature that enters in $\tau_i$. According to [1], $T_i$ should be 2.3 times larger than the average energy $\varepsilon_{inj}$ of the injected ions. The reason is that in Budker's model we have $Q \sim x^{-2}$ as $x \to 0$; i.e., the ion source is strongly enriched in the low-energy region. When a more realistic source is considered and account is taken of the ion cooling of the electron, the value of $T_i/\varepsilon_{inj}$ turns out, as a rule, to be lower by a factor 2-3, and this leads to a considerable decrease of $\tau_{\parallel}{}^i$ compared with the Budker model.

Without specifying the form of $F(x, \mu, t)$, we consider some general integral equations that characterize the balance of the number of ions and of their energy in a trap. To avoid excessively cumbersome expressions, we neglect the influence of the ambipolar potential, the last term in (3.2) which describes the ion diffusion by electrons, and the correction of order $m_eT_i/m_iT_e$ in the penultimate term of (3.2). These effects do not affect qualitatively the results that follow, and can only change the numerical coefficient by 20-30%. Integrating (3.2) over the volume of the confinement region (3.6) with weights 1 and $x^2$, we obtain the balance equations for the number of ions and for the energy:

$$dn/dt = -\nu n/\tau_i + I, \tag{3.20}$$

$$\frac{3}{2} n \frac{dT_i}{dt} = T_i \frac{n}{\tau_i} \left( \frac{3}{2} \nu - \gamma - 3\rho \right) + I \left( \varepsilon_{\text{inj}} - \frac{3}{2} T_i \right). \tag{3.21}$$

The dimensionless coefficients $\nu(t)$, $\gamma(t)$, and $\rho(t)$ are given by

$$\nu = -2\pi \int_0 \frac{dx}{x} \left( \frac{\partial G}{\partial x} + \frac{\mu}{x} \frac{\partial G}{\partial \mu} + \frac{1}{x} \frac{\partial}{\partial \mu} (1-\mu^2) \frac{\partial G}{\partial \mu} \right) (1-\mu)^2 \frac{\partial F}{\partial \mu} \bigg|_{\mu=\mu_b}, \tag{3.22}$$

$$\gamma = -2\pi \int_0^\infty x\, dx \left( \frac{\partial G}{\partial x} + \frac{\mu}{x} \frac{\partial G}{\partial \mu} + \frac{1}{x} \frac{\partial}{\partial \mu} (1-\mu^2) \frac{\partial G}{\partial \mu} \right) (1-\mu)^2 \frac{\partial F}{\partial \mu} \bigg|_{\mu=\mu_b}, \tag{3.23}$$

$$\rho = \frac{4}{3\sqrt{\pi}} \frac{\Lambda_{il}^{\varepsilon}}{\Lambda_i} \left( \frac{m_e T_i^3}{m_i T_e^3} \right)^{1/2}. \tag{3.24}$$

$I = \int S dv$ is the number of ions entering per unit time into a unit volume; $\varepsilon_{\text{inj}}$ is the average energy of the injected particles. The coefficients $\nu$ and $\gamma$ characterize the particle and energy losses due to scattering into the loss cone, and $\rho$ accounts for the ion cooling by the electrons. At an arbitrary magnetic-field profile the expression for $\rho$ must be multiplied by the coefficient (3.8). Within the framework of the model in which it is assumed that G can be replaced by $\bar{G}$, and that the angular dependence of the function $F(x, \mu, t)$ is determined by lowest eigenfunction $M_0(\mu)$, the coefficients $\nu$ and $\gamma$ take the simplest form:

$$\nu = \frac{\lambda_0(\lambda_0+1)}{2} \int_0^\infty \frac{dx}{x} \bar{F}(x, t) \frac{\partial \bar{G}(x, t)}{\partial x}, \tag{3.25}$$

$$\gamma = \frac{\lambda_0(\lambda_0+1)}{2} \int_0^\infty x\, dx \bar{F}(x, t) \frac{\partial \bar{G}(x, t)}{\partial x}, \tag{3.26}$$

where $\bar{G}$ and $\bar{F}$ are defined by (3.9).

In the derivation of (3.20) and (3.21), the normalization conditions used for $F(x, \mu, t)$ corresponded to the chosen method of reducing (3.2) to nondimensional form:

$$\int_0^\infty \bar{F}(x,\ t)\, x^2 dx = 1, \quad \int_0^\infty \bar{F}(x,\ t)\, x^4 dx = 3/2. \tag{3.27}$$

It can be proved with the aid of (3.27) that $3\nu/2 > \gamma$. This inequality follows, if $\nu$ and $\gamma$ are defined in expressions (3.25) and (3.26), from the following relations:

$$\frac{3}{2}\nu - \gamma = -\frac{\lambda_0(\lambda_0+1)}{2}\int_0^\infty dx\left[\frac{\partial}{\partial x}\left(\frac{1}{x^3}\frac{\partial \bar{G}}{\partial x}\right)\right]\int_0^x \xi^2\left(\frac{3}{2}-\xi^2\right)\bar{F}(\xi)\, d\xi, \tag{3.28}$$

since the expression in the square brackets in (3.28) is always negative according to (3.9), and the inner integral in (3.28) is always positive when (3.27) is taken into account. The indicated inequality means that the average energy of the ions that leave the trap is lower than the thermal energy. This is most strongly manifested in the Budker model, where $\gamma = 0.43\nu$. In the other limiting case, when $F(x,\ \mu,\ t)$ is close to a $\delta$ function in energy, we have $\gamma \to 3\nu/2$.

In accordance with (3.20) and (3.21), the density and the average energy of the ions should relax at $I \neq 0$ to a stationary state:

$$n = I\tau_i/\nu, \quad T_i = \varepsilon_{\text{inj}}\frac{\nu}{\gamma + 3\rho}. \tag{3.29}$$

In the stationary state, $\tau_{\parallel}{}^{i}$ takes, by definition, the form

$$\tau_{\parallel}^{i} = n/I = \tau_i/\nu. \tag{3.30}$$

From (3.30), in view of (3.25) and (3.18), it follows, in particular, that $\tau_{\parallel}{}^{i}$ has a logarithmic dependence on R. Equation (3.29) shows, furthermore, that ion cooling by electrons influences strongly the value of $T_i/\varepsilon_{\text{inj}}$. Without allowance for the cooling ($\rho = 0$) the average ion energy in the trap should be approximately twice as large as $\varepsilon_{\text{inj}}$. However, even for classical electron-energy loss, as follows from the results of Section 2, we have $\rho \approx 0.15\text{-}0.2$ and $3T_i/2 \lesssim \varepsilon_{\text{inj}}$. The appearance of additional channels for electron-energy loss increases the parameter $\rho$ and accordingly decreases $T_i/\varepsilon_{\text{inj}}$ even more. The stationary density of the ions can be expressed with the aid of (3.29) in terms of the source parameters:

$$n = \left(\frac{I\sqrt{m_i}\,\varepsilon_{\text{inj}}^{3/2}}{\sqrt{2}\pi e^4\Lambda_i}\right)^{1/2} \frac{\nu^{1/4}}{(\gamma+3\rho)^{3/4}}. \tag{3.31}$$

We also present for $(n\tau_\varepsilon)_\parallel^i$ an expression that characterizes the confining properties of the trap when used for thermonuclear applications:

$$(n\tau_e)_\parallel^i = \frac{3T_i}{2\varepsilon_{\text{inj}}} n\tau_\parallel^i = \frac{3\sqrt{m_i}\,\varepsilon_{\text{inj}}^{3/2}}{2\sqrt{2}\pi e^4\Lambda_i} \frac{\nu^{3/2}}{(\gamma+3\rho)^{5/2}}. \tag{3.32}$$

In the absence of an ion source, Eqs. (3.20) and (3.21) describe the free decay of the plasma ionic component. The main features of this process can be traced most simply and lucidly under the condition that $T_i/T_e$ is independent of t. $T_i/T_e$ should be constant in those cases when $T_e$ is determined by heat from ions, particularly for a classical energy balance of the electronic component (see Section 2). At $T_i/T_e$ = const, Eq. (3.2) for $F(x, \mu, t)$ has a stationary solution ($\partial F/\partial t = 0$) corresponding to time-independent values of $\nu$, $\gamma$, and $\rho$. Since (3.2) is an equation of parabolic type, the function $F(x, \mu, t)$, irrespective of its initial form, should relax within a time $\lesssim \tau_i$ to this stationary state. The equation for the stationary function $F(x, \mu)$, with allowance for the simplifying assumptions made above, is

$$\widehat{C}_i(F) + \nu F + \frac{1}{3}\left(\frac{3}{2}\nu - \gamma\right)\frac{1}{x^2}\frac{\partial}{\partial x}(x^3 F) = 0. \tag{3.33}$$

The considered stationary state of the dimensionless function $F(x, \mu)$ corresponds to a self-similar evolution of the total ion distribution function $f_i(v, \mu, t)$, where the entire dependence of $f_i$ on t reduces to time dependences of only the ion density and the average ion energy. The onset of a self-similar regime was first observed in [18] in numerical calculations of the evolution of $f_i$.

Multiplying (3.21) by $\tau_i/nT_i$, and (3.20) by $\tau_i/n$, and subtracting one from the other, we get at I = 0 the relation

$$\frac{d\tau_i}{dt} = \frac{5}{2}\nu - \gamma - 3\rho, \tag{3.34}$$

which shows that in the self-similar regime $\tau_i$ is linear in time:

$$\tau_i(t) = \tau_{i0} + \left(\frac{5}{2}\nu - \gamma - 3\rho\right)t. \tag{3.35}$$

Depending on the sign of $(5\nu/2 - \gamma - 3\rho)$, the value of $\tau_i(t)$ either increases without limit or vanishes after a finite time. It is this which determines the character of the plasma decay. If $5\nu/2 \neq \gamma + 3\rho$, then n and $T_i$ vary either in accordance with a power law, or vanish after a finite time:

$$n = n_0\left[1 + \left(\frac{5}{2}\nu - \gamma - 3\rho\right)\frac{t}{\tau_{i0}}\right]^{-\zeta_n}, \quad \zeta_n = \frac{\nu}{\frac{5}{2}\nu - \gamma - 3\rho},$$
$$T_i = T_{i0}\left[1 + \left(\frac{5}{2}\nu - \gamma - 3\rho\right)\frac{t}{\tau_{i0}}\right]^{\zeta_T}, \quad \zeta_T = \frac{2\left(\frac{3}{2}\nu - \gamma - 3\rho\right)}{3\left(\frac{5}{2}\nu - \gamma - 3\rho\right)}. \tag{3.36}$$

The value of n always decreases with time, while $T_i$ either increases or decreases, depending on the sign of $(3\nu/2 - \gamma - 3\rho)$. If $5\nu/2 = \gamma + 3\rho$, then n and $T_i$ vary exponentially:

$$n = n_0 \exp\left\{-\frac{\nu}{\tau_{i0}}t\right\}, \quad T_i = T_{i0}\exp\left\{-\frac{2\nu}{3\tau_{i0}}t\right\}. \tag{3.37}$$

It should be noted that if no account is taken of the cooling of the ions by the electrons ($\rho = 0$), the classical decay of the plasma should always follow, in accordance with (3.36), a power law with $T_i$ a growing function of the time.

Thus, both the stationary state and the free decay of the ion component of a plasma depend substantially on the relative magnitudes of the three dimensionless parameters $\nu$, $\gamma$, and $\rho$. The parameter $\rho$ is determined by the energy balance of the electronic component. The two other parameters $\nu$ and $\gamma$ are determined by the form of the function F and depend principally on the mirror ratio. In the general case the parameters $\nu$ and $\gamma$ can be obtained only by numerical solution of the corresponding equation for the function F. In many cases, nonetheless, we can simulate the function $F(x, \mu)$ quite accurately by some analytic expression, and then obtain, with the aid of (3.21), (3.22) or (3.25), (3.26), approximate expressions for $\nu$ and $\gamma$. Owing to the integral dependences of $\nu$ and $\gamma$ on F, the values of $\nu$ and $\gamma$ obtained in this manner are much more accurate than the initial model for $F(x, \mu)$.

The function $F(x, \mu)$ can be simulated relatively simply in the self-similar decay regime. As noted above, in this regime the angular dependence of $F(x, \mu)$ is described well enough by the lowest eigenfunction $M_0(\mu)$. In other words, $F(x, \mu)$ can be represented in the form

$$F(x, \mu) \approx \overline{F}(x) M_0(\mu) \Big/ 4\pi \int_0^{\mu_b} M_0(\mu)\, d\mu. \tag{3.38}$$

We analyze next the form of $\overline{F}(x)$ at $x^2 \ll 1$ and $x^2 \gg 1$. With the aid of (3.9) and (3.10), it is easy to verify that at $x^2 \ll 1$ the principal role in Eq. (3.33) is played by scattering into the loss cone and by diffusion along x, so that the approximate form of $\overline{F}(x)$ can be obtained from the equation

$$\frac{1}{x^2} \frac{\partial}{\partial x} x^2 \frac{\partial \overline{F}}{\partial x} - \frac{\lambda_0(\lambda_0+1)}{x^2} \overline{F} \approx 0, \tag{3.39}$$

according to which $\overline{F}$ should be proportional to $x^{\lambda_0}$. At $x^2 \gg 1$, the angle scattering ceases to play a dominant role, and the form of the function $\overline{F}(x)$ is determined mainly by the balance between the diffusion along x and the effective dynamic friction:

$$\frac{1}{x^2} \frac{\partial}{\partial x} \left\{ \left[ 1 + \frac{1}{3} \left( \frac{3}{2} \nu - \gamma \right) x^3 \right] \overline{F} + \frac{1}{2x} \frac{\partial \overline{F}}{\partial x} \right\} + \nu \overline{F} \approx 0. \tag{3.40}$$

As $\lambda_0 \to 0$, the parameters $\nu$ and $\gamma$ should also tend to zero. In this case, Eq. (3.40) determines the Maxwellian tail of the distribution function: $\overline{F} \sim \exp\{-x^2\}$. When $\nu \neq 0$ and $\gamma \neq 0$, the solution of Eq. (3.40) at $x^2 \gg 1$ has the following asymptotic form:

$$\overline{F} \sim \left[ 1 + \frac{(3\nu - 2\gamma)}{6} x^3 \right]^{\frac{2\nu}{9\nu - 6\gamma}} \exp\left\{ -x^2 - \frac{3\nu - 2\gamma}{15} x^5 \right\}. \tag{3.41}$$

Thus, $\overline{F}(x)$ should increase at low energies in proportion to $x^{\lambda_0}$, and should decrease somewhat faster than $\exp\{-x^2\}$ at high energies. It is easy to verify that the relatively simple model function

$$\overline{F}(x) = 2\left(1 + \frac{\lambda_0}{3}\right)^{(3+\lambda_0)/2} x^{\lambda_0} \exp\left\{ -\left(1 + \frac{\lambda_0}{3}\right) x^2 \right\} \Big/ \Gamma\left(\frac{3+\lambda_0}{2}\right) \tag{3.42}$$

satisfies both the foregoing qualitative requirements and the normalization conditions (3.27). As expected, Eq. (3.42) goes over as $\lambda_0 \to 0$ into a Maxwellian function, and as $\lambda_0 \to \infty$ into a $\delta$ function. Moreover, if the parameters $\nu$ and $\gamma$ are calculated with the aid of (3.42) and substituted in (3.33), the normalized solution of (3.33), obtained by numerical methods, differs from (3.42) by less than 10% in the entire interval $0 < x^2 \lesssim 4$. At $x^2 > 4$, the relative deviation of (3.42) from the exact solution of (3.33) becomes quite appreciable, but this region makes no contribution of any significance to the basic integral relations. Thus, (3.42) is a good enough analytic model of the distribution function and can be used to obtain various integral relations. In particular, it may turn out to be quite useful in the analysis of the possible kinetic instabilities of the plasma in simple mirror traps.

Substituting the model function (3.42) in (3.25) and (3.26) and integrating, we easily obtain expressions for $\nu(\lambda_0)$ and $\gamma(\lambda_0)$. These expressions are rather unwieldy. In the interval $0.4 \lesssim \gamma_0 \lesssim 2$, however, which corresponds to the most interesting range of mirror ratios $1.5 \lesssim R \lesssim 60$, we can approximate $\nu(\lambda_0)$ and $\gamma(\lambda_0)$ with an error up to 5% by the following simple relations:

$$\nu \approx 1.28\frac{\lambda_0(\lambda_0+1)}{2+\lambda_0} \approx \frac{1.91}{\ln R + \sqrt{\ln R}}, \tag{3.43}$$

$$\gamma \approx 0.314\lambda_0(\lambda_0+1) \approx 0.723/\ln R. \tag{3.44}$$

In the presence of an ion source the equation for the stationary function $F(x, \mu)$ takes the form

$$\widehat{C}_i(F) + \frac{\rho}{x^2}\frac{\partial}{\partial x}(x^3 F) + Q(x, \mu) = 0. \tag{3.45}$$

If the angular distribution of the source is smooth enough, the dependence of F on $\mu$ does not differ greatly from $M_0(\mu)$. In that case, just as under conditions of self-similar plasma decay, the form of the function $\overline{F}(x)$ at $x^2 \ll 1$ is determined by Eq. (3.39), so that $\overline{F}(x) \sim x^{\lambda_0}$ in the low-energy region. At $x^2 \gg 1$, the form of the function $\overline{F}(x)$ can be found from the approximate equation

$$\frac{1}{x^2}\frac{\partial}{\partial x}\left[(1+\rho x^3)\overline{F} + \frac{1}{2x}\frac{\partial \overline{F}}{\partial x}\right] \approx 0, \tag{3.46}$$

according to which we have, at high energies,

$$\bar{F} \sim \exp\left\{-x^2 - \frac{2}{5}\rho x^5\right\}. \tag{3.47}$$

Comparison of (3.47) with (3.41) shows that at $\rho \lesssim 0.3\lambda_0$ the asymptotic form of $\bar{F}(x)$ in the stationary state does not differ strongly from the asymptotic form of $\bar{F}(x)$ in the self-similar regime. Therefore, the parameters $\nu$ and $\gamma$ in the stationary state with $\rho \leq 0.3\lambda_0$ and with a smooth angular distribution of the source are well enough approximated by expressions (3.43) and (3.44). At a narrow angular distribution of the source, close to $\delta(\mu)$, expressions (3.43) and (3.44) also provide a correct estimate of the parameters $\nu$ and $\gamma$, provided R is replaced in them by 1.5R. An increase of $\rho$ has relatively little effect on the values of $\nu$ and $\gamma$. Thus, at $\rho$ = 3-4, the parameters $\nu$ and $\gamma$ decrease by an approximate factor 1.5 compared with (3.43) and (3.44). With allowance for the indicated corrections, expressions (3.43) and (3.44) provide thus a sufficiently accurate estimate of the parameters $\nu$ and $\gamma$ for most cases of practical interest.

It was assumed everywhere above that the ions stem from ionization of fast neutral atoms injected into the trap. This is correct under conditions of possible thermonuclear applications, when the energy of the injected atoms is 100 keV. At an injection energy lower than 100 keV, it is necessary to take into account charge exchange on top of the ionization. Although charge exchange does not change the total number of ions in the trap, it nevertheless influences actively the evolution of the distribution function, since it returns to their birthplace the ions that "run away" as a result of diffusion. Charge exchange plays a particularly important role in laboratory plasma with moderate injection energy on the order of 10-20 keV, for in this energy range the cross-sections for charge exchange of hydrogen and deuterium are about 4-5 times larger than the cross-sections for ionization. Without dwelling in detail on effects involved in allowance for charge exchange, we note that this process can be easily incorporated in the scheme described above, by replacing the source function Q(x, $\mu$) in (3.2) by the expression

$$\left(1+\frac{\tau_{\text{ion}}}{\tau_{\text{ce}}}\right)Q-\frac{\tau_i}{\tau_{\text{ce}}}F(x,\ \mu),$$

where $\tau_{ion}$ is the ionization time and $\tau_{ce}$ is the time of charge exchange of the plasma ions with neutral atoms.

## 4. Traps with Improved Longitudinal Plasma Confinement

The study of the main laws governing plasma confinement in simple mirror traps has shown that the prospects of using such a trap as a basis for a thermonuclear reactor are quite doubtful, since even for classical longitudinal losses the value of $n\tau_{\varepsilon}$ turns out to be insufficient to obtain a satisfactory reactor-power gain. Much attention is therefore being paid to searches for effective ways of lowering the longitudinal plasma losses in open traps. The most attractive from this point of view are traps in which a potential barrier for the ions can be produced between the bulk of the plasma and the wall to which the force lines go. By analogy with electron confinement, the ion lifetime in such a system should increase exponentially with increasing height of the barrier, and it is this which ensures the necessary amplification $n\tau_{\varepsilon}$. At present there are two known types of adiabatic trap in which a potential barrier is used to improve the longitudinal ion confinement. These are tandem-mirror traps and traps with rotating plasma.

Tandem-Mirror Traps. The idea of the tandem-mirror (or ambipolar) trap was independently advanced by two groups of workers [31, 32]. The magnetic system of a tandem-mirror trap consists of a long central solenoid with a uniform field $B_0$, at the ends of which are located dual mirrors constituting small end traps (Fig. 7). To ensure magnetohydrodynamic stability of the entire trap, an additional quadrupole field in the outermost traps produces a magnetic well in the transverse direction. It is assumed further that neutral injection maintains in the outermost traps a plasma with density $n_{out}$ and an average ion energy $\varepsilon_{out}$, while the central solenoid is filled with plasma having $n_0 \ll n_{out}$ and $T_i \sim T_e \ll \varepsilon_{out}$. In view of the positive potential of the plasma and the high longitudinal thermal conductivity of the electrons, their distribution, just as in a simple mirror trap, should be close to Maxwellian with a temperature that is approximately constant over its length. In this case, taking into account the quasineutrality and the Boltzmann distribution of the electron density along the magnetic field, the profile of the potential, as shown in Fig. 7, should duplicate the density profile.

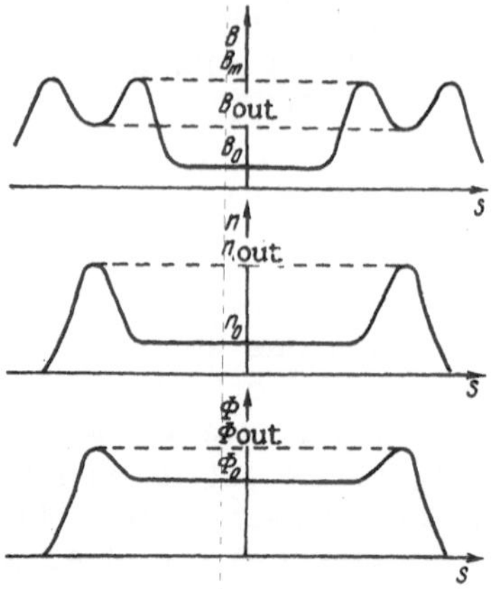


Fig. 7. Longitudinal profiles of the magnetic field B(s), of the density n(s), and of the electric potential Φ(s) in a tandem-mirror trap.

The high-energy ions are confined in the outermost traps, as in a simple mirror trap, by the presence of magnetic mirrors. Since $\varepsilon_{out} \gg T_e$, the longitudinal electric field has little effect on the confinement of the high-energy ions. It is this which essentially makes it possible to maintain the required plasma-density profile. In contrast to high-energy ions, the "warm" ions that fill the central section are confined mainly by the potential difference between the outermost trap and the center:

$$e\Delta\Phi \equiv e(\Phi_{out} - \Phi_0) = T_e \ln (n_{out}/n_0) \gg T_i. \quad (4.1)$$

The region of warm-ion confinement in velocity space takes, therefore, the form of a hyperboloid

$$\mu^2 \leqslant \frac{R_0 - R_{out}}{R_0} + \frac{R_{out}}{R_0} \frac{2e\Delta\Phi}{m_i v^2}, \quad (4.2)$$

similar to the confinement region for the electrons

$$\mu^2 \leqslant \frac{R_0 - 1}{R_0} + \frac{1}{R_0} \frac{2e\Phi_0}{m_e v^2}, \quad (4.3)$$

where $R_0 = B_m/B_0$, $R_{out} = B_m/B_{out}$. A more detailed analysis shows that (4.2) determines the confinement region of ions having an energy lower than $e\Delta\Phi R_{out}/(R_{out} - 1)$. Ions with higher energy are reflected by the first mirror without reaching the center of the outermost trap, so that their confinement region is of the form $\mu^2 \leq (R_0 - 1)/R_0$. The difference between the confinement region for high-energy ions and (4.2) does not alter noticeably the lifetime of the plasma of the central section, inasmuch as $R_0 \gg R_{out}$ under typical conditions.

The presence of a high potential barrier makes the problem of the ion loss from the central section perfectly analogous to the problem of electron loss in a simple mirror trap (see Section 2). Therefore, the change of the ion density and of their energy in the central section is described by expressions similar in form to (2.18) and (2.19), with $T_e$ replaced by $T_i$, $\Phi$ by $\Delta\Phi$, $\tau_e$ by $\tau_i$, and R by $R_0/R_{out}$. The coefficient $\alpha$ for ions is equal to 1/2 as against 1 for electrons, since the oblique scattering of the ions by electrons can be neglected at $T_i \sim T_e$.

It was shown in [13] that in a multicomponent plasma in which the different ion components are in thermodynamic equilibrium with one another ($T_i = T_j$), expressions similar to (2.18) and (2.19) can be written for each ion component. The quantities $\tau_i$ and $\alpha_i$ that enter in these expressions then take the form

$$\frac{1}{\tau_i} = \sqrt{2}\,\pi e^4 \sum_j \frac{Z_i^2 Z_j^2 n_j \Lambda^{i/j} \sqrt{m_i}}{m_j T_i^{3/2}}, \tag{4.4}$$

$$\alpha_i = \frac{1}{2} \sum_j Z_j^2 n_j \Lambda^{i/j} \Big/ \sum_j \left(\frac{m_i}{m_j}\right) Z_j^2 n_j \Lambda^{i/j}, \tag{4.5}$$

where the summation is over all ion species. For an arbitrary magnetic-field profile, the quantities $\tau_i$ and $\alpha_i$ are averaged in analogy with the procedure described in Section 2 for electrons:

$$\bar{\tau}_i^{-1} = \oint \frac{\Delta\Phi ds}{[\Phi_{out} - \Phi(s)]\,\tau_i(s)} \Big/ \oint \sqrt{\frac{\Delta\Phi}{\Phi_{out} - \Phi(s)}}\,ds, \tag{4.6}$$

$$\bar{\alpha}_i = \oint \frac{\alpha_i(s) B_0 \Delta\Phi \bar{\tau}_i ds}{B(s)\,[\Phi_{out} - \Phi(s)]\,\tau_i(s)} \Big/ \oint \sqrt{\frac{\Delta\Phi}{\Phi_{out} - \Phi(s)}}\,ds. \tag{4.7}$$

The longitudinal electron loss in a tandem-mirror trap, just as in a simple mirror trap, is determined by expressions (2.18) and (2.19). The generalizations of these expressions to allow for a multicomponent plasma and for an arbitrary magnetic-field profile remain in force here. The total energy balance of the electronic components includes the heating of the electrons by the hot ions confined in the outermost traps, and the energy exchange between the electrons and the warm ions of the central sections. These two processes can be described by relation (2.34). In contrast to a simple mirror trap, in a tandem-mirror trap the heating by hot ions is a small factor in the electron energy balance $n_{out} \times L_{out}R_{out}/n_0 L_0 R_0$, where $L_{out}$ and $L_0$ are the effective lengths of the outermost trap and of the central section. In the absence of additional electron heating, this lowers considerably the ratio $T_e/\varepsilon_{out}$ compared with the case of a simple mirror trap. Another distinctive feature of the behavior of the electronic component in a tandem-mirror trap is the presence of a group of electrons captured in the outermost traps. The appearance of this group is due to the fact that the potential barrier for the ions is simultaneously a potential barrier for electrons. Since the trapped electrons are in the outermost traps all the time, they are subject to more intense heating by the hot ions than the untrapped electrons. As a result, notwithstanding the comparatively rapid heat exchange within the electronic component, the temperature of the trapped electrons should be somewhat higher than that of the bulk of the electrons. This improves the confinement of the central-section ions, since the potential barrier for the ions is proportional precisely to the temperature of the trapped electrons. To enhance this effect, a "thermal barrier" was suggested in [33], meaning an artificial lowering of the potential between the central section and the outer trap so as to reduce the heat exchange between the bulk of the electrons and the electrons trapped in tho outer traps.

The confinement of the hot ions in the outer traps does not differ from the confinement in a simple mirror trap, considered in Section 3, except that the parameter $\rho$ in the outer trap can exceed by more than an order of magnitude the value typical of a classical scheme of a simple mirror trap. According to [31], in a tandem-mirror trap without a "thermal barrier" and without additional heating we have $\rho \gtrsim 1$-$2$. The high value of the parameter $\rho$ affects very adversely the energy balance of the outer trap. In view of their small volume, however, the outermost traps should not influence strongly the power gain of a tandem-mirror thermonuclear reactor [31, 32]. From this point of view, the outer traps are only complicated mirrors that permit a substantial improvement of the ion confinement in the central section.

We have discussed above longitudinal losses of electrons and of warm ions of the central section, corresponding to the case of infrequent collisions, when time of particle travel along the trap is much shorter than the time to fill the loss hyperboloid. For warm ions and electrons, the weak-collision condition is of the form

$$\tau_i \frac{R_{\text{out}}}{R_0}\left(\frac{e\Delta\Phi}{T_i}\right)^{3/2} \gg L_0 \sqrt{\frac{m_i}{2e\Delta\Phi}}, \tag{4.8}$$

$$\tau_e \frac{1}{R_0}\left(\frac{e\Phi_0}{T_e}\right)^{3/2} \gg L_0 \sqrt{\frac{m_e}{2e\Phi_0}}. \tag{4.9}$$

At plasma parameters typical of the tandem-mirror reactor, the conditions (4.8) and (4.9) are met with a large margin [31, 32]. In experiments with relatively low particle energy ($T_i \sim T_e \sim$ 100-200 eV), however, conditions (4.8) and (4.9) may not be met. In this case, relations (2.18), (2.19) and the analogous expression for ion loss are no longer valid.

In the limiting case of frequent collisions, when conditions inverse to (4.8) and (4.9) are met, the particle distribution function is close to Maxwellian not only in the confinement region, but also inside the loss hyperboloid. The longitudinal losses from the central section are determined by the gasdynamic fluxes of the particles escaping through the corresponding potential barriers in the outer traps. By calculating these fluxes it is easy to obtain the following expressions for the lifetimes of the thermal ions and of the electrons:

$$\tau_\parallel^i = \frac{R_0}{R_{\text{out}}} L_i \sqrt{\frac{\pi m_i}{2T_i}} \exp\left\{\frac{e\Delta\Phi}{T_i}\right\}, \tag{4.10}$$

$$\tau_\parallel^e = R_0 L_e \sqrt{\frac{\pi m_e}{2T_e}} \exp\left\{\frac{e\Phi_0}{T_e}\right\}. \tag{4.11}$$

where $L_i$ and $L_e$ are the effective trap lengths for the ions and electrons, respectively:

$$L_{i,e} = \int (B_0 n_{i,e}(s)/B(s)\, n_0)\, ds. \tag{4.12}$$

Calculating also the energy flux, we obtain the average energy carried away by each ion and electron from the plasma:

$$\langle \varepsilon \rangle_i = e\Delta\Phi + 2T_i, \quad \langle \varepsilon \rangle_e = e\Phi_0 + 2T_e. \tag{4.13}$$

We note that in the limit of frequent collisions the above-barrier energy is 2T, and not T as in the infrequent-collision limit (2.20).

The transition region between the infrequent and frequent collisions was investigated in [34]. Calculations by the Monte Carlo method have shown that the lifetime in the transition region is well described by the simple interpolation formula

$$\tau_{\parallel} \approx \tau_{\parallel \mathrm{infr}} + \tau_{\parallel \mathrm{freq}}, \tag{4.14}$$

where $\tau_{\parallel \mathrm{infr}}$ and $\tau_{\parallel \mathrm{freq}}$ are the lifetimes in the limit of infrequent and frequent collisions, respectively. An equally simple interpolation formula can describe also the average energy carried away by the particles from the trap:

$$\langle \varepsilon \rangle = e\Phi + \left( \frac{2\tau_{\parallel \mathrm{freq}} + \tau_{\parallel \mathrm{infr}}}{\tau_{\parallel \mathrm{freq}} + \tau_{\parallel \mathrm{infr}}} \right) T. \tag{4.15}$$

Traps with Rotating Plasma. By traps with rotating plasma are usually meant axisymmetric open traps containing besides the longitudinal magnetic field also a strong radial electric field E that causes the plasma to rotate at a speed $cE_{\perp}/B$. The simplest rotating plasma trap, shown schematically in Fig. 8, is an axisymmetric mirror trap having on each end a system of coaxial electrodes that are in contact with the plasma. The required radial profile of the electric field is produced in the plasma by applying independently to each electrode an appropriate high voltage.

The basic principles of plasma confinement in rotating plasma traps, as well as the results of the initial stage of experiments on traps of this type, are described in detail in the review [35]. Further development of research into this class of traps, including an analysis of the prospects for use in reactors, is reflected in [36]. Besides the tra-

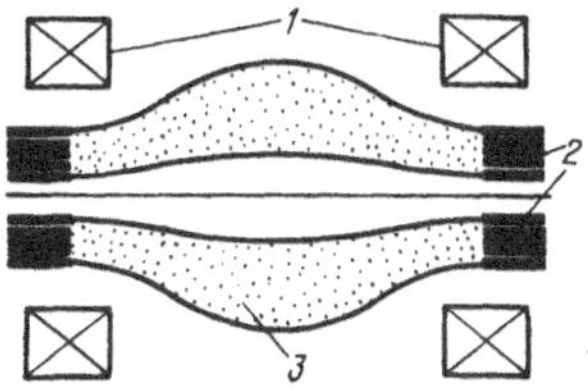


Fig. 8. Simplest scheme of a trap with rotating plasma: 1) current-carrying coils; 2) coaxial electrodes; 3) plasma.

ditional trap scheme with rotating plasma, a modified variant was also considered, viz., a centrifugal trap [37], which offers a number of advantages over the traditional scheme.

We shall dwell briefly on the main principles of longitudinal confinement of particles in rotating plasma traps. Owing to the axial symmetry of the trap, the magnetic force lines and the particle drift trajectories lie on surfaces on which the magnetic flux is constant, $\Psi(r, z)$ = const (see Section 1). Since the longitudinal profile of the plasma density changes little on going from one surface $\Psi$ = const to another, and the longitudinal profile of the potential is connected with the density by the Boltzmann relation (1.5), the potential difference between neighboring constant-$\Psi$ surfaces in the bulk of the plasma remains approximately constant along the force line:

$$\left.\frac{\partial\Phi}{\partial\Psi}\right|_{\Psi=\text{const}} = \text{const}. \qquad (4.16)$$

From (4.16) follows the so-called "isorotation law" [35], according to which the frequency $\Omega_E$ of plasma rotation about the magnetic axis of the trap is constant on the drift surface:

$$\Omega_E = (cE_\perp/rB)_{\Psi=\text{const}} \approx \text{const}. \qquad (4.17)$$

It is therefore convenient to consider the plasma confinement in a rotating coordinate frame, in which the particles are acted on by a centrifugal force $mr\Omega_E^2$ that tends to push them out into a region where the drift surface has a maximum.

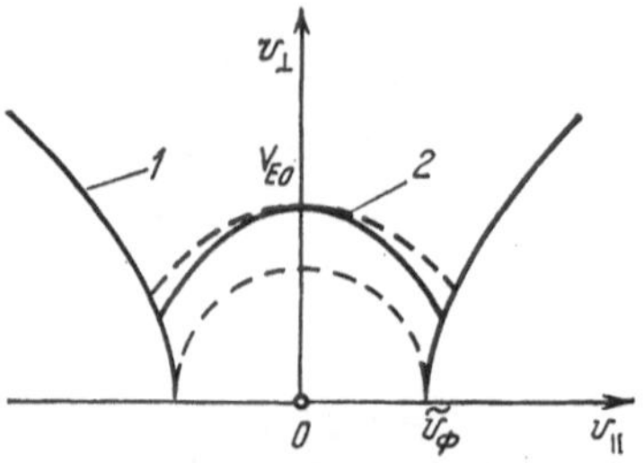


Fig. 9. Ion-confinement region in a rotating plasma trap: 1) boundary of confinement region; 2) "source surface."

With allowance for the conservation of the energy and of $J_\perp$, the ion-containment region takes, in the rotating frame, the form

$$v_{\parallel 0}^2 \leqslant v_{\perp 0}^2 (R-1) + V_{E0}^2 \left(1 - \frac{r_m^2}{r_0^2}\right) - \frac{2e\Phi_0}{m_i}, \qquad (4.18)$$

where $V_{E_0} = cE_0/B_0$; $r_m$ is the radius of the constant-$\Psi$ surface in the mirror. The right-hand side of (4.18) contains one more term than (1.3), and at $r_m < r_0$ this term can be called the centrifugal barrier. In the simplest trap scheme shown in Fig. 8 we have $r_m^2 = r_0^2 B_0/B_m$, so that $r_0^2/r_m^2 = R$.

In the general case, the axisymmetric magnetic-field configuration can be chosen such that the parameters $r_m^2/r_0^2$ and $B_m/B_0$ are independent. In particular, $r_m^2/r_0^2$ can be made small enough even at $B_m \leq B_0$. The centrifugal trap of [37] was based on this circumstance. We confine ourselves for brevity to the simplest rotating plasma trap system, which describes nevertheless quite fully the features of longitudinal plasma confinement in traps of this class.

To improve substantially the ion confinement compared with a simple mirror trap, the quantity $V_{E_0}^2$ must according to (4.18) satisfy the condition

$$V_{E0}^2 > \frac{2R}{m_i (R-1)} (e\Phi_0 + T_i). \qquad (4.19)$$

In this case, the ion-confinement-region boundary is the surface of a two-cavity hyperboloid (Fig. 9) with vertex at the point $v_\perp = 0$, $v_\parallel = \tilde{v}_\Phi > \sqrt{2T_i/m_i}$, where

$$\tilde{v}_\Phi = \left(\frac{R-1}{R} V_{E0}^2 - \frac{2e\Phi_0}{m_i}\right)^{1/2}. \quad (4.20)$$

As $m_i \tilde{v}_\Phi^2 / 2T_i$ is increased, the ion lifetime, just as in the central section of a tandem-mirror trap, should increase exponentially.

The trap considered solves, in a rather original manner, the problem of ion heating. It suffices for this purpose to introduce neutral atoms of negligibly low energy into the trap. In the rotating frame these atoms rotate with frequency $\Omega_E$ in a direction opposite to the plasma rotation in the laboratory frame. Ionization of the neutral atoms produces ions with $v_\perp = V_E$, $v_\parallel = 0$, which is equivalent to injection of high-energy neutral atoms in ordinary open traps. With allowance for the conservation of the energy and of $J_\perp$, an ion produced at some point on a force line where the magnetic field is equal to $B^*$ has at the central section of the trap the following values of $v_{\perp 0}$ and $v_0$:

$$v_{\perp 0} = V_{E0}\frac{B_0}{B^*}, \quad v_0^2 = V_{E0}^2 - \frac{2e}{m_i}(\Phi_0 - \Phi^*) > \tilde{v}_\Phi^2. \quad (4.21)$$

Thus, the ion sources referred to the central sections are located in velocity space on a certain "source surface" shown in Fig. 9. The "injection" energy exceeds the height of the effective potential barrier for ions, thereby ensuring maintenance of a stationary state without additional heating of the plasma.

Since $V_{E_0}^2 \ll v_{T_e}^2$, the action of the centrifugal forces on the electrons can be neglected. Therefore, electron confinement in a rotating plasma trap does not differ qualitatively from electron confinement in a simple mirror trap, apart from some increase of the parameter $e\Phi_0/T_e$, which is necessary to maintain the particle balance under conditions when the ion loss is lowered.

The presence of an effective potential barrier for the ions points to the possibility of an approximate description of the ion loss by relations (2.18) and (2.19), in which the electron parameters are replaced by the corresponding ion parameters. Such a description, however, is not quite correct under conditions of a rotating plasma trap, since the presence of a source distorts noticeably the ion distribution function in the region $v \geq \tilde{v}_\Phi$. Numerical calculations carried out with a fully conservative difference scheme developed for this case [29, 37] show that in the presence of a source the ion loss should exceed by 1.5-2 times the estimate obtained with the aid of the modified expressions (2.18), (2.19).

The specific plasma-heating mechanism due to the plasma rotation imposes certain constraints on the values of $V_{E_0}$, R, and the principal parameters of the plasma. In fact, setting, in the stationary state, the energy carried away from the trap by the electron–ion pair equal to the injection energy, we arrive with the aid of (2.20), (4.20), and (4.21) at the relation

$$T_i + T_e \approx m_i V_{E0}^2/2R. \tag{4.22}$$

Furthermore, according to numerical-calculation results given in [36], $T_e \approx 0.035 m_i V_{E_0}^2$ for all $R \geq 2$, so that at $R \leq 6$ the quantity $T_e$ in (4.22) can be neglected. For the same reason we can neglect the quantity $2e\Phi_0/m_i$ in (4.20) and obtain the following expression for the relative height of the ion potential barrier:

$$(m_i \tilde{v}_\Phi^2/2T_i) \approx (R-1), \tag{4.23}$$

which shows that at constant $T_i$ the value of $n\tau_{\parallel}^{e}$ in the rotating plasma trap should increase exponentially with increase of R:

$$n\tau_{\parallel}^{e} = \frac{3n\tau_{\parallel}^{i}}{2R} \sim n\tau_i e^R \ln R. \tag{4.24}$$

Thus, from the viewpoint of classical longitudinal losses, rotating plasma traps seem quite attractive and can, in principle, ensure a sufficiently large power gain in a thermonuclear reactor. Nonetheless, research into such systems still faces a number of complicated and quite peculiar problems, principal among which is assurance of magnetohydrodynamic stability with allowance for the symmetry and rotation, and maintaining in the plasma the necessary profile of a radial electric field of intensity about 100 kV/cm.

Multimirror Traps. To conclude this section, we shall dwell briefly also on one possibility of lowering the dipole losses in a plasma in an open adiabatic trap. We consider, in particular, sufficiently long dense-plasma traps operating in the regime of frequent collisions, when the trap length L

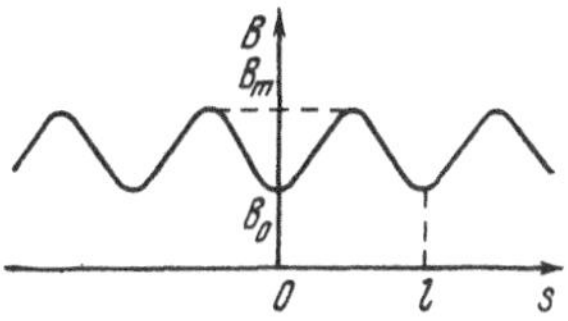


Fig. 10. Longitudinal profile of magnetic field in a multimirror trap.

satisfies the condition $L \gg v_{Ti}\tau_i/R$. Since such systems were not considered in detail in this review, we confine ourselves only to very simple estimates that explain the gist of the matter. It must nevertheless be noted that the discussion of multimirror traps in this review is not fortuitous, since the principle of longitudinal plasma confinement in these traps makes use of certain properties of mirror traps operating in the infrequent-collision regime.

Consider a long system constituting a chain of tandem-connected mirror traps (Fig. 10), each of which has a length $\ell$ and a mirror ratio $R = B_m/B_0 \gtrsim 2$. The total length of the system is $L = N\ell$, where $N \gg 1$. Let the system considered be filled with plasma with ion density and temperature, such that the following conditions hold:

$$l \lesssim \frac{v_{Ti}\tau_i}{R} \ll L. \qquad (4.25)$$

Each individual trap contains both trapped ions that oscillate between mirrors by virtue of energy and $J_\perp$ conservation, and freely escaping ones. The trapped ions should execute, according to (4.25), at least several oscillations between the mirrors before they become free as a result of collisions.

The particles and the energy are longitudinally transported only by the free (untrapped) particles which, in contrast to the trapped ones, can also have a macroscopic velocity $V_\parallel$ directed from the center toward the end of the system. During a time of scale $\tau_i/R$, much shorter than the passage time of the entire system, the untrapped ion has a probability of about 1 of being captured in one of the local traps, and will be replaced by one of the previously trapped ions. As a result of this exchange, the longitudinal momentum of the untrapped particles, per unit volume, decreases by an amount of the order of $nm_iV_{\|}/R$. This momentum is transferred to the trapped ions, and is in turn transferred after a time $\ell/v_{Ti}$ by reflection from the mirrors, to the magnetic field. The momentum transfer can be interpreted as friction of the untrapped ions against the trapped ones, and in final analysis against the magnetic field. The presence of friction alters the character of the plasma motion – the plasma longitudinal transport is determined not by the gasdynamic flow in a uniform magnetic field, but by diffusion "seepage" from one local trap to another.

The macroscopic velocity $V_{\|}$ can be found by equating the friction force $\sim m_i n V_{\|}/\tau_i$ to the pressure gradient $\sim nT_i/RL$:

$$v_{\|} \sim v_{Ti}^2 \tau_i / RL. \tag{4.26}$$

As a result, we obtain the following estimate of the longitudinal lifetime in a multimirror system:

$$\tau_{\|} \sim L^2 R^2 / v_{Ti}^2 \tau_i. \tag{4.27}$$

As expected, the diffusion character of the longitudinal loss leads to a quadratic dependence of $\tau_{\|}$ on L. According to (4.27), the longitudinal lifetime in a multimirror system exceeds considerably both the plasma lifetime $\tau_{\|}{}^L \sim LR/v_{Ti}$ in a long trap with simple mirrors at the ends, and the lifetime $\tau_{\|}{}^L \sim \tau_i \ln R$ in an individual short trap.

The initial idea of multimirror confinement was set forth in [38], where the estimates cited above were obtained. A detailed theoretical investigation of the longitudinal plasma loss in multimirror systems, based on a solution of the Boltzmann equations for ions and electrons, was carried out in [39-41]. The results of these papers confirmed on the whole the validity of the estimates above.

## Conclusion

Summarizing, we can state that as of now the theory of longitudinal classical plasma loss in open adiabatic traps has been sufficiently fully developed and includes both a qualitative understanding of the principal physical processes and well-developed methods for their description. This permits not only a detailed analysis of the already known trap types, but also an estimate of the prospects of newly proposed systems. It must be borne in mind, of course, that classical longitudinal losses constitute only one of the possible channels of plasma loss and may in final analysis not play the dominant role. Nonetheless, a reliable understanding of the classical mechanism will provide an investigation of any system for plasma confinement (at least during the initial stage) with some "margin of optimism." The present review, of course, could not include a detailed analysis of all the questions connected with classical longitudinal plasma loss in open traps of various types. One can hope, however, that the ideas and methods expounded here will be useful in the solution of many specific problems.

## REFERENCES


1. G. I. Budker, in: Plasma Physics and the Problem of Controlled Thermonuclear Reactions [in Russian], Izd. Akad. Nauk SSSR, Moscow (1958), Vol. III, p. 3.
2. A. S. Bishop, Project Sherwood, Addison-Wesley, Reading, Massachusetts (1958).
3. D. V. Sivukhin, in: Reviews of Plasma Physics, Vol. 1, M. A. Leontovich (ed.), Consultants Bureau, New York (1965).
4. V. V. Mirnov and D. D. Ryutov, Pis'ma Zh. Tekh. Fiz., 5, 678 (1979).
5. E. E. Yushmanov, Zh. Eksp. Teor. Fiz., 49, 588 (1965).
6. D. D. Ryutov, Fiz. Plazmy, 5, 1189 (1979).
7. L. D. Landau, Zh. Eksp. Teor. Fiz., 7, 203 (1937).
8. M. N. Rosenbluth, W. McDonald, and D. Judd, Phys. Rev., 107, 1 (1957).
9. B. A. Trubnikov, Zh. Eksp. Teor. Fiz., 34, 1341 (1958).
10. V. A. Trubnikov, in: Reviews of Plasma Physics, Vol. 1, M. A. Leontovich (ed.), Consultants Bureau, New York (1965).
11. D. V. Sivukhin, in: Reviews of Plasma Physics, Vol. 4, M. A. Leontovich (ed.), Consultants Bureau, New York (1966).
12. V. P. Pastukhov, Nucl. Fusion, 14, 3 (1974).
13. R. H. Cohen, M. E. Rensink, T. A. Cutler, and A. A. Mirrin, Nucl. Fusion, 18, 1229 (1974)
14. L. Spitzer, Physics of Fully Ionized Gases, Wiley, New York (1962).
15. A. H. Futch, J. P. Holdren, J. Killeen, and A. A. Mirrin, Plasma Phys., 14, 211 (1971).

16. I. K. Konkashbaev, I. S. Landman, and F. R. Ulinich, Zh. Eksp. Teor. Fiz., 74, 956 (1978).
17. G. F. Bing and J. E. Roberts, Phys. Fluids, 4, 1039 (1961).
18. D. J. Ben Daniel and W. P. Allis, Plasma Phys. (J. Nucl. Energ., Part C), 4, 31, 79 (1962).
19. T. K. Fowler and M. Rankin, Plasma Phys. (J. Nucl. Energ., Part C), 8, 121 (1966).
20. L. G. Kuo-Petravic, M. Petravic, and C. J. H. Watson, Proc. Intern. Conf. on Nucl. Fusion Reactors, Culham, England (1969), paper 2.4.
21. K. D. Marx, Phys. Fluids, 13, 1355 (1970).
22. J. E. Roberts and M. L. Carr, End Losses from Mirror Machines, Rep. UCRL 5651 (1960).
23. J. Killin and K. D. Marx, in: Computational Methods in Plasma Physics [Russian translation], Mir, Moscow (1974), p. 417.
24. J. Killin, A. Mirrin, and M. Rensink, in: Computational Methods in Physics, Controlled Thermonuclear Fusion [Russian translation], Mir, Moscow (1980), p. 419.
25. A. V. Bobylev and V. A. Chuyanov, Zh. Vychisl. Mat. Mat. Fiz., 16, 407 (1976).
26. I. F. Potapenko and V. A. Chuyanov, Zh. Vychisl. Mat. Mat. Fiz., 19, 458 (1979).
27. A. V. Bobylev, I. F. Potapenko, and V. A. Chuyanov, Dokl. Akad. Nauk SSSR, 255, 1348 (1980).
28. A. V. Bobylev, I. F. Potapenko, and V. A. Chuyanov, Zh. Vychisl. Mat. Mat. Fiz., 20, 513 (1980).
29. V. I. Volosov and M. S. Pekker, Zh. Vychisl. Mat. Mat. Fiz., 20, 1341 (1980).
30. I. F. Potapenko and V. A. Chuyanov, Zh. Vychisl. Mat. Mat. Fiz., 22, 751 (1982).
31. G. I. Dimov, V. V. Zakaidakov, and M. E. Kishinevskii, Fiz. Plazmy, 2, 597 (1976).
32. T. K. Fowler and B. G. Logan, Comments Plasma Phys. Controlled Fusion, 2, 167 (1977).
33. D. E. Baldwin and B. G. Logan, Phys. Rev. Lett., 43, 1318 (1979).
34. T. G. Rognlien and T. A. Catler, Nucl. Fusion, 20, 1009 (1980).
35. B. Lehnert, Nucl. Fusion, 11, 485 (1971).
36. A. A. Bekhtenev et al., Nucl. Fusion, 20, 579 (1980).
37. V. I. Volosov and M. S. Pekker, Nucl. Fusion, 21, 1275 (1981).
38. G. I. Budker, V. V. Mirnov, and D. D. Ryutov, Pis'ma Zh. Eksp. Teor. Fiz., 14, 320 (1971).

39. V. V. Mirnov and D. D. Ryutov, Nucl. Fusion, 12, 627 (1972).
40. A. Makhijani, A. J. Lichtenberg, M. A. Lieberman, and B. G. Logan, Phys. Fluids, 17, 1291 (1974).
41. B. G. Logan, I. G. Brown, A. J. Lichtenberg, and M. A. Lieberman, Phys. Fluids, 17, 1302 (1974).
42. O. B. Firsov, in: Plasma Physics and the Problem of Controlled Thermonuclear Reactions [in Russian], Izd. Akad. Nauk SSSR, Moscow (1958), Vol. III, p. 327.
43. I. J. Spalding, Nucl. Fusion, 8, 161 (1968).
44. M. G. Haines, Nucl. Fusion, 17, 811 (1977).
45. V. A. Sizonenko and K. N. Stepanov, Zh. Tekh. Fiz., 45, 741 (1975).
46. E. E. Yushmanov, Fiz. Plazmy, 4, 23 (1978).

# SPECTRAL-LINE BROADENING IN A PLASMA

V. I. Kogan, V. S. Lisitsa, and G. V. Sholin

## Introduction

The gist of phenomena of spectral-line broadening is that the plasma line in question is emitted by atoms not in a narrow spectral band on the order of the natural width (as would be the case for one atom in vacuum), but in a much wider frequency interval $\Delta\omega$. The causes of line broadening are, first, that the observed line is emitted not by an individual atom but by an entire ensemble of atoms whose emission frequencies can differ for various reasons and, second, that the emitting atom is acted upon by external fields produced by the particles and waves in the plasma. The actual cause of the broadening governs both the half-width of the line and its profile, i.e., the intensity distribution in frequency. The intensity $I(\omega)$ of a given line can be represented as the product of the total (integrated over the spectrum) intensity $I_0$ by a factor $J(\omega)$ that determines the line profile, so that

$$I(\omega) = I_0 J(\omega), \qquad \int_{-\infty}^{\infty} J(\omega)\, d\omega = 1. \tag{1.1}$$

The value of $I_0$ is determined by balance of the excitation and deexcitation of the atom levels, whereas the profile $J(\omega)$ depends on more subtle details of the interaction between the atom and the plasma; it is these details which are the subject of broadening theory.

We note that of interest to the broadening processes is a narrow frequency band $\Delta\omega$ of frequencies that are low compared with the unperturbed line frequency $\omega_0$. Therefore, the external broadening actions on an atom are weak compared with the intra-atomic parameters.

This weakness of the broadening action notwithstanding, study of the line contour is quite important for at least two reasons. First, such a study yields values of plasma parameters such as temperature, density, electric and magnetic field strengths, the presence of oscillations, and others. For astrophysical objects, for example, this is frequently the only source of information. Second, the line width and profile influence strongly radiation-transport processes inside the plasma, the emergence of the radiation from the system, establishment of equilibrium between the radiation and the medium, and others.

The sensitivity of the interaction between the radiation and the medium to details of the broadened-line contours is due to the resonant character of this process. Even a small detuning $\Delta\omega = \omega - \omega_0$ from resonance causes an abrupt decrease of the radiation-absorption coefficients.

Broadening processes can be qualitatively visualized by regarding the perturbation of the atom as a perturbation of the oscillations of a classical oscillator (a "tuning fork" of sorts), placed in the plasma. Any perturbation that upsets the monochromaticity of the oscillator broadens its spectrum. The broadening problem consists of two parts: 1) investigation of the statistical properties of the broadening perturbation; 2) investigation of the character of the action of this perturbation on the atom. In Section 2 we consider the statistical properties of the electric fields that act on an atom in a plasma. In Section 3 are analyzed the character of the interaction of the atom with the plasma microfield, and the line-broadening mechanism. In Section 4 is considered the static theory of line broadening and deviations from the theory on account of thermal motion of the particles and the inhomogeneity of the perturbing field. In Section 5 is developed the impact theory of line broadening by the electric fields of the waves and the particles. In Section 6 is considered the transition between the impact and the static broadening mechanisms. Comparison of the theory with experiment is the subject of the concluding Section 7.

## 1. General Equations for the Intensity Distribution in a Line

Consider a system consisting of a radiating atom, perturbing particles, and a radiation field. If the motion of the perturbing particles can be described classically, the Hamiltonian of the system can be written in the form

$$H = H_1 + H_2 + H_R + V_{12} + V_{1R}, \tag{1.2}$$

where $H_1$ is the Hamiltonian of the radiating atom; $H_R$ is the Hamiltonian of the radiation field; $H_2$ is the Hamiltonian of the perturbing particles; $V_{1R}$ is the potential of the interaction between the atom and the radiation field; $V_{12}$ is the potential of the interaction between the radiator and the surrounding particles, and is a function of the time. Since we are interested in the distribution of the line intensity near its unperturbed frequency $\omega_0$, the interaction of the radiation field with the perturbing particles is disregarded.

The sum of the first three terms of (1.2) is the unperturbed Hamiltonian of the system $H_0 = H_1 + H_2 + H_R$. Corresponding to it is a complete set of orthonormal time-independent eigenfunctions of the form $\Psi = \Psi_1\Psi_2\Psi(\ldots n_k \ldots)$, where $\Psi_1$ and $\Psi_2$ are the eigenfunctions of the radiator and of the perturbing particles, respectively, while $\Psi(\ldots n_k \ldots)$ describes the state of the radiation field in the space of the occupation numbers $n_k$. Let the system be, at the instant $t = 0$, in a state A specified by a wave function $\Psi(0) = \varphi_a(0)\Psi_R(n_k)$. Here $\varphi_a$ denotes, for brevity, the product of $\varphi_1$ and $\varphi_2$, while $\Psi_R$ is labeled by only one index $n_k$ pertaining to that species of photons which will be absorbed. In the absence of the interaction $V_{12}(t)$, the absorption problem reduces to finding the probability $P_{AB}$ that the system will go over from a state $\varphi_a\Psi_R(n_k)$ into a state $\varphi_b\Psi_R(n_k - 1)$ that is the stationary state of the Hamiltonian $H_0$ at a later instant of time $t = T$ [1]. The argument $n_k - 1$ indicates that one photon of species k has been absorbed. The presence of the interaction $V_{12}(t)$ no longer permits the state of the system to be described by the product $\varphi_b\Psi_R(n_k - 1)$. The interaction $V_{12}(t)$ of the atom with the surrounding particles is stronger than the interaction $V_{1R}$ of the atom with the radiation field. It is therefore necessary first to find the state of the system at the instant $t = T$ determined by the interaction potential $V_{12}(t)$. If we denote by $\varphi_n(t)$ the wave function that coincided at the instant $t = 0$ with $\varphi_n(0)$

and is now subject to a Schrödinger equation with the Hamiltonian

$$H_m = H_1 + H_2 + V_{12}(t), \tag{1.3}$$

it is easy to verify that

$$\varphi_n(t) = U_m(t)\,\varphi_n(0), \tag{1.4}$$

where the evolution operator $U_m(t)$ satisfies the equation

$$i\hbar\dot{U}_m(t) = H_m(t)\,U_m(t), \qquad U_m(0) = 1. \tag{1.5}$$

The probability of transition of the system into the state $\Psi\,t) = \varphi_b\Psi_R(n_k - 1)$ can now be obtained by using the evolution operator that corresponds to the total Hamiltonian:

$$P_{BA} = |\langle\Psi(T)\,|\,\hat{U}\,|\,\Psi(0)\rangle|^2. \tag{1.6}$$

Since the interaction of the radiation field with the rest of the system is neglected, the evolution operator can be represented in first-order approximation as the product of $U_m(t)$ by $U_R(t)$, where $U_R(t)$ is the evolution operator of the radiation field, with $U_mU_R = U_RU_m$. Putting $U_{0c} = U_m(t)U_R(t)$, applying now the operator $U_{0c}^{-1}(t)$ to Eq. (1.6), and using the successive approximations

$$U_{0c}^{-1}U = 1 + \frac{1}{i\hbar}\int_0^T U_{0c}^{-1}H_{1R}U_{0c}dt, \tag{1.7}$$

we obtain

$$P_{BA} = \frac{1}{\hbar^2}\left|\int_0^T dte^{+i\omega t}\langle\varphi_b(0)\,\Psi_R(n_k \pm 1)\,|\,U_m^{-1}H_{1R}U_m\,|\,\varphi_a(0)\,\Psi_R(n_k)\rangle\right|^2, \tag{1.8}$$

where $\hbar\omega$ is the energy of the radiated (or absorbed) photon.

Using next the standard expressions [1] for the matrix elements of the interaction of the atom with the radiation field and for the density of states of the radiation field, we obtain in the dipole approximation ($kr \ll 1$) the following expression for the energy radiated per unit time by an individual atom:

$$I(\omega) = \frac{\omega^4(n_k \pm 1)}{4\pi^2c^3T}\left|\int_0^T dte^{i\omega t}\langle\varphi_b(0)\,|\,U_m^{-1}\,\mathbf{d}\mathbf{e}_k U_m\,|\,\varphi_a(0)\rangle\right|^2, \tag{1.9}$$

where $\mathbf{d} = e\mathbf{r}$ is the dipole moment of the atom, and $\mathbf{e}_k$ is the polarization vector of a photon with momentum $\mathbf{k}$.

Summing finally over all finite states and averaging over all the initial states of the atom, we obtain the basic equation for the analysis of spectral-line profiles:

$$I(\omega) = \frac{\omega^4 (n_k + 1)}{4\pi^2 c^3 T} \sum_{e_k} \int_0^T dt \int_0^T dt' e^{i\omega(t-t')} \operatorname{Spur} \{\rho_0 U_m^{-1}(t) \mathbf{d}\mathbf{e}_k U_m(t) \times U_m^{-1}(t') \mathbf{d}\mathbf{e}_k U_m(t')\}. \quad (1.10)$$

Here $\rho_0$ is the density matrix. An important assumption in what follows is that $(\rho(t) = \rho_0)$ is constant, and that this matrix is diagonal in the internal coordinates of the atom, i.e.,

$$U_m(t)\rho_0 = \rho_0 U_m(t), \qquad (\rho_0)_{\alpha\alpha'} = P_\alpha \delta_{\alpha\alpha'}. \quad (1.11)$$

Using the cyclic permutation under the Spur sign, we can reduce the evolution operators in (1.10) to the operator

$$U_m(t \to t') = U_m^{-1}(t') U_m(t),$$

which describes the system evolution during the time interval from t to t'. At a fixed $\tau = t' - t$ the integrand averaged over the collisions should not depend on the initial and final instants of time, and is determined only by the duration $\tau$. The double integral in (1.10) is therefore proportional to T, and the intensity distribution in the line or, more accurately, the radiation power per atom, is given by

$$I(\omega) = \frac{\omega^4 (n_k + 1)}{4\pi^2 c^3} \int_0^\infty d\tau e^{i\omega\tau} \operatorname{Spur} \{\rho_0 (\mathbf{d}\mathbf{e}_k) U_m^{-1}(\tau) (\mathbf{d}\mathbf{e}_k) U_m(\tau)\}. \quad (1.12)$$

Integration of (1.12) with respect to the frequency $\omega$ yields, obviously, the total intensity $I_0$ of the atom's dipole radiation. The evolution operator $U_m(t)$ describes thus only the redistribution of the intensity over the frequency, i.e., the line contour, and does not affect the total intensity $I_0$. It is therefore convenient to separate the line contour $J(\omega)$ in explicit form. In the case of an isotropic medium $J(\omega)$ is independent of the radiation polarization, so that (1.12) can be averaged over the directions of the vector $\mathbf{e}_k$. As a result we get

$$J(\omega) = \frac{1}{\pi} \operatorname{Re} \int_0^\infty d\tau e^{i\omega\tau} \Phi(\tau), \tag{1.13}$$

$$\Phi(\tau) = \frac{1}{|d_{ab}|^2} \operatorname{Spur} [\rho_0 dU_m^{-1}(\tau) dU_m(\tau)]. \tag{1.14}$$

The function $\Phi(\tau)$, which is the Fourier transform of $J(\omega)$, is called the autocorrelation function of the dipole moments of the atom. It describes, obviously, the time evolution of the atomic states. The physical meaning of $\Phi(\tau)$ can be easily understood from the following simple reasoning: If the action of the medium on the atom is strong enough, the atom rapidly "loses memory" of the initial state – the correlation function attenuates rapidly, and this leads obviously to a large width of the line contour $J(\omega)$.

Strictly speaking, the averaging should be both over the states of the atoms and over the states of the perturbing particles. In our classical-trajectory approximation, however, the Spur operation in (1.14) is replaced simply by averaging over the coordinates and velocities of all the perturbing particles, an operation which we shall denote hereafter by the symbol {...}. The states $a$ and b of the atom are, generally speaking, degenerate, so that the summation in (1.3) is still quite complicated. By virtue of the already noted weakness of the broadening interactions, however, the states $a$ and b are not intermixed by the perturbation, so that the summations over the sublevels $\alpha$, $\alpha'$ and $\beta$, $\beta'$ pertaining to $a$ or b proceed independently.

Of importance for the understanding of the physical fundamentals of the broadening processes is the case when the perturbation causes no transitions between the states, i.e., the evolution operator is diagonal:

$$(U_m)_{\alpha\alpha'} = \delta_{\alpha\alpha'} \exp(iE_\alpha t/\hbar) \exp\left[-i\int^t \varkappa_\alpha(\tau)\, d\tau\right], \tag{1.15}$$

where $\varkappa_\alpha(t)$ is the frequency shift produced in the state $\alpha$ by the perturbation (e.g., the Stark shift of a level in an electric field).

The only broadening mechanisms in this case are frequency (phase) shifts caused by the perturbation (frequency modulation). The atom can be regarded here as a classical oscillator with modulated frequency. This model is called the classical adiabatic oscillator model [1]. The line contour $J(\omega)$ takes in this model a particularly simple form:

$$J(\omega) = \frac{1}{\pi} \operatorname{Re} \int_0^\infty d\tau \exp(i\Delta\omega\tau)\, \Phi(\tau), \tag{1.16}$$

$$\Phi(\tau) = \left\{ \exp\left[ -i \int_0^\tau \varkappa(t)\, dt \right] \right\}, \tag{1.17}$$

where $\Delta\omega = \omega - \omega_0$ [$\omega_0 = (E_a - E_b)/\hbar$] is the shift of the observed frequency $\omega$ relative to the unperturbed transition frequency $\omega_0$; $\varkappa(t) = \varkappa_a - \varkappa_b$ is the difference phase shift of the upper (a) and lower (b) states of the atomic oscillator.

We confine ourselves here to the simplest formulation of the broadening problem, which nonetheless gave good account of itself from the standpoint of obtaining specific results of practical importance. The connection between broadening theory and the general kinetic theory of electromagnetic processes, and a discussion of problems connected with the formulation of the line-broadening problem, can be found in the monographs [2, 4, 5]. We limit ourselves below, of necessity, only to a limited citation of the literature, an exhaustive list of which is given in [2, 3].

## 2. Plasma Microfield

The Holtsmark Distribution. The electric field produced by the plasma ions and electrons near a radiating atom is called the plasma microfield. A fundamental role in microfield theory (and simultaneously in static line-broadening theory; see Section 4) is played by the Holtsmark model [6]. It deals with the distribution (at a given point of space and at a given instant of time) of the values of the microfield $F \equiv |\mathbf{F}|$ produced by an aggregate $\mathcal{N}$ of perfectly randomly disposed immobile ions:

$$\mathbf{F} = \sum_{k=1}^{\mathcal{N}} \mathbf{F}_k = e \sum_{k=1}^{\mathcal{N}} \mathbf{r}_k / r_k^3. \tag{2.1}$$

The probability $W(\mathbf{F})d\mathbf{F}$ is obviously equal to that fraction of configuration space $\mathbf{r}_1, \mathbf{r}_2, \ldots, \mathbf{r}_{\mathcal{N}}$ in which Eq. (2.1) is valid. To separate this fraction automatically, we average over the configuration space the following $\delta$ function:

$$W(\mathbf{F}) = \left\langle \delta\left(\mathbf{F} - e\sum_{k=1}^{\mathcal{N}} \frac{\mathbf{r}_k}{r_k^3}\right)\right\rangle = \frac{1}{(2\pi)^3}\int d\boldsymbol{\rho}\, e^{i\boldsymbol{\rho}\mathbf{F}} \left\langle \exp\left(-ie\boldsymbol{\rho}\sum_{k=1}^{\mathcal{N}} \frac{\mathbf{r}_k}{r_k^3}\right)\right\rangle, \tag{2.2}$$

where the expansion of the $\delta$ function in a Fourier integral is used.

We agree to carry out the averaging in (2.2) over the distribution of mutually uncorrelated ions $\prod_{k=1}^{\mathcal{N}}\left(\frac{d\mathbf{r}_k}{V}\right)$ (V is volume of the system). Going next to the limit $\mathcal{N} \to \infty$, $V \to \infty$ at $\mathcal{N}/V \equiv N = \text{const}$ (N is the ion-number density), we arrive at the chain of equations

$$\left\langle \exp\left(-ie\boldsymbol{\rho}\sum_{k=1}^{\mathcal{N}} \frac{\mathbf{r}_k}{r_k^3}\right)\right\rangle = \int \ldots \int \exp\left(-ie\boldsymbol{\rho}\sum_{k}^{\mathcal{N}} \frac{\mathbf{r}_k}{r_k^3}\right)\prod_{k=1}^{\mathcal{N}} \frac{d\mathbf{r}_k}{V}$$
$$= \left(\int \exp\left(-ie\frac{\boldsymbol{\rho}\mathbf{r}}{r^3}\right)\frac{d\mathbf{r}}{V}\right)^{\mathcal{N}} \equiv \left\{1 - \int \frac{d\mathbf{r}}{V}\left[1 - \exp\left[-ie\frac{\boldsymbol{\rho}\mathbf{r}}{r^3}\right]\right]\right\}^{\mathcal{N}} \to \exp\left\{-N\int d\mathbf{r}\left[1 - \exp\left(-ie\frac{\boldsymbol{\rho}\mathbf{r}}{r^3}\right)\right]\right\}$$
$$= \exp\left[-N(\lambda e\rho)^{3/2}\right], \tag{2.3}$$

where $\lambda = 2\pi(4/15)^{2/3} = 2.603$. Substituting (2.3) in (2.2), integrating over the angles of the vector $\boldsymbol{\rho}$, and using the (automatically ensuing) independence of $W(\mathbf{F})$ of the direction of $\mathbf{F}$, we get ultimately [6, 7]

$$W(\mathbf{F})\,d\mathbf{F} \equiv W(F)\,4\pi F^2 dF = \mathcal{H}\left(\frac{F}{F_0}\right)\frac{dF}{F_0}, \tag{2.4}$$

where we have introduced a "natural" scale for the ion microfield $F_0$ and the Holtsmark function $\mathcal{H}$:

$$F_0 = \lambda e N^{2/3} \quad (\lambda \approx 2.603), \tag{2.5)*$$

$$\mathcal{H}(\beta) = \frac{2}{\pi}\beta \int_0^\infty \sin(\beta x) \exp\left(-x^{3/2}\right) dx. \tag{2.6}$$

The function $\mathcal{H}(\beta)$ is normalized to $\int_0^\infty \mathcal{H}(\beta)\, d\beta = 1$. Its maximum is at $\beta = 1.607$, and its asymptote is

$$\mathcal{H}(\beta) \approx \begin{cases} 1{,}496\beta^{-5/2}\left(1 + 5{,}107\beta^{-3/2}\right), & \beta \gg 1, \\ \frac{4}{3\pi}\beta^2\left(1 - 0{,}463\beta^2\right), & \beta \ll 1. \end{cases} \tag{2.7}$$

The asymptote $\beta^{-5/2} \propto F^{-5/2}$ for strong fields is due, as should be the case, to only one ion (the closest to the observation point). This asymptote follows actually from the distribution $W(r)dr \propto r^2 dr$ of the distances r to the nearest ion and from the relations $F(r) \propto r^{-2}$, $W(r)dr = W(F)dF$. The asymptote $B^2 \propto F^2$ for weak fields (produced by a large number of ions) coincides with the asymptotic Gauss (more accurately, Gauss–Rayleigh) distribution. The Holtsmark distribution (2.4)-(2.6) describes thus the transition from a Gaussian distribution of weak fields to a "binary" distribution of strong fields.

Some Generalizations of the Holtsmark Distribution. Any departure from the simplifying assumptions inherent in the Holtsmark model necessitates a corresponding generalization of the model. Thus, allowance for the effects of mutual correlation of the ions and the screening of their electric field due to the interaction with the electrons complicates greatly the structure of the W(**F**) distribution compared with (2.1)-(2.3), etc. Next, when the microfield distribution is used to analyze line broadening, the fact that the radiating atom has proper finite space-time scales (dimensions, lifetime, etc.), in conjunction with the thermal motion of the broadening ions, makes the W(**F**) distribution itself insufficient at one space-time point (or, equivalently, the distribution of the microfield **F** alone without its deriv-

*Note that $F_0$ is numerically very close to the field $F_{av} = er_0^{-2}$ at the average internuclear distance $r_0$ defined by the relation $(4\pi/3)Nr_0^3 = 1$.

atives $\dot{\mathbf{F}}, \ddot{\mathbf{F}}, \ldots$). Let us examine briefly the corresponding generalizations of the Holtsmark distribution.

A. The broadening ions interact in the plasma with one another via screened Coulomb potentials

$$V(\mathbf{r}_1, \mathbf{r}_2, \ldots, \mathbf{r}_{\mathcal{N}}) = \sum_{1=k<l}^{\mathcal{N}} \frac{e^2}{r_{kl}} \exp(-r_{kl}/\rho_D), \tag{2.8}$$

where $\rho_D$ is the Debye radius. In an equilibrium plasma of temperature T, the probability of a given ion configuration is described by the product of the volume element $\prod_{k=1}^{\mathcal{N}} \left(\frac{d\mathbf{r}_k}{V}\right)$ (see above) by the "Boltzmann" factor $\exp[-V(\mathbf{r}_1, \ldots, \mathcal{N})/kt]$ that takes the ion interaction into account. Allowance for this factor decreases the probability of configurations with large values of V, i.e., with close distances between ions, which correspond to large values of the microfield F. Therefore, the real microfield distribution should give higher probabilities of weak fields and lower probabilities of strong ones compared with the Holtsmark distribution (Fig. 1 [8]).

The deviation from this distribution is determined, naturally, by the deviation $\mathcal{N}_D^{-1}$ of the plasma from ideal, where $\mathcal{N}_D \equiv (4\pi/3)N\rho_D^3$ is the average number of ions in a Debye-radius sphere; in specific calculations (including those illustrated in Fig. 1), one used frequently the parameter $\alpha = N_D^{-1/3} \propto N^{1/6}T^{-1/2}$, which is the ratio of the average distance between ions $r_0$ $[N(4\pi/3)r_0^3 = 1]$ to the Debye radius. The limit $\alpha = 0$ corresponds to the Holtsmark distribution; with increasing $\alpha$ the deviations from it increase and become quite noticeable, for example, already at $\alpha =$ 0.4 (i.e., $\mathcal{N}_D \cong 16$).

The criterion $\mathcal{N}_D \gg 1$ for the applicability of the Holtsmark model limits, strictly speaking, also the applicability of the published more general calculations of the plasma-microfield distribution, making it difficult to estimate the accuracy of the latter in the region $\mathcal{N}_D \sim 1$. A detailed exposition of the effects connected with ion correlation in a plasma is contained in [2, 9, 10].

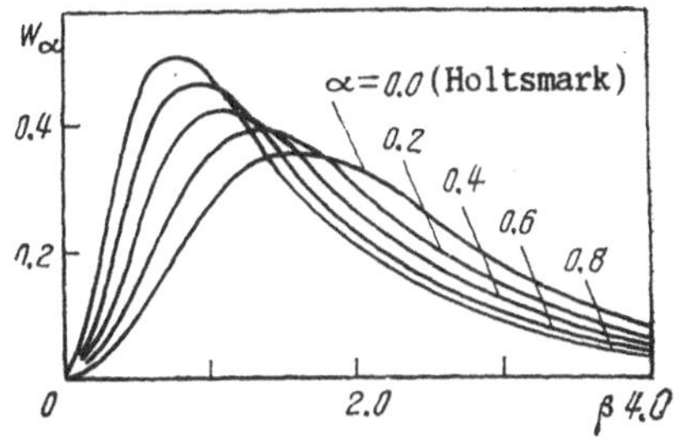


Fig. 1. Microfield distribution in a weakly nonideal plasma.

B. A "dynamic" generalization of the Holtsmark distribution, in which account is taken of the motion of the "field" particles, was given (as applied to stellar-dynamics problems) by Chandrasekhar and von Neumann [7] in several close forms. Namely the following were constructed: 1) the distributions $W(\mathbf{F}, \dot{\mathbf{F}})$ and $W(\mathbf{F}, \dot{\mathbf{F}}, \ddot{\mathbf{F}})$; 2) the two-time distribution $W(\mathbf{F}_1, \mathbf{F}_2)$ that determines the probability of realizing the values $\mathbf{F}_1$ and $\mathbf{F}_2$ of a microfield at a given point at the instant of time $t = 0$ and $t = \tau$; 3) the distribution $W(\mathbf{F}_1, \mathbf{F}_2)$ of a microfield at a given instant of time but at two points of space $\mathbf{r}_1$ and $\mathbf{r}_2$. Distributions of type 1 and 2 make possible a description of the ion-microfield temporal dynamics governed by the thermal motion of the ions in terms of the moments of the corresponding distributions [7].* Some results of this analysis will be used in Section 4.

To develop an adiabatic "plural" broadening theory based on vector addition of the moving-ion fields [13], it is necessary to use a many-time distribution $W(\mathbf{F}_1, \mathbf{F}_2, \ldots, \mathbf{F}_M)$ of the microfield and successive infinitely close M ($M \to \infty$) instants of time.

C. We recall, finally, that both generalizations A and B of the Holtsmark distribution can be "synthesized" into a single scheme [14].

*A distribution of type 2 can be calculated in explicit form for small and large values of $\tau$ [11]. This makes it possible to track in greater detail the dynamics of the destruction and "induction" of a correlation between $\mathbf{F}(0)$ and $\mathbf{F}(\tau)$ [12].

Distribution of Ion-Electron Microfields with Allowance for Collective Effects (of Turbulent Noise). We proceed to consider the distribution of the microfield that is important for many types of a real laboratory plasma. The electric microfield in a plasma always has two components – low-frequency $F_{ion}$, due to the ions, and high-frequency $F_{el}$, due to the electrons. In turn, by virtue of the long-range character of the Coulomb forces, one can separate in each of these microfield components collective and individual components that can be regarded as statistically independent. Since the "high-frequency" field acts nonadiabatically on the atom, the statistics of the distribution of its amplitudes hardly manifests itself in the line profile. It will be shown below that the principal rôle in the high-frequency component is played by the autocorrelation function, or the energy density of the noise of the Langmuir waves.

In the case of the linear Stark effect, when the static approximation can be used for the description of the broadening (Section 4), the line profile is proportional to the distribution function $W(F_{ion})dF_{ion}$ of the low-frequency component of the field. It is important here that the static description implies a possibility of choosing a time interval $\Delta t$ that is long enough compared with the period $\tau_e = 2\pi/\omega_{pe}$ of the electron Langmuir oscillations and, at the same time, small compared with the characteristic time $\tau_i \sim N^{-1/3}/v_{Ti}$ of the change of the ion configuration. Moreover, since the description of the nonadiabatic effects calls for averaging over a time interval $2\pi/\omega_{pe} \lesssim \Delta t_e \ll \gamma_e^{-1}$, the characteristic Hamiltonian that describes the interaction of the emitter with the low-frequency field component $F_{ion}$ should include the fields of the Coulomb centers that are already averaged over the electron motion. Since time averaging is equivalent to ensemble averaging, it is natural to regard the resultant quantity $F_{ion}$ as the vector sum of the fields produced by ion charges screened at a radius determined by the electron component of the plasma: $r_{De} = v_{Te}/\omega_{pe}$. This can be easily verified by considering, in the Debye–Hückel approximation, the screening of the ion field for the case $T_e \gg T_i$, and, furthermore, when all other plasma ions can be regarded at rest ($\Delta t \ll N^{-1/3}/v_{Ti}$). The last assumption is equivalent to taking the limit as $T_i \to 0$ in the Boltzmann factor $\exp(-e\varphi_i/T_i)$ for the ion-density distribution about the considered charged center. As a result, $\exp(-e\varphi_i/T_i) \to 0$, and the potential is determined by a Debye–Hückel function in which the screening is due only to electrons.

The calculation of the distribution function of the low-frequency fields reduces thus to finding the probability distribution of the resultant field of like-charged screened Coulomb centers:

$$\mathbf{F} = \sum_j \mathbf{F}_j(\mathbf{r}_j), \qquad \mathbf{F}_j = \frac{e}{r_j^3}\mathbf{r}_j(1 + \varkappa_{De} r)\exp(-\varkappa_{De} r), \tag{2.9}$$

to which is vectorially added the resultant field of the collective low-frequency plasma oscillations. That these two contributions are statistically independent can be demonstrated by following the method of Bohm and Pines [15] for separating collective coordinates.

The need for resorting to the method of collective variables is due to the known circumstance that the partition function of a system of Coulomb-interacting charged particles diverges. If no artificial procedures of cutting off the interaction are resorted to, this fact must be regarded as evidence of impossibility of ascribing the properties of a Gibbs susbsystem to an individual charged particle. On the other hand, the procedure of separating the collective coordinates can be treated just as a method of finding Gibbs ensembles that describe correctly the Hamiltonian of a system of charged particles:

$$H = \sum_{\alpha, i} \frac{\mathbf{p}_{i\alpha}^2}{2m_\alpha} + \frac{1}{2}\sum_{\alpha, \beta, i \neq j} \frac{e_\alpha e_\beta}{|\mathbf{r}_i - \mathbf{r}_j|}. \tag{2.10}$$

Following [15], we can use the appropriate canonical transformations to write the Hamiltonian (2.10) in the form

$$H = \sum_{\alpha, i} \frac{\mathbf{p}_i^{''2}}{2m_\alpha} + \frac{1}{2}\sum_{\alpha, \beta, i \neq j} \frac{e_\alpha e_\beta \exp(-k_c|\mathbf{r}_i'' - \mathbf{r}_j''|)}{|\mathbf{r}_i'' - \mathbf{r}_j''|} + \frac{1}{2}\sum(|Q_k''|^2 + \omega_k^2|P_k''|). \tag{2.11}$$

Here $\omega_k$ satisfies the dispersion equation

$$1 = \frac{4\pi}{V}\sum_{i, \alpha} \frac{e_\alpha^2}{\left[\omega_k - \frac{(\mathbf{k}\mathbf{p}_i'')}{m_\alpha}\right]^2 m_\alpha}. \tag{2.12}$$

Averaging now over the electron component, it is easy to verify that Eq. (2.12) describes the low-frequency ion-sound branch of the plasma oscillations, while the parameter $k_c$ in the Hamiltonian (2.11) corresponds to an electron Debye radius $k_c = \rho_{De}^{-1}$. Thus, two weakly interacting subsystems, screened Coulomb centers and collective oscillations $Q_k$ and $P_k$, turn out to be the analog of one system of strongly interacting charged particles. It is important that, whereas the partition function, and hence all the thermodynamic quantities, for a system of charged particles is not defined, the partition function and all the thermodynamic quantities of a Hamiltonian transformed by introducing collective variables are all finite. The condition that ensures validity of this substitution is, in the case of the low-frequency branch, the inequality $k < \rho_{De}^{-1}$ at $\omega \approx \omega_{pi}$, or

$$\left\langle \left( \frac{kp''}{m\omega_k} \right)^2 \right\rangle_{av} \simeq \frac{T_i}{T_e} \ll 1 \qquad (2.13)$$

where $\omega \approx k\sqrt{T_e/m_i} \ll \omega_{pi}$.

The distribution of a low-frequency microfield is thus defined by the general equation

$$W(\mathbf{F}) = \int \delta\left(\mathbf{F} - \sum_i \mathbf{F}_i\right) P_N(H)\, d\mathbf{r}_i'' \ldots d\mathbf{P}_k'' \ldots, \qquad (2.14)$$

where the integration is over the new canonical variables $\mathbf{r}_i''$, $\mathbf{p}_i''$, $\mathbf{Q}_k''$, $\mathbf{P}_k''$, and the Gibbs factor is given by

$$P_N(H) = \frac{\exp(-H/T_i)}{\int \exp\{-H/T_i\}\, d\mathbf{r}_i'' \ldots d\mathbf{P}_k'' \ldots}. \qquad (2.15)$$

The resultant microfield can be represented in the form

$$\sum_i \mathbf{F}_i = \sum_i \frac{e}{r_i''^3} \mathbf{r}_i'' (1 + \varkappa_{De} r_i'') \exp(-\varkappa_{De} r_i'') + \left( \frac{4\pi}{V} \right)^{1/2} \sum_{k<k_c} \mathbf{Q}_k'' \exp(i\,\mathbf{kr}''), \qquad (2.16)$$

where $\mathbf{r}''$ is the coordinate of the neutral observation point. Taking the Fourier transform of (2.14), it is easy to verify that the spectral functions for the individual and collective components of the microfield are factorizable:

$$W(\mathbf{q}) = \int \exp(i\,\mathbf{qF})\,W(\mathbf{F})\,d\mathbf{F} = W_{\text{ind}}(\mathbf{q})\,W_{\text{coll}}(\mathbf{q}), \tag{2.17}$$

where

$$W_{\text{ind}}(\mathbf{q}) = \frac{\displaystyle\int \exp\left\{-i\mathbf{q}\sum_i \frac{e}{r_i^3}\mathbf{r}_i(1+\varkappa_{De} r_i)\,e^{-\varkappa_D r_i} - \frac{e^2}{2T_i}\sum \frac{e^{-\varkappa_D|\mathbf{r}_i-\mathbf{r}_j|}}{|\mathbf{r}_i-\mathbf{r}_j|}\right\} d\mathbf{r}_1\ldots d\mathbf{r}_N}{\displaystyle\int \exp\left\{-\sum \frac{e^2}{2T_i}\frac{\exp[-\varkappa_D|\mathbf{r}_i-\mathbf{r}_j|]}{|\mathbf{r}_i-\mathbf{r}_j|}\right\}\times d\mathbf{r}_1\ldots d\mathbf{r}_N} \tag{2.18}$$

(the integration is carried out here only over the configuration space of the ions);

$$W_{\text{coll}}(\mathbf{q}) = \frac{\displaystyle\int \exp\left\{-i\mathbf{q}\left(\frac{4\pi}{V}\right)^{1/2}\sum \mathbf{Q}_k e^{i\mathbf{kr}} - \frac{1}{2T_i}\sum_{k<\varkappa_D}\mathbf{Q}_k^2\right\} d\mathbf{Q}_{k_1}\ldots d\mathbf{Q}_{k_0}}{\displaystyle\int \exp\left\{-\frac{1}{2T_i}\sum_{k<\varkappa_D}\mathbf{Q}_k^2\right\} d\mathbf{Q}_{k_1}\ldots d\mathbf{Q}_{k_0}}. \tag{2.19}$$

(Here the integration is over the collective coordinates that describe the amplitudes of the low-frequency oscillations.) Calculation of the collective part $W_{\text{coll}}$ leads to the Rayleigh distribution:

$$W_{\text{coll}}(F) = \frac{3F^2}{F_R^3}\sqrt{\frac{6}{\pi}}\exp\left[-\frac{3F^2}{2F_R^2}\right], \tag{2.20}$$

where it is easy to obtain for the ion-sound spectrum

$$\frac{F_R^2}{8\pi} = \frac{1}{8\pi}\int_0^\infty F^2 W_{\text{coll}}(F)\,dF \simeq \frac{1}{4}T_i\frac{1}{(2\pi)^2}\int_0^{\varkappa_{De}} k^2\,dk \sim \left(\frac{T_i}{T_e}\right) f(\varkappa_{De}).$$

The energy density of the ion-sound oscillations in a two-temperature plasma is thus proportional to $T_i/T_e$, i.e., is smaller the more the electron temperature exceeds that of the ions. At the same time, the role of the charge screening decreases with increasing $T_e$. In a strongly nonequilibrium plasma, the energy density of the oscillations can differ by many orders of magnitude from its thermodynamic equilibrium value. In this case, $F_R$ must be regarded as a certain external parameter that determines the degree of disequilibrium of the medium.

An individual component of a microfield can be calculated with the aid of the equation of $W_{ind}(\mathbf{q})$ if the function

$$P_{\mathcal{N}} = \prod_{i,j} \exp\left\{-\frac{e^2}{2T_i}\frac{\exp[-\varkappa_D|\mathbf{r}_i-\mathbf{r}_j|]}{|\mathbf{r}_i-\mathbf{r}_j|}\right\}$$

is represented in terms of correlation functions of second $g_{ij}^s = \exp[-V_{ij}/T_i] - 1$ and higher orders

$$P_s = \prod_{k(1)} P_1(\mathbf{r}_j) \prod_{k(2)} (1+g_{je}^s) \prod_{h(3)} (1+g_{iem}^s) \;\ldots$$

and $W_{ind}(\mathbf{q})$ is expressed in terms of the partial distribution functions

$$W_{ind}(\mathbf{q}) = \sum_{s=0}^{N} W_{ind}^s(\mathbf{q})$$

where

$$W_{ind}^s(\mathbf{q}) = \frac{\mathcal{N}!}{(\mathcal{N}-s!)\,s!}\int \ldots \int \sum \varepsilon_j \ldots \varepsilon_s P_{\mathcal{N}} d\mathbf{r}_1 \ldots d\mathbf{r}_s. \quad (2.21)$$

Here $\varepsilon_k \equiv e^{iqF_k} - 1$.

Since the ions repel one another, in the limit as $T_i \to 0$ the exponential factors should vanish, so that the pair correlation function $g_{ij}^2 \to -1$ and, hence, all the $P_s$ for $s \geq 2$, are exponentially small in this limit.

Using next the assumption that the distribution of the particles in space $P_1 = 1/V$ is random, which corresponds to a "gaseous" system, we arrive at the Ecker–Müller distribution [17]

$$W_{ind}(\mathbf{q}) = \exp\left\{\frac{\mathcal{N}}{V}\int [e^{-iqF} - 1]\, dr\right\}, \quad (2.22)$$

where $\mathbf{F} = -(e\mathbf{r}/r^3)(1 + \varkappa_{De}r)e^{-\varkappa_{De}r}$; $\mathcal{N}$ is the number of screened particles; V is the volume of the system. The known Holtsmark distribution follows from (2.22) in the limit $T_e \to \infty$ or $\varkappa_{De} \to 0$. Moreover, it is easy to verify that the resultant distribution, which is a convolution of $W_{ind}(F)$ and $W_{coll}(F)$, also goes in the limit $T_e \to \infty$, $(T_i/T_e) \to 0$ into a Holtsmark distribution if the low-frequency noise is at the thermal level corresponding to the ion temperature.

The Holtsmark distribution thus corresponds indeed to neglect of the thermal motion of the ion. Naturally, as $T_i \to 0$, owing to the exponential smallness of the Boltzmann factor in the pair correlation function, ion–ion correlations are likewise absent from $W_{ind}$. The principal effect of these correlations is the existence of a nonzero thermal (or above-thermal) level of the ion-sound noise, which influences the resultant field distribution. This influence can be taken into account quite simply in the case of a two-temperature plasma with ion-sound noise determined by the temperature $T_i$. It is easy to verify that the ratio $\alpha = F_H/F_R$ of the amplitude scales of the individual and collective fields has, in this case, a rather large value

$$\alpha = \left(\frac{T_e}{T_i}\right)^{1/2}\left(\frac{\pi^5}{36}\right)^{1/12}\left(\frac{T_e}{e^2N^{1/3}}\right)^{1/4} \gg 1, \tag{2.23}$$

so that the relation $\alpha\beta \gg 1$, where $\beta = F/F_H$, can be assumed satisfied over the greater part of the distribution. The resultant microfield distribution, which is a convolution of $W_{ind}(F)$ and $W_{coll}(F)$ at $\beta \gg \alpha^{-1}$, can then be represented in the form

$$W(\delta) \simeq W_{ind}(\beta)\left\{1 - \left(\frac{4}{3}\right)^{1/3}\frac{1}{\beta}\left(\frac{1}{\pi}\right)^{11/12}\left(\frac{T_i}{T_e}\right)^{1/2}\left(\frac{e^2N^{1/3}}{T_e}\right)^{1/4} \times\left[1 - \beta\frac{d}{d\beta}\ln W_{ind}(\beta)\right]\right\}. \tag{2.24}$$

It can be seen from this relation that the most important change occurs in the region of sufficiently small $1/\alpha \ll \beta \lesssim 1$, where the correction term is essentially positive. Both the distribution and its maximum are then shifted somewhat into the region $\beta \lesssim 1$. This is indeed to be expected, since the mutual correlations of the ions tend primarily to decrease the probability of strong field. It is remarkable that in the limiting case $(T_i/T_e) \to 0$ the correction necessitated by the

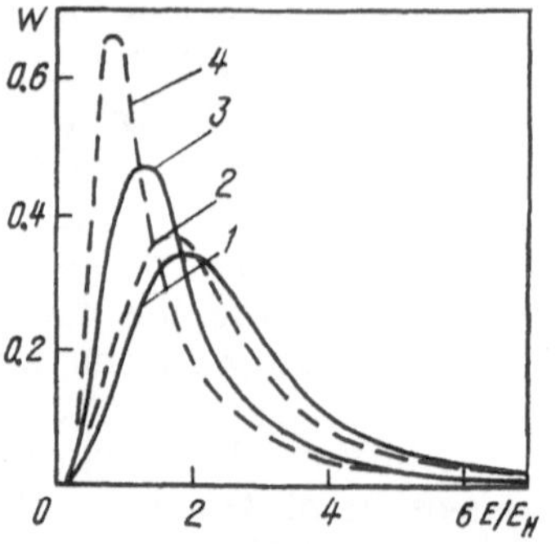


Fig. 2. Distribution of a microfield in a turbulent plasma (Holtsmark–Rayleigh distribution).

ion–ion correlations also vanishes. The inverse limiting case $(T_i/T_e) \to \infty$ (at finite but not too small $T_e$, so that the number of particles in a Debye-radius sphere is large, $N\rho_{De}^3 \gg 1$) can be readily seen, by examining the properties of the convolution integral

$$W_{RH} = \alpha\beta \int_0^\infty \left\{\exp\left[-\frac{3}{2}\alpha^2(\beta-u)^2\right] - \exp\left[-\frac{3}{2}\alpha^2(\beta+u)^2\right]\right\} \frac{W_{ind}(u)}{u} du, \tag{2.25}$$

where $\alpha = F_H/F_R$ and $\beta = F/F_H$, to lead to a distribution of Gaussian type, namely to a Rayleigh distribution on the section $F \lesssim F_R$:

$$W_{RH}(F)\,dF = 3\sqrt{\frac{6}{\pi}} \exp\left(-\frac{3}{2}\frac{F^2}{F_R^2}\right)\frac{F^2}{F_R^3}\,dF. \tag{2.26}$$

A two-temperature analysis yields, therefore, the conditions under which a Gaussian-type distribution can be realized in a thermal plasma. It suffices for this purpose to change the order of the limiting transitions in the functions that describe the ion–ion correlations and the charge screening. In the one-temperature model $T = T_e = T_i$, the limiting transition $T \to \infty$, as shown in [11] and as follows from (2.22), leads to a Holtsmark distribution.

It must be recognized, however, that in the two-temperature model the case $T_i > T_e$ corresponds to effective damping of the ion-sound oscillations. One can therefore hardly expect realization of such a distribution in an equilibrium plasma.

The microfield distributions for isotropic turbulence with high ion-sound level were analyzed and calculated in [18-22]. Figure 2 shows a typical resultant microfield distribution for an isotropic case with $\alpha = E_H/E_R = 1$. The distribution $W(\beta)$ corresponds to convolution of the Rayleigh and Holtsmark functions $W_R$ and $W_H$ (curves 1), and also to a Rayleigh function with nearest-neighbor distribution $W_B$ (curve 2). Plots of the functions $W_H(\beta)$ and $W_B(\beta)$ (curves 3 and 4, respectively) are also shown for comparison. Usually, $\alpha \ll 1$ in experiments with turbulent plasma, so that the corresponding convolution practically coincides with the Rayleigh function everywhere except in the far "wing" $\beta \gg 1$. The use of $W_R$ as the initial field distribution in a turbulent plasma is therefore not merely a convenient and simple model, but is actually a good approximation of the true picture of the phenomena [80].

To obtain specific results, it is necessary to make definite assumptions concerning the axisymmetric normalized distribution function $W(F, \cos\theta)$ of the static electric fields in the plasma. We use as a model a superposition of a one-dimensional spectrum with wave vector along the system axis, and a two-dimensional spectrum with wave vectors in the perpendicular plane (for details see Section 4).

As a rule, interest attaches in experiment to the case when the field amplitude $\langle F \rangle$ of the electrostatic oscillations is considerably larger than the average fields $F_0 \approx 2.6eN^{2/3}$ produced by individual ions. In this case, the contribution of the individual particles to the distribution of the static fields can be regarded as negligibly small, and the distribution of the total field $\mathbf{F} = \mathbf{F}_\parallel + \mathbf{F}_\perp$ is then given by

$$W(\mathbf{F}, \cos\theta)\, d\mathbf{F} d(\cos\theta) = \sqrt{\frac{2}{\pi \langle F_\parallel^2 \rangle}}\, \frac{F^2}{\langle F_\perp^2 \rangle} \exp\left\{ -F^2 \left[ \frac{1}{\langle F_\perp^2 \rangle} + \left( \frac{1}{2\langle F_\parallel^2 \rangle} - \frac{1}{\langle F_\perp^2 \rangle} \right) \cos^2\theta \right] \right\} dFd(\cos\theta). \quad (2.27)$$

The directivity pattern corresponding to the distribution is an ellipsoid of revolution with axis along z. As the degree of anisotropy $\eta = \sqrt{\frac{2\langle E_\parallel^2 \rangle}{\langle E_\perp^2 \rangle}}$ increases from 0 to $\infty$

this ellipsoid changes in shape from a strongly oblate "lentil" to a very prolate "sausage," and turns into a sphere at $\eta = 1$.

The actual calculation of the anisotropic static contour will be carried out in Section 4.

## 3. Line-Broadening Mechanisms in a Plasma. Character of Atom Interaction with a Plasma Microfield

The mechanisms responsible for the line broadening can be classified either in accordance with the types of the physical effects that cause the broadening, or in accordance with the character of their action on the radiating atom. In the latter case the different broadening mechanisms exhibit common features.

The most important physical effects responsible for line broadening in a plasma are those of Doppler, Stark, and Zeeman. The Doppler effect shifts the frequency radiated by the atom by an amount

$$\Delta\omega_D = \mathbf{kv}, \quad |\mathbf{k}| = \omega_0/c, \tag{3.1}$$

where $\mathbf{k}$ is the wave vector of a wave of frequency $\omega_0$ and $\mathbf{v}$ is the atom velocity (c is the speed of light).

The Stark effect shifts and splits the atomic levels in an external field F by an amount

$$\Delta\omega_S = C_k \, |\mathbf{F}|^{k/2}, \tag{3.2}$$

where $C_k$ is called the Stark-effect constant; for hydrogen-like levels (k = 2), we have

$$C_2 = \frac{3}{2} n (n_1 - n_2) \hbar/me, \tag{3.3}$$

where n, $n_1$, and $n_2$ are the principal and parabolic quantum numbers of the atom.

The constant $C_4$ of the quadratic Stark effect does not have a universal structure and is calculated numerically for each atom (see [1]).

The Zeeman effect splits the atomic levels in a magnetic field B by an amount

$$\Delta\omega_3 = M\mu_0 B, \quad -l \leqslant M \leqslant l, \tag{3.4}$$

where M is the component of the orbital angular momentum $\ell$ of the atom in the direction of B, and $\mu_0 = e\hbar/2mc$ is the Bohr magneton (we present the simplest result for a zero-spin atom, see [1]).

To complete the picture, we mention the so-called electrodynamic broadening, due to the Stark splitting of the atom levels in an electric Lorentz field $\mathbf{F} = [\mathbf{vB}]/c$ which takes place in the reference frame of the atom as it moves across a magnetic field.

Regardless of the physical effect, the ensuing broadening depends on the character of the action of the effect on the atom. A distinction is therefore made between two principal broadening mechanisms, static and impact.

These two broadening mechanisms are demarcated by the following parameter. Let $\Delta\omega_i$ denote the characteristic shift of the atom frequency governed by one of the physical effects (3.1)-(3.4). Furthermore, let $\tau_c{}^i$ be the correlation time of the physical quantities that cause the broadening. For the Doppler effect, e.g., $\tau_c{}^D \sim L/v$ is the free-path time of the atom (L is the mean free path); for the Stark broadening $\tau_c{}^S \sim T_F$ is the lifetime of the field (equal, for pair collisions, simply to the time of flight $\rho/v$ of the broadening particle), etc. The product of these two parameters $\Delta\omega_i\tau_c{}^i$, called in radiophysics the modulation index [23], determines in fact the character of the broadening.

The static broadening mechanism is realized if the modulation index is large:

$$\Delta\omega_i\tau_c^i \gg 1, \tag{3.5}$$

meaning, obviously, deep and slow frequency shifts.

The static line contour is obtained under the assumption that the frequency shifts $\Delta\omega_i$ are independent of the time (i.e., are static). An individual atom emits, in this case, an infinitely narrow line:

$$J = \delta(\Delta\omega - \Delta\omega_i) \tag{3.6}$$

($\Delta\omega = \omega - \omega_0$ is the shift of the observed frequency $\omega$ relative to the unperturbed one $\omega_0$).

Stark broadening proper appears after (3.6) is averaged with the distribution function $W(\Delta\omega_i)$ of the shifts, which yields

$$J(\omega) = \int W(\Delta\omega_i)\,\delta(\Delta\omega - \Delta\omega_i)\,d\Delta\omega_i = W(\Delta\omega). \qquad (3.7)$$

This averaging takes into account the fact that the frequency shifts of the different atoms are different, and radiation is observed not from one but from an entire ensemble of atoms.* The differences between the atom shifts for the Doppler broadening (3.1) are due to the difference between the atom velocities, for the Stark broadening (3.2) to the difference between the values of the field **F** at different points of space, and for the Zeeman broadening (3.4) to the spatial inhomogeneity of the magnetic field.

For a Maxwellian distribution of the atoms in velocity, the static Doppler contour takes, in accordance with (3.7), the form

$$J_D(\omega) = \frac{1}{kv_0}\exp\left[-\left(\frac{\Delta\omega}{kv_0}\right)^2\right],$$

$$v_0 = \sqrt{2T/M}, \qquad (3.8)$$

where T and M are the temperature and mass of the atoms.

If the electric fields in the plasma have a distribution $W_H(F)$ given by the Holtsmark function (see Section 2), the static Stark contour of the line is of the form

$$J_S(\omega) = \frac{1}{C_k F_0^{k/2}}\,\mathscr{H}\left[\left(\frac{\Delta\omega}{C_k F_0^{k/2}}\right)^{2/k}\right], \quad F_0 = 2{,}603eN^{2/3}, \qquad (3.9)$$

where N is the density of the broadening particles and $\mathscr{H}(x)$ is the Holtsmark function [6] (see Section 2).

Validity criteria for the static approximation are obtained by substituting in (3.6) the characteristic mean values $\Delta\omega_i$ and $\tau_c^i$. Thus, for Doppler broadening, substituting $\Delta\omega_i \sim (v_0/c)\omega_0$ and $\tau_c^i \sim L/v_0$, we get

*This circumstance is the reason why this type of distribution was previously called "statistical" [1]. In laser physics it is called "inhomogeneous" because the frequency shifts of different atoms are different [24].

$$L \gg \lambda; \tag{3.10}$$

i.e., the mean free path should be large compared with the wavelength of light.

For Stark broadening, analogously, putting $\Delta\omega_i \sim C_k \times (eN^{2/3})^{k/2}$ (splitting in the field $eN^{2/3}$, produced by a particle at the average interparticle distance) and $\tau_c^i \sim N^{-1/3}/v_0$, we obtain for the linear ($C_2$) and quadratic ($C_4$) Stark effects

$$N(C_2/v_0)^3 \equiv g_2 \gg 1 \text{ and } N(C_4/v_0) = g_4 \gg 1; \tag{3.11}$$

i.e., the static approximation is valid for high density and low velocity of the plasma particles.

We consider now the impact mechanism of broadening, which is realized at the condition alternate to (3.5)

$$\Delta\omega_i \tau_c^i \ll 1, \tag{3.12}$$

and corresponding obviously to small and rapidly varying frequency shifts.

It is intuitively obvious that a weak rapidly fluctuating perturbation should lead, just as interaction of an atom with zero oscillations of the electric field, to a Lorentzian line profile:

$$J(\omega) = \frac{1}{\pi} \frac{\gamma_i}{\Delta\omega^2 + \gamma_i^2}, \tag{3.13}$$

where the parameter $\gamma_i$, called the impact line width,* depends on the specific broadening mechanism. The impact broadening mechanism is realized under conditions inverse to (3.10) and (3.11).

We consider the transition from the static to the impact mechanism and the structure of the impact width $\gamma_i$ for specific examples of Doppler and Stark line broadening. We use the classically adiabatic broadening model, in which the radiating atom is regarded as a classical oscillator whose spectrum broadening is due only to modulation of its frequency as a result of a Doppler or Stark shift.

*In laser physics a broadening of type (3.13) is called "homogeneous" in view of the fact that the width $\gamma_i$ is the same for all atoms.

The line profile $J(\omega)$ takes in this model the form

$$J(\omega)=\frac{1}{\pi}\,\mathrm{Re}\int_{0}^{\infty}dt\exp(-i\Delta\omega t)\,\Phi(t), \tag{3.14}$$

where $\Phi(t)$ is the correlation function of the oscillator amplitudes [1, 73]:

$$\Phi(t)=\left\langle \exp\left[i\int_{0}^{t}\Delta\omega_i(\tau)\,d\tau\right]\right\rangle. \tag{3.15}$$

Here $\langle\ldots\rangle$ means averaging over the random variables that determine the phase shift $\Delta\omega_i$.

We begin with the case of Doppler broadening. We assume, for the change of the atom velocity, following [25], the Brownian-motion model. Then, expanding the exponential in (3.15) in a series and averaging term by term, we can verify that $\Phi(t)$ can be expressed in terms of the mean square $\langle r^2(t)\rangle$ of the displacement of a Brownian particle (the odd terms of the expansion average out to zero), and this yields

$$\Phi(t)=\exp[-k^2\langle r^2(t)\rangle],\quad \langle r^2(t)\rangle=\frac{v_0^2}{\nu^2}\,[1-e^{-\nu t}-\nu t], \tag{3.16}$$

where $\nu$ is the collision frequency in the Brownian-motion model and determines the mean free path $L = v_0/\nu$.

At small $\nu$ (more accurately at $kv_0/\nu = L/\lambda \gg 1$), the main contribution to (3.14) is made by the region of small t ($\nu t \ll 1$). Expanding (3.16) at $\nu t \ll 1$ and substituting the result in (3.14), we arrive at the Doppler contour (3.8).

At large $\nu$ ($L/\lambda \ll 1$) the main contribution to (3.14) is made by the region $\nu t \gg 1$, where the shift $\langle r^2(t)\rangle \approx v_0^2 t/\nu$ is diffusive with a diffusion coefficient $D = v_0^2/\nu = v_0 L$. Substitution of this value of $\langle r^2(t)\rangle$ in (3.14) leads to the Lorentz profile (3.13), where the impact width $\gamma_i$ is equal to

$$\gamma_i\equiv\gamma_d=(kv_0)^2/\nu\equiv k^2 D. \tag{3.17}$$

The width $\gamma_d$ turns out to be smaller by a factor $kv_0/\nu \ll 1$ than the Doppler width; i.e., collisions lead to an effective narrowing of this line. This effect was discovered by Dicke [1] (see also [24]). For example, for Coulomb collisions between a radiating ion of charge Z = 10 and plasma ions having an average Z ~ 10 at a temperature T ~ 1 keV, narrowing of a line of $\lambda \sim 100$ Å should be observed already at a plasma–ion density $N_Z \sim 10^{20}$ cm$^{-1}$ [26].

Consider now Stark broadening. Assume that the overall Stark shift $\Delta\omega_S$ of the atom frequency consists of the shifts $\Delta\omega_{Sk}$ produced by the individual broadening particles (the model of scalar addition of perturbations [27]):

$$\Delta\omega_{Sk} = \sum_{k=1}^{\mathcal{N}} \Delta\omega_{Sk}, \quad \Delta\omega_{Sk} = \frac{C_{2i}}{R_k^{2i}(t)} \quad (i = 1,\ 2), \tag{3.18}$$

where the trajectory $R_k^2 = \rho_k^2 + v^2t^2$ of the broadening particle is assumed to be a straight line (with an impact parameter $\rho_k$), and $\mathcal{N}$ is the total number of particles in the volume V ($N = \mathcal{N}/V$ is the particle density).

In this model, the correlation function (3.15) breaks up into a product of single-particle functions, so that [27]

$$\Phi(t) = \left\langle \prod_{l=1}^{\mathcal{N}} \exp\left[ i \int_0^t \Delta\omega_{S l}(\tau)\, d\tau \right] \right\rangle_{\mathcal{N}}$$

$$= \left\langle 1 - \frac{1}{V} \int d\mathbf{r} \left\{ 1 - \exp\left[ i \int_0^t \Delta\omega_{S 1} d\tau \right] \right\} \right\rangle_1^{\mathcal{N}=NV}$$

$$= \exp[-NV(t)];$$

$$V(t) = 2\pi \int_0^\infty \rho d\rho \int_{-\infty}^{\infty} dZ \left\{ 1 - \exp\left[ i \int_z^{z+vt} \Delta\omega_S(u)\, du \right] \right\}. \tag{3.19}$$

The symbol $\langle\ldots\rangle_{\mathcal{N}}$ denotes here averaging over an ensemble of $\mathcal{N} = NV$ particles with density N in a volume V, and $\langle\ldots\rangle_1$ denotes averaging over the phase volume of one particle. In the derivation of (3.19) we have taken the limit $\mathcal{N} \to \infty$, $V \to \infty$, $N = \mathcal{N}/N = \text{const}$ and introduced the cylindrical coordinates $d\mathbf{r} = 2\pi\rho d\rho dZ$ with axis $0Z \parallel v$ (here and elsewhere, the velocity will be assumed for simplicity to be fixed and equal to a certain characteristic velocity of the

Maxwellian distribution). The quantity V(t) is called the "collisional" volume.

We now introduce the characteristic parameters of broadening theory. It can be seen from (3.19) that the collision efficiency is characterized by the change of the phase shift $\Delta\varphi$ of the atomic oscillator. The phase shift $\Delta\varphi$ acquired in one collision equals

$$\Delta\varphi=\int_{-\infty}^{\infty}\Delta\omega s_i(t)\,dt=C_n\int_{-\infty}^{\infty}\frac{du}{(\rho^2+v^2)^{n/2}}=\pi\frac{C_n}{\rho^{n-1}v}. \qquad (3.20)$$

If $\Delta\varphi \sim \pi$, the oscillator phase collapses completely, and the atom "forgets" after such a collision the initial value of the phase. The value $\rho_B$ of the impact parameter at which $\Delta\varphi \sim \pi$ is called the Weisskopf radius:

$$\rho_B=(C_n/v)^{1/(n-1)}. \qquad (3.21)$$

This radius determines the effective phase-collapse cross-section $\sigma_B \sim \pi\rho^2_B$, called the optical-collision cross-section. An important role is played by the Weisskopf frequency (or reciprocal time) of the collision:

$$\Omega_B=v/\rho_B=v^{\frac{n}{n-1}}/C_n^{1/(n-1)}. \qquad (3.22)$$

The Weisskopf radius determines also the interaction effective volume, of order $\rho_B^3$, which contains, on the average,

$$N\rho_B^3\equiv g \qquad (3.23)$$

particles. Thus, the atom interacts simultaneously with a large number of particles if $g \gg 1$ and with one (nearest) particle if $g \ll 1$. The parameter $g^{-1}$ is consequently the pairing (binarity) parameter of the broadening collisions.

The characteristic time between collisions is determined by the frequency (or by the reciprocal of the free path time):

$$\gamma\sim Nv\sigma_B. \qquad (3.24)$$

We now examine the character of the evolution of the correlation function (3.19) [12, 28]. It is easily seen that the only scale of V(t) is $\Omega_B^{-1}$. There are therefore two characteristic time regions $\Omega_B\tau \ll 1$ and $\Omega_B\tau \gg 1$. At $\Omega_B\tau \ll 1$ we obtain from (3.19)

$$\mathrm{Re}\,V(\tau)=\int_0^\infty 4\pi r_0^2 dr_0\,[1-\cos(C_n\tau/r_0^n)]$$

$$=\frac{4\pi}{3}(C_n\tau)^{3/n}\Gamma\left(\frac{n-3}{n}\right)\sin\left(\frac{n-3}{n}\right)\frac{\pi}{2},\quad \Omega_B\tau\ll 1. \qquad (3.25)$$

The time evolution of the collision volume is determined here completely by the form of the static interaction potential $C_nR^{-n}$. This region, which corresponds to low velocities (or to a short time $\tau$) is customarily called static.

For long times $\Omega_B\tau \gg 1$ we have

$$\mathrm{Re}\,V(\tau)=\mathrm{Re}\,V'(\infty)\,\tau=v\sigma_B\tau,\quad \Omega_B\tau\gg 1, \qquad (3.26)$$

where $V'(\infty)$ is the derivative of the collision volume, equal to

$$\mathrm{Re}\,V'(\infty)=v2\pi\int_0^{\rho_m}\rho d\rho\left[1-\cos\left(\frac{C_n}{v}\int_{-\infty}^{\infty}\frac{du}{(\rho^2+u^2)^{n/2}}\right)\right]. \qquad (3.27)$$

It can be seen from (3.26) that in the region $\Omega_B\tau \gg 1$ the evolution of $V(\tau)$ is connected with the cross-section for isolated collisions. This is usually called the impact volume.

The collision volume $V(\tau)$ is thus a monotonically increasing function of $\tau$ with scale $\Omega_B^{-1}$ and with two characteristic asymptotes, static (3.25) and impact (3.26).

It follows from the foregoing analysis that the line-broadening problem involves two basic tasks. The first is an investigation of the statistics of the fluctuating plasma microfield, and the other is the dynamics of its action on the atom. The statistical distributions were considered in Section 2. We shall dwell below on the main problems of atom interaction with a plasma microfield.

In the investigation of broadening processes we used above the adiabatic model, in which only the change of the energy of the atomic states (the phase modulation) is taken into account. This model is valid if the perturbations are slow enough and cause no transitions between the different levels of the atom. It is applicable, in particular, for broadening of the lines of nonhydrogenlike atoms by plasma ions via the quadratic Stark effect.

Let us examine the dynamics of the interaction of an atom with a perturbing electric field **F**(t) using the following very simple example. Let the upper (radiating) level $\alpha$ of the atom be perturbed by interaction with one adjacent sublevel $\alpha'$ and let the field **F**(t) be produced by one traveling charged plasma particle. We are interested in the profile of the line produced in transition from level $\alpha$ to a lower level $\beta$, the perturbation of which is neglected. The width of the radiated line (as well as the magnitude of the perturbation) is assumed small compared with the energy difference $\Delta E_{\alpha\alpha'} = E_\alpha - E_{\alpha'} = \hbar\omega_{\alpha\alpha'}$ of the levels $\alpha$ and $\alpha'$ (this is called the case of an isolated line, which is realized, e.g., for lower levels of helium).

The Hamiltonian $\hat{H}$ of this system is

$$\hat{H} = \hat{H}_0 + \hat{V}(t) = \hat{H}_0 + e\mathbf{r}_A\mathbf{F}(t) = \hat{H}_0 + \frac{e^2\mathbf{r}_A\mathbf{r}(t)}{r^3(t)}, \qquad (3.28)$$

with $\hat{H}_0$ the unperturbed Hamiltonian of the atom (with levels $\alpha$, $\alpha'$, and $\beta$), $e\mathbf{r}_A$ the dipole moment of the atom, and $\mathbf{r}(t) = \boldsymbol{\rho} + \mathbf{v}t$ the coordinate of the perturbing particle.

It is convenient to describe the time variation of an atomic state by means of an evolution operator $\hat{T}(t, t_0)$ that satisfies the Schrödinger equation (in the interaction representation):

$$i\hbar\frac{d\hat{T}(t, t_0)}{dt} = \exp(iH_0t/\hbar)V(t)\exp(-iH_0t/\hbar)\hat{T}(t, t_0). \qquad (3.29)$$

Using the operator T, the wave function $\Psi(t)$ of the atom can be written as

$$\Psi_\alpha(t) = \sum_{\alpha'} T_{\alpha\alpha'}(t, t_0)\exp(-iE_{\alpha'}t/\hbar)\psi_{\alpha'};$$
$$T_{\alpha\alpha'}(t_0, t_0) = \delta_{\alpha\alpha'}, \qquad (3.30)$$

where $\psi_\alpha$ is an unperturbed function corresponding to the Hamiltonian $\hat{H}_0$.

The evolution of the atom depends on the ratio of the change of the perturbation $v/\rho_{ef}$ (the reciprocal travel time of a particle over an effective distance $\rho_{ef}$) and the frequency $\omega_{\alpha\alpha'}$ that separates the radiating and the perturbing levels. If

$$\rho_{ef}\,\omega_{\alpha\alpha'}/v \gg 1, \tag{3.31}$$

the perturbation is adiabatic, i.e., does not cause inelastic transitions between the states $\alpha$ and $\alpha'$. In this case, the evolution operator T is diagonal:

$$T_{\alpha\alpha'} = \delta_{\alpha\alpha'} \exp\left[-i/\hbar \int_{t_0}^{t} \frac{C_4}{r^4(\tau)}\, d\tau\right], \tag{3.32}$$

where the constant of the quadratic Stark effect is

$$C_4 = \frac{1}{3}\left(\frac{\hbar}{m}\right)^2 \frac{|\mathbf{r}_{A\alpha\alpha'}|^2}{\omega_{\alpha\alpha'} a_0^2} \left(a_0 = \frac{\hbar^2}{me^2}\right). \tag{3.33}$$

The use of the "adiabatic" evolution operator (3.32) is obviously equivalent to the adiabatic model (3.19) of the oscillator. A criterion for the applicability of this model is obtained from (3.31), in which $\rho_{ef}$ is replaced by the Weisskopf radius $\rho_B \sim (C_4/v)^{1/4}$; this yields

$$\frac{\hbar\omega_{\alpha\alpha'}}{mv^2} \frac{|r_{\alpha\alpha'}|}{a_0} \gg 1. \tag{3.34}$$

We emphasize that in the adiabatic model the perturbation frequency $v/\rho_B$ is considered small only compared with the level spacing $\omega_{\alpha\alpha'}$, but not compared with the frequency detuning $|\Delta\omega| \ll \omega_{\alpha\alpha'}$. With respect to the latter, the frequency $v/\rho_B$ can be either low (static limit) or high (impact limit); see above.

Under a condition inverse to (3.34), the perturbation causes inelastic (nonadiabatic) transitions between the levels $\alpha$ and $\alpha'$. These transitions are, in fact, the main cause of broadening in this case. The entire frequency region $|\omega| < \omega_{\alpha\alpha'}$ lies then in the impact-broadening region.

Calculations of the evolution operator T for the impact mechanism of broadening (in both the adiabatic and nonadiabatic cases) are based on the general principles of impact theory. In essence, the collisions are assumed to be very rapid, so that the characteristic times within which the evolution operator changes noticeably turn out to be much longer than the collision times. This makes it possible to consider for the operator T, in place of the exact equations (3.29), equations averaged over the collision operators. Indeed, we choose a time interval $\Delta t$ that is long compared with the collision time $\rho_{ef}/v$, but is still short compared with the characteristic variation time of the operator T. We have then for the average increment $\Delta T$ of the operator T during the time interval from t to t + $\Delta t$:

$$\Delta\{T\}=\{T(t+\Delta t,\ 0)-T(t,\ 0)\}=\{[T(t+\Delta t,\ t)-1]\,T(t,\ 0)\}, \quad (3.35)$$

where the braces {...} denote averaging over the ensemble of the broadening particles.

By virtue of the choice of $\Delta t$, the two factors in the right-hand side of (3.35) are statistically independent, so that they can be broken up into a product of mean values. Next, the evolution operator T(t + $\Delta t$, t), regarded as instantaneous, can be replaced for the time $\Delta t$ by the operator $T(-\infty, +\infty)$, which obviously coincides with the scattering matrix (S matrix). Finally, since the individual collisions are independent, the result reduces simply to multiplying by the number of collisions during the time $\Delta t$, viz., $Nvf(v)dv2\pi\rho d\rho\Delta t$.

Taking in (3.35) the limit as $\Delta t \to 0$ and solving the resultant differential equation, we obtain an equation for the evolution operator in the impact approximation:

$$\{T(t)\}=\exp(+i\widehat{H}_0 t/\hbar)\exp(-i\widehat{H}_0 t/\hbar+\widehat{\Phi}t), \quad (3.36)$$

where the time-independent operator $\hat{\Phi}$ is called the impact-broadening operator and is connected in obvious fashion with a scattering matrix $\hat{S}$:

$$\widehat{\Phi}=N\int_0^\infty vf(v)\,dv\int_0^\infty 2\pi\rho d\rho\,[\widehat{S}(\rho,\ v)-1]. \quad (3.37)$$

Thus, in the impact approximation the calculation of the evolution operator T(t) reduces to the substantially simpler calculation of the operator $\Phi$.

Nonetheless, the problem of determining the $\hat{S}$ matrix in (3.37) remains serious. Its solution is obvious in the adiabatic limit (3.32), where the $\hat{S}$ matrix reduces simply to the factor $e^{i\eta(\rho v)}$, where $\eta(\rho, v)$ is the phase shift of an atomic oscillator after one collision (see the oscillator model above).

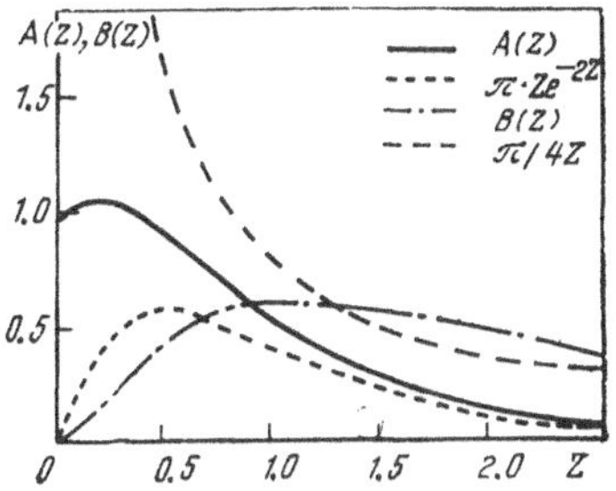


Fig. 3. The functions A(Z) and B(Z) that determine the contribution of inelastic and elastic processes to the broadening.

For fast nonadiabatic collisions, the $\hat{S}$ matrix can be calculated by perturbation theory. The use of this theory is justified by specific features of the long-range Coulomb perturbation that acts on the atom, whereby the main contribution to the broadening is made by long-range (weak) collisions.

Confining ourselves in the solution of (3.29) to second-order perturbation theory and substituting the perturbation in the form (3.28), we get

$$\{\langle\alpha\mid\hat{S}-1\mid\alpha\rangle\}=-\Bigg\{\frac{e^2}{\hbar^2}\sum_{\sigma,\nu}\langle\alpha\mid r_{A\sigma}\mid\alpha'\rangle\langle\alpha'\mid r_{A\nu}\mid\alpha\rangle$$

$$\times\int\limits_{-\infty}^{\infty}dt_1\int\limits_{-\infty}^{t_1}dt_2\exp\left[i\left(\omega_{\alpha\alpha'}t_1+\omega_{\alpha'\alpha}t_2\right)\right]F_\sigma(t_1)F_\nu(t_2)\Bigg\}. \qquad (3.38)$$

Using the explicit form $\mathbf{F} = e\mathbf{r}\cdot r^{-3}$ ($r_\sigma(t) = \rho_\sigma + v_\sigma t$), averaging in (3.38) over the angle variables, and expressing the integrals with respect to $t_1$ and $t_2$ in (3.38) in terms of the Bessel functions $K_0(Z)$ and $K_1(Z)$, we get [29]

$$\{\langle\alpha\mid\hat{S}-1\mid\alpha\rangle\}$$

$$=-\frac{2}{3}\left(\frac{e^2}{\hbar\rho v}\right)\mid\langle\alpha\mid\mathbf{r}\mid\alpha'\rangle\mid^2\left[A\left(\frac{\rho\omega_{\alpha\alpha'}}{v}\right)+iB\left(\frac{\rho\omega_{\alpha\alpha'}}{v}\right)\right], \qquad (3.39)$$

$$A(Z)=Z^2\left[K_0^2(\mid Z\mid)+K_1^2(\mid Z\mid)\right], \qquad (3.40)$$

$$B(Z) = \frac{2Z}{\pi}\,\mathcal{P}\int_0^\infty \frac{A(Z')}{Z^2 - Z'^2}\,dZ'. \tag{3.41}$$

The functions A and B in (3.39) determine, respectively, the inelastic and elastic parts of the scattering phase $\delta = \Gamma + i\eta$ (this can be easily verified by putting $S = e^{\delta}$). Figure 3 shows plots of the functions A and B together with their asymptotic values and makes clear the relative roles of the elastic and inelastic processes in the broadening. In fact, in the adiabatic region $Z_{ef} = (\rho_{ef}\omega_{\alpha\alpha'}/v) \gg 1$ the main contribution to the broadening is made by the elastic scattering [in this case, $B(Z) \approx \pi/4Z$], and the phase advance $\mathrm{Re}\, i\delta = \eta(\rho, v)$ coincides here with the results of the classically adiabatic model. At $Z_{ef} = (\rho_{ef}\omega_{\alpha\alpha'}/v) \ll 1$, on the contrary, the principal role is played by the inelastic processes (here $A \approx 1$).

In the nonadiabatic region $(\rho_{ef}\omega_{\alpha\alpha'}/v) \ll 1$ an expression for the diagonal elements of the operator $\Phi_{\alpha\alpha'}$ is obtained by substituting (3.39) (with $A \approx 1$ and $B \cong 0$) in (3.37) and integrating:

$$\mathrm{Re}\,\Phi_{\alpha\alpha} = -\frac{4\pi}{3} N \left(\frac{\hbar}{m}\right)^2 \frac{|\mathbf{r}_{\alpha\alpha'}|^2}{a_0^2} \left\langle \frac{1}{v} \right\rangle \ln \frac{v}{\rho_{\min}\omega_{\alpha\alpha'}}, \tag{3.42}$$

where we introduced lower ($\rho_{\min}$) cutoff parameters in the logarithmically diverging integrals with respect to $\rho$. We choose $\rho_{\min}$ to satisfy the condition for applicability of perturbation theory:

$$|\{\langle \alpha | S(\rho_{\min}) - 1 | \alpha \rangle\}| \sim \left(\frac{\hbar}{mv\rho_{\min}}\right)^2 \frac{|\mathbf{r}_{\alpha\alpha'}|^2}{a_0^2} \approx 1. \tag{3.43}$$

It can be seen from (3.42) that the contribution of the "weak" collisions $\rho > \rho_{\min}$ is logarithmically large compared with that of the "strong" ones, and this justifies the use of perturbation theory.

The foregoing covers in practice the problems of binary theory of broadening by particles for the case of isolated line. We proceed now to the case of overlapping lines: $\omega_{\alpha\alpha'} \to 0$. We confine ourselves exclusively to hydrogenlike levels corresponding to total overlap of the lines ($\omega_{\alpha\alpha'} \to 0$).

A specific feature of hydrogen-line broadening is primarily the absence of the parameter $\omega_{\alpha\alpha'}$ that separates the adiabatic from the nonadiabatic collisions, and hence an inherent redefinition of the very concept of "adiabaticity." Indeed, action of a constant electric field transforms the "regular" (i.e., diagonalizable by perturbations) wave functions of the hydrogen atom from spherical (as above) into parabolic. If the frequency of the perturbation is small enough compared with the frequency of the Stark-level splitting $C_2F/e \sim C_2\rho^{-2}$, there are no transitions between the Stark states; i.e., the atom "follows" the field adiabatically. The adiabaticity criterion follows from (3.39) if $\omega_{\alpha\alpha'}$ is replaced by $C_2\rho^{-2}$, so that

$$C_2/\rho v \gg 1 ; \tag{3.44}$$

i.e., the close and slow particle travels become adiabatic, whereas the remote (weak) and fast ones become nonadiabatic.

For a more detailed consideration of this question we consider the wave function $\Psi(t)$ not in the laboratory frame XYZ, but in a frame X'Y'Z' whose axis 0Z' is directed at each instant of time along the field $\mathbf{F}(t)$:

$$\psi(t) = \exp[i L_X \varphi(t)] \psi'(t). \tag{3.45}$$

Here $\psi'$ is the wave function of the atom in the rotating frame, $\exp[iL_X\varphi(t)]$ is the operator of rotation through an angle $\varphi(t)$ and separates the frames XYZ and X'Y'Z' ($L_X$ is the X component of the orbital angular momentum of the atom).

Substituting (3.45) in the Schrödinger equation, we get

$$i\hbar \frac{\partial \psi'}{\partial t} = [\hat{H}_0 + d_Z F(t) + \hbar L_X \dot{\varphi}(t)] \psi' \equiv [\hat{H}_0 + \hat{V}] \psi'. \tag{3.46}$$

Thus, in the rotating frame the atom effects, besides the interaction $d_ZF(t)$ with the electric field, also a "magnetic" interaction $L_X\dot{\varphi}$ connected with the instantaneous angular velocity $\dot{\varphi}$ of the rotation. The designation "magnetic" is justified in view of the full analogy between the interaction $L_X\dot{\varphi}$ and the interaction $L_X\mu_0B$ of an atom with an external magnetic field.

The magnetic perturbation due to the rotation causes nonadiabatic transitions between the Stark states. If the change of the field is slow, the rotation can be taken into account by perturbation theory, with the quantity (3.44) as the parameter.

For the case of pair collisions, there exists an interesting possibility of finding the exact wave-function evolution of a hydrogen level with a fixed principal quantum number n in the field of a passing charged particle. This possibility is due to the fact that both the electric $d_Z F$ and the magnetic $L_X \dot{\varphi}$ perturbations have the same time dependence, so that their ratio remains constant during the collision. Indeed, the value of $\dot{\varphi}$ can be easily found from the conservation law for the angular momentum of the broadening particle:

$$Mr^2(t)\,\dot{\varphi}(t) = Mv\rho, \quad \dot{\varphi}(t) = v\rho/r^2(t), \tag{3.47}$$

from which it is indeed clear that F(t) and $\dot{\varphi}(t)$ have the same dependence on r(t).

Thus, a hydrogen atom colliding with a particle is located in an electric field perpendicular to an effective (proportional to $\dot{\varphi}$) magnetic field; both fields vary synchronously in time, so that the angle $\beta$ that their vector sum or difference makes with the OX axis remains constant. We introduce the vectors

$$\boldsymbol{\omega}_{1,2}(t) = \dot{\boldsymbol{\varphi}}(t) \mp \frac{B}{e}\,\mathbf{F}(t) \quad \left(B = \frac{3}{2}\,n\,\frac{\hbar}{m}\right), \tag{3.48}$$

whose directions remain constant in the course of the collision. Thus, for the angle $\beta_2$ between $\boldsymbol{\omega}_2$ and the OX$\|$**F** axis we have, taking (3.47) into account,

$$\tan \beta_2 = \rho v/B. \tag{3.49}$$

The atom's evolution can be determined if it is possible to construct states that correspond to quantization of the atom along the directions $\boldsymbol{\omega}_{1,2}$. We construct such states by using the Coulomb-field symmetry properties that are manifested in the presence of another integral of the motion, viz., the Runge–Lenz vector [30, 31]:

$$\mathbf{A} = \frac{1}{2m}\left([\mathbf{pL}] - [\mathbf{Lp}]\right) - e^2\mathbf{r}/r \tag{3.50}$$

(**p** is the electron momentum). The matrix elements of the operator **A** for states with fixed n coincide with the matrix elements of the atom's dipole moment: $\mathbf{d} = -\frac{3}{2}\frac{ea_0}{\hbar}\mathbf{A}$.

Introducing new "momentum operators"

$$\mathbf{J}_{1,2} = (\mathbf{L} \pm \mathbf{A})/2, \tag{3.51}$$

we rewrite the perturbation potential V(t) in (3.46) in the form

$$V(t) = \hbar\,[\mathbf{J}_1\omega_1(t) + \mathbf{J}_2\omega_2(t)]. \tag{3.52}$$

It can be seen from (3.52) that the atom states of interest to us [which diagonalize the perturbation V(t)] correspond to a definite projection of $\mathbf{J}_1$ on $\omega_1$ (designated below by the quantum number n') and of $\mathbf{J}_2$ on $\omega_2$ (designated n"). The possibility of choosing independently the quantized quantities $\mathbf{J}_1$ and $\mathbf{J}_2$ is due to the presence of two independent integrals of the motion in the Coulomb field (L and A) and is based on the fact that the operators $\mathbf{J}_1$ and $\mathbf{J}_2$ obey the usual angular-momentum commutation rules, and the projections of $\mathbf{J}_1$ and $\mathbf{J}_2$ commute fully with one another.

The functions $u_{n'n''}$ that diagonalize the perturbation (3.52) can be obtained from the ordinary parabolic functions $u_{i_1 i_2}$ ($i_1$ and $i_2$ are the quantum numbers of the projections of the vectors $\mathbf{J}_1, \mathbf{J}_2$ along the axis OX∥F) by simple rotation through the constant angles $\beta_1$ and $\beta_2$, so that [31, 32]

$$u_{n'n''} = \sum_{i_1, i_2} D^{(j)}_{n'i_1}(\beta_1)\, D^{(j)}_{n''i_2}(\beta_2)\, u_{i_1 i_2}, \tag{3.53}$$

where $D_{k\ell}^{(j)}$ is the Wigner rotation function [33]; here j = (n − 1)/2 and the quantum numbers of the projections n', n", $i_1$, and $i_2$ change, as usual, from −j to +j.

Choosing the wave functions $u_{n'n''}$, we obtain from (3.52) for the change ΔE(t) of the atom energy in the rotating coordinate frame

$$\Delta E(t) = (n' + n'')\,\hbar\,|\omega_{1,2}(t)| = \hbar\,(n' + n'')\,\sigma\,\frac{\rho v}{e}F(t); \quad \sigma = \sqrt{1 + (B/\rho v)^2}. \tag{3.54}$$

The wave function ψ'(t) thus takes in the rotating system the form [34, 12]

$$\psi'(t) = u_{n'n''} \exp\left[-i(n'+n'')\sigma\frac{\rho v}{e}\int^{t} F(\tau)\,d\tau\right]. \tag{3.55}$$

It follows from (3.55) that the wave-function evolution is connected (just as in the adiabatic model of an oscillator!) only with the modulus of the electric field F(t). This makes the problem similar to adiabatic theory with specially defined "components" $u_{n'n''}$. The nonadiabaticity effects reduce to a dependence of the amplitudes of these components on the particle-travel parameters, and also make the phase factor somewhat more complicated. This circumstance permits many results of the adiabatic model to be used in the general case. It is clear beforehand that the general theory (3.55) contains just as many parameters as the adiabatic model, so that one can speak only of numerical differences, furthermore, small in a number of cases.

The final results can also be represented in a more general form, by introducing the evolution operator $\hat{T}(t)$ of the hydrogen level [32]:

$$\langle i_1' i_2' | \hat{T}(t) | i_1 i_2 \rangle = \sum_{n', n''=-j}^{+j} D^{(j)}_{i_1' n'}(\beta) D^{(j)}_{i_2' n''}(\pi-\beta) D^{(j)}_{i_1 n'}(\beta) D^{(j)}_{i_2 n''}(\pi-\beta)$$

$$\times \exp\left[-i(n'+n'')\hbar\sigma\frac{\rho v}{e}\int^{t} F(\tau)\,d\tau\right], \tag{3.56}$$

where $D_{ik}^{(j)}$ are rotation operators.

All the foregoing results can be readily used in the case of hydrogenlike-ion line broadening, for which it is important to take the bending of the broadening-particle trajectory into account [35]. Indeed, by virtue of the conservation of the angular momentum of the particle in a Coulomb field, we have for the angular velocity $\dot{\varphi}$ the previous expression (3.47), and for the rotation angle $\varphi(t)$ of the field **F**(t) (which determines the phase of the wave function) we have the expression

$$\varphi(t) = \varphi_0 + \rho v_\infty \int_0^t \frac{d\tau}{r^2(\tau)}, \tag{3.57}$$

where $\varphi_0$ is the initial angle and $v_\infty$ is the electron velocity at infinity.

In contrast to the rectilinear motion, the value of $\varphi(t)$ as $t \to \infty$ is bounded by the value $\varphi(\infty) = \arccos \varepsilon^{-1}$, where $\varepsilon = 1 + M^2\rho^2 v_\infty^4/(Z-1)^2 e^4)^{1/2}$ is the eccentricity of the hyperbola (Z is the charge of the radiating atom). The degree of deviation from linearity is obviously determined by the ratio of the radius $\rho$ to the Landau radius $\rho_\Lambda = Ze^2/Mv_\infty^2$.

It is clear from the foregoing that to calculate the evolution of the wave function of a hydrogen atom there is no need to resort to the impact- or static-theory approximation. The corresponding limiting results follow automatically from the general solution after suitably averaging over the collision parameters. Thus, at low velocities ($v \to 0$) the atom states $u_{n'n''}$ go over into parabolic states with the OX axis along the field $\mathbf{F}(t)$, and the energy changes (3.54) go over into Stark frequency shifts. By the same token, the transition to the static adiabatic model is traced.

The results of the impact theory for hydrogen should be obtained from the general equations (3.42) with $\omega_{\alpha\alpha'} \to 0$. It can be seen from (3.42), however, that the impact line width $\mathrm{Re}\,\Phi_{\alpha\alpha}$ diverges logarithmically at the upper limit $\rho_{max} \sim v/\omega_{\alpha\alpha'} \to \infty$.

This circumstance is fundamental and the reason for it is that the impact theory, based on replacing the evolution operator by the $\hat{S}$ matrix, does not apply, strictly speaking, to a Coulomb interaction. Indeed, the evolution operator $\hat{T}(t)$ calculated for finite times t in second-order perturbation theory is obviously connected with an integral of the pair correlation function of the electric fields [36, 37]

$$\langle \hat{T}_{\alpha\alpha}(t) \rangle \backsim \int_0^t dt' \int_0^{t'} dt'' \langle F(t') F(t'') \rangle \backsim t \ln \frac{vt}{\rho_{min}}. \qquad (3.58)$$

The last result shows that the derivative $d\hat{T}/dt$ is actually not expressed in terms of the constant operator $\Phi$, but contains a logarithmic distorting factor. This factor indicates that distant travel paths with $\rho \sim vt$ cannot be regarded as terminated [36, 37]. In a plasma the values of $\rho_{max}$ are actually bounded by $\rho_D$, so that at $t_{ef} \gtrsim \rho_D/v \sim \omega_{pe}^{-1}$ the logarithm must be cut off at $\rho_{max} \sim \rho_D$, which obviously reaffirms the validity of the impact theory. In broadening theory, however, the effective times $t_{ef}$ can be determined by values of order $\Delta\omega^{-1}$, which are smaller than $\omega_{pe}^{-1}$, and this should lead to logarithmic distortions of the Lorentzian (impact) line profile [37].

## 4. Static Theory and Its Generalizations

The initial assumption of the static theory is that the line shift (more the shift of any of its component $\alpha \to \beta$) is a function of a certain physical parameter $\gamma$ (of the radiating-atom velocity, of the external magnetic field strength, of the electric microfield, etc.) whose distribution $W(\gamma)$ is assumed known. In this scheme each atom "sees" its value of the parameter $\gamma$, and the emission spectrum of an individual atomic-line component takes the form

$$I_{\alpha\beta}^{(\gamma)}(\omega) = I_{\alpha\beta}\delta[\Delta\omega - f(\gamma)]. \tag{4.1}$$

The observed intensity is obtained by summing the intensities from the individual atoms or, equivalently, by averaging (4.1) over the distribution $W(\gamma)$. The corresponding integration yields, obviously, the following result for the static intensity distribution in the component $\alpha \to \beta$ :

$$I_{\alpha\beta}(\omega) = I_{\alpha\beta}W[f^{-1}(\Delta\omega)]\frac{df^{-1}(\Delta\omega)}{d\Delta\omega}, \tag{4.2}$$

where $f^{-1}$ is a function inverse to f.

If, for example, the role of $\gamma$ is assumed by the atom-velocity component $v_i$, then $f(\gamma) = \omega_0 v_i/c$, and if the distribution $W(v_i)$ is Maxwellian, Eq. (4.2) yields the known Doppler line contour. For each case of the linear Stark effect in an ion microfield **F** we have $\gamma = |\mathbf{F}| = F$, $f(\gamma) = C_{\alpha\beta}F$ ($C_{\alpha\beta}$ is the Stark constant of the considered component), and $W(\gamma)$ is the Holtsmark distribution of the microfield (2.4), or its distribution shown in Fig. 1, so that

$$I_{\alpha\beta}(\omega) = \frac{I_{\alpha\beta}}{C_{\alpha\beta}} W\left(\frac{\Delta\omega}{C_{\alpha\beta}}\right). \tag{4.3}$$

To obtain the static contour $I_{ab}(\omega)$ of the line as a whole it is necessary to substitute in (4.3) the distribution W(F) indicated above, sum over all the Stark components $\alpha$ and $\beta$ of the upper and lower levels, and normalize by dividing by the total line intensity $I_0 = \sum_{\alpha,\beta} I_{\alpha\beta}$. In the limit of an ideal plasma ($\mathcal{N}_D = \infty$), this yields [see (2.4)]

$$I_{ab}(\omega) = \frac{1}{I_0} \sum_{\alpha,\beta} \frac{I_{\alpha\beta}}{C_{\alpha\beta} F_0} \mathcal{H}\left(\frac{\Delta\omega}{C_{\alpha\beta} F_0}\right). \tag{4.4}$$

This formula was used in [38] to calculate a number of hydrogen lines.

We emphasize that the Holtsmark line profile is highly nonbinary (this is, as a matter of fact, the advantage of the statistical approximation). A transition to the binary case takes place according to (2.7) only in the line wing – at large values of $\Delta\omega$:

$$I_{ab} \sim \frac{2\pi N}{\Delta\omega^{5/2}} \frac{1}{I_0} \sum_{\alpha,\beta} C_{\alpha\beta}^{3/2} I_{\alpha\beta} \equiv \frac{2\pi N C^{3/2}}{\Delta\omega^{5/2}}, \tag{4.5}$$

where the effective Stark constant C for the line as a whole was introduced. An approximate estimate of this constant is [12, 39]

$$C = \left(\frac{3}{8}\right)^{2/3} \frac{\hbar}{m} (n_a^2 - n_b^2), \tag{4.6}$$

with $n_a$ and $n_b$ the principal quantum numbers of the levels a and b.

Expression (4.5) is already linear in the density N of the broadening particles. The transition to this binary result occurs at

$$\Delta\omega \gg (C/e) F_0 = \lambda C N^{2/3} \equiv \Delta\omega_0. \tag{4.7}$$

The quantity $\Delta\omega_0$ determines the characteristic scale of Holtsmark broadening of a hydrogen line. The line profile (4.4) satisfies the normalization condition

$$\int_0^\infty I_{ab}(\omega)\, d\omega = 1, \tag{4.8}$$

which follows from the normalizability of the Holtsmark function.

Recall that some hydrogen lines have central Stark components (e.g., $L_\alpha$, $H_\alpha$, $H_\gamma$) and others do not ($H_\beta$, $H_\delta$); static broadening, by its very meaning, pertains only to noncentral (sideband) Stark components, while broadening of the central (unshifted) component is via the impact mechanism (see Section 5).

Noise Anisotropy. It turns out that measurement of light polarization inside the Stark profile of a spectral line yields information on the electrostatic-noise spectrum of a turbulent plasma [40, 41]. The influence of homogeneous but anisotropic high-frequency electrostatic noise on the Stark profiles of hydrogen lines can equally well be described within the framework of the static approximation [41]. The polarization procedure is based on a general property of the polarization of Stark components in a constant electric field, viz., the preferred bunching of circularly polarized $\sigma$ components near the line center and the departure into the wing by the linearly polarized $\pi$ components.

We consider the most typical organization of an experiment: appearance of turbulent noise is expected in apparatus having a symmetry axis along OZ. Naturally, the directivity pattern of the noise must have axial symmetry, but the turbulence level and the character of the anisotropy of the noise remain unknown.

Spectral line profiles $S_1(x)$ and $S_2(x)$ are recorded at two positions of the polarizer (1 – parallel to the system axis, 2 – perpendicular to it). To reveal distinctly the anisotropy effect, we investigate the difference $D(x) = S_1 - S_2$ of these two profiles (here, $x = \lambda - \lambda_0$ is the shift from the line center in angstroms).

In the static approximation the line profile is described by the expression

$$S_i(x) = \sum_{k,\nu} \int_{F=0}^{\infty} \int_{-1}^{+1} \int_{u=-\infty}^{\infty} W(F, \cos\theta)\, I_{k\nu} f_{k\nu}(\cos\theta)\, \delta(x - F\Delta_{k\nu} - ud)\, M(u)\, du\, dF\, d(\cos\theta). \tag{4.9}$$

The subscript $i = 1, 2$ conforms here to the position of the polarizer, and the subscript $\nu = \pi, \sigma$ reflects the character of the polarization of the component. The second term in the argument of the $\delta$ function is the Stark shift, where

$\Delta_{k\nu} = (3ea_0\lambda_0^2/4\pi c\hbar)[n(n_1 - n_2) - n'(n_1' - n_2')]$ is the reduced Stark constant ($a_0$ is the Bohr radius, n the principal quantum number, and $n_1$ and $n_2$ the electric quantum numbers). The third term is the shift due to the Doppler effect, with $d = v_0\lambda_0/c$ the Doppler width ($v_0 = \sqrt{2T/M}$ is the average modulus of the thermal velocity of the atoms in the direction of the line of sight). The velocity averaging is over the Maxwellian distribution

$$M(u)\,du = \frac{1}{\sqrt{\pi}}\,e^{-u^2}du\;(u = v_i/v_0).$$

The set of four functions $f_{i\nu}(\cos\theta)$ gives the angular dependence of the observed Stark-component intensity. These functions are readily determined by considering the components, along the direction of polarizer 1 or 2, of a radiating dipole either oriented along the field **F** ($\pi$ component) or rotating in a plane perpendicular to **F** ($\sigma$ component):

$$f_{1\pi} = 3\cos^2\theta, \qquad f_{2\pi} = \frac{3}{2}(1-\cos^2\theta),$$
$$f_{1\sigma} = \frac{3}{2}(1-\cos^2\theta), \qquad f_{2\sigma} = \frac{3}{4}(1+\cos^2\theta). \tag{4.10}$$

The functions $f_{i\nu}$ are so normalized that the areas under the profiles are equal to unity: $\int S_1(x)dx = \int S_2(x)dx = 1$. Use is made here of a general property of the relative intensities $I_{k\nu}$ of Stark components:

$$\sum_k I_{k\pi} = \frac{1}{2}\sum_k I_{k\sigma} = 1/3. \tag{4.11}$$

The difference contour D(x) contains only the linear combinations

$$f_{1\pi} - f_{2\pi} = \frac{3}{2}(3\cos^2\theta - 1)$$

and

$$f_{1\sigma} - f_{2\sigma} = \frac{3}{4}(1 - 3\cos^2\theta).$$

Introducing the notation

$$\tilde{I}_{k\pi} = 3I_{k\pi}, \qquad \tilde{I}_{k\pi} = -\frac{3}{2}I_{k\sigma}, \tag{4.12}$$

we can omit the subscript $\nu$ from the summation in (4.9), since the angle factor does not depend here on $\nu$:

$$D(x) = \sum_k \tilde{I}_k \int_0^\infty dF \mathcal{F}(F) \int_{-\infty}^{\infty} du M(u) \delta(x - F\Delta_k - ud), \tag{4.13}$$

where

$$\mathcal{F}(F) = \int_{-1}^{+1} W(F, \cos\theta) \frac{1}{2} (3\cos^2\theta - 1) d(\cos\theta). \tag{4.14}$$

We consider the case when the Doppler broadening is small compared with the Stark shift for all the sideband components. We then have, accurate to terms of second order in the smallness parameter $\varepsilon = d / \Delta_k \langle F \rangle$,

$$D(x) = \sum_{k \neq 0} \frac{\tilde{I}_k}{|\Delta_k|} \mathcal{F}\left(\frac{x}{\Delta_k}\right) + \frac{\tilde{I}_0}{d} M\left(\frac{x}{d}\right) \int_0^\infty \mathcal{F}(F) dF. \tag{4.15}$$

This profile, apart from a narrow central part, is the usual pattern of Stark broadening of a spectral line by static field. The only difference is that the signs of Stark-component intensities alternate, in accordance with (4.12), and the role of the "isotropic" distribution function is assumed by $\mathcal{F}(F)$. This function has a simple physical meaning. It shows whether fields of given amplitude F have a preferred direction, and the extent to which this direction is more probable than all others. $\mathcal{F}(F) \equiv 0$ for isotropic noise distribution. On deviation from isotropy $\mathcal{F}(F)$ differs from zero, and this difference increases with increasing anisotropy of the noise distribution. $\mathcal{F}(F) > 0$ if the directivity pattern of the fields is prolate along the system axis, and $\mathcal{F}(F) < 0$ if it is oblate.

We put $S_F = \int_0^\infty \mathcal{F}(F) dF$. This "norm" $S_F$ of the function $\mathcal{F}(F)$ depends only on the degree $\eta = \langle F_F \rangle / \langle F_x \rangle$ of the anisotropy, but not on the scale of the fields. The function $S_F(\eta)$ increases monotonically from $-1/2$ to $+1$, with $S_F(1) = 0$.*

*For simplicity, we shall use below the notation

$$\langle F_\parallel^2 \rangle \equiv F_z^2, \quad \frac{1}{2} \langle F_\perp^2 \rangle \equiv F_x^2.$$

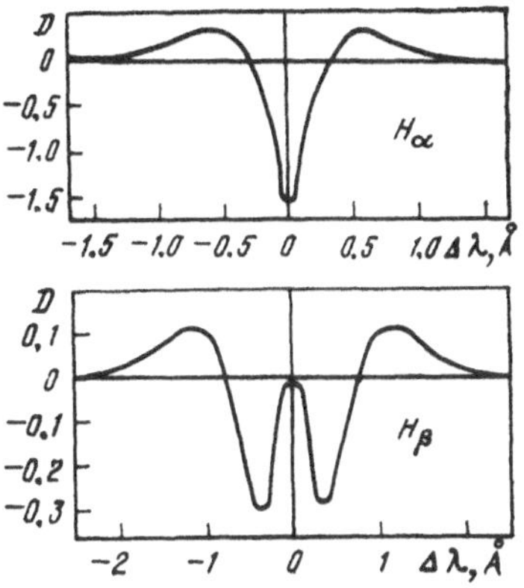


Fig. 4. Difference profiles of the lines $H_\alpha$ and $H_\beta$ in a plasma with isotropic noise distribution

Within the framework of this model, the function $\mathcal{F}(F)$, and hence also the difference contour D(x), can be expressed analytically:

$$\mathcal{F}(F) = \frac{F^2}{\sqrt{8\pi}\, F_x F_x^2}\, e^{-\frac{F^2}{2F_x^2}} \left\{ \Phi\left[\frac{3}{2}, \frac{5}{2}, (F_x^{-2} - F_z^{-2})\, F^2/2\right] - \Phi\left[\frac{1}{2}, \frac{3}{2}, (F_x^{-2} - F_z^{-2})\, F^2/2\right]\right., \qquad (4.16)$$

where $\Phi(\alpha, \gamma, z)$ are confluent hypergeometric functions.

The function

$$S_F(\eta) = \frac{\eta^2 + 1/2}{\eta^2 - 1} - \frac{3}{2}\, \frac{\eta^2}{|\eta^2 - 1|^{3/2}} \begin{cases} \arcsin\sqrt{1 - \eta^{-2}}; & \eta > 1, \\ \ln(\eta^{-1} + \sqrt{\eta^{-2} - 1}), & \eta < 1 \end{cases} \qquad (4.17)$$

has likewise a simple analytic form.

A typical difference profile for the lines $H_\alpha$ and $H_\beta$ is shown in Fig. 4, drawn for "sausage" type distributions. For anisotropy of opposite sign ("lentil" type distributions), the picture should be inverted relative to the abscissa axis. Thus, from only the form of the difference profile we can directly determine qualitatively the spatial directivity of the noise.

From the difference profile we can determine not only the ratio $F_zF_x$, but each of these fields separately. The quantity $F_{max} = \max(F_z, F_x)$ can be estimated from the position $x_{extr}$ of the far extremum of the difference profile

$$F_{max} \approx \frac{x_{extr}}{\Delta_\pi}, \quad \left(\Delta_\pi^2 = \frac{\sum_k I_{k\pi}/\Delta_k^p}{\sum_k I_{k\pi}/\Delta_k^{p+1}}, \quad p = \begin{cases}1, & \eta > 1\\ 2, & \eta < 1\end{cases}\right).$$

The value of $F_{max}$ can be determined more accurately from the variation of the intensity in the wing

$$D_{AS}(x) \backsim \begin{cases}\exp[-x^2/2F_{max}^2\Delta_\pi^2], & D_{AS}(x) > 0,\\ -x\exp[-x^2/2F_{max}^2\Delta_\pi^2], & D_{AS}(x) < 0.\end{cases} \tag{4.18}$$

The polarization method of measuring noise anisotropy can be used even when the Doppler broadening exceeds the Stark broadening. In that case, expanding the $\delta$ function in (4.13) in powers of the smallness parameter $\Delta_k\langle F\rangle/d \ll 1$, we obtain

$$D(x) = \sum_k \frac{\tilde{I}_k}{d}\int_0^\infty dF\mathcal{F}(F)\left\{1 + \frac{\Delta_k}{d}F\frac{d}{du}\right.$$
$$\left. + \frac{1}{2}\left(\frac{\Delta_k}{d}\right)^2 F^2\frac{d}{du^2} + \ldots\right\}M(u)\Big|_{u=x/d}. \tag{4.18a}$$

This expression contains sums of the form $\sum_k \tilde{I}_k(\Delta_k)^q$, $q = 0, 1, 2$. Since $\tilde{I}_{-k} = +\tilde{I}_k$, and $\Delta_{-k} = -\Delta_k$, these sums are nonzero only for even q. Further, from (4.11) and (4.12) we have $\sum_k \tilde{I}_k = 0$ and, accurate to fourth order in $\Delta_k\langle F\rangle/d$:

$$D(x) - \left(2\frac{x^2}{d^2} - 1\right)\frac{e^{-x^2/d^2}}{\sqrt{\pi}\, d}\left[\sum_k \frac{1}{2}\tilde{I}_k\frac{\Delta_k^2}{d^2}\int_0^\infty \mathcal{F}(F)F^2\,dF\right]. \tag{4.19}$$

At $d \gg \Delta_k\langle F\rangle$ the difference profile is thus small compared with the polarization profiles $S_i(x)$.

Influence of Thermal Motion of the Broadening Ions on the Line Profile. Criteria for Validity of the Static Approximation. The static theory is one of the basic postulates of the description of the profiles of hydrogen lines in a plasma.

It is therefore important to specify the conditions for the action of a microfield **F** on a plasma to be static, meaning for neglect of the time variation of **F**. It was already noted in Section 3 that favoring the validity of the static approximation is a large value of the parameter $g_2 = N(C_2/v)^3$ (of the modulation index). The criterion $g_2 \gg 1$ is, however, only a certain condition that is integral in the spectrum and does not take into account the more subtle criteria that are differential in the spectrum.

We note in this connection that the very existence of a static limit in broadening theory (a limit is apparently unrelated to the Fourier expansion that characterizes this theory) is by far not obvious beforehand. This was reflected in Pauli's skeptical attitude toward the Holtsmark theory (1927; see [42]). Pauli's main countering argument was the undisputed fact that the ion microfield is not static. The direct approach to a resolution of this problem is obviously to calculate the corrections to the static line profile for the thermal motion of the ions. The requirement that these corrections be relatively small is indeed the criterion of static broadening by ions. This was done in [43] in the binary case, and in [13] for the multiple (Holtsmark) case within the framework of the phase-modulation model. An illustrative physical picture of the influence of thermal motion of ions on a radiating model (within the framework of the very same model) is given in [44], and we shall mainly follow this picture. A complete calculation of the corrections, with allowance for nonadiabaticity, was made in [45]. Thermal corrections with account taken of motion of the radiating atom are analyzed in [2].

Allowance for a nonstatic microfield **F**(t) is made difficult by the complexity of its static properties. In fact, at $g_2 \gg 1$ the atom (according to Section 3) interacts with a large number of broadening particles (usually ions), making the temporal evolution of resultant multiparticle field **F**(t) quite intricate. Naturally, this prevents also, in the general case, a determination of the character of the atom's evolution and, in turn, of its spectrum. We confine ourselves therefore to the case when the ion field F(t) varies slowly enough in time.

In a static field **F**(0) the states of the hydrogen atom are oriented along the 0Z axis, which is parallel to the vector **F**(0). When **F**(t) changes with time, both its magnitude and direction change. The change of the magnitude (modulus) of $\mathbf{F}(0)$ is obviously governed by those components whose time derivatives are parallel to $\mathbf{F}(0)$, while the change of its direction is governed by the components $\dot{\mathbf{F}}_{\perp}$, perpendicular to $\mathbf{F}(0)$ ($\dot{\mathbf{F}}^2=\dot{\mathbf{F}}_{\parallel}^2+\dot{\mathbf{F}}_{\perp}^2$). Let us examine the influence of both types of change on the radiation by the atom.

Variation of the field's modulus leads to a time variation of the Stark frequency shifts, so that the change of the phase of the atomic oscillator takes the form

$$C_2\int_0^{\tau} F(t)\,dt \simeq C_2F(0)\,\tau + C_2\dot{F}(0)\frac{\tau^2}{2}.$$

The additional phase shift $\Delta\varphi(\tau) = C^2\dot{F}(0)(\tau^2/2)$ due to the thermal motion causes loss of phase coherence of the radiation. In fact, at $\Delta\varphi(\tau) \sim 1$, the atom "forgets" its initial phase – the coherence becomes disordered. As a result of this disorder, the atom's radiation spectrum in the given field f(0) is not infinitely narrow ($\propto\delta[\Delta\omega - C_2F(0)]$) but is spread out to a value $|C_2\dot{F}|^{-1/2}$. The order of magnitude of the time $T_\Phi$ of phase-coherence loss is

$$T_\Phi(F) \sim \langle |C_2\dot{F}|^{-1/2}\rangle_F \sim \sqrt{T_F/C_2F}. \tag{4.20}$$

The symbol $\langle\ldots\rangle_F$ stands here for averaging over all values of $\dot{F}$ at a fixed modulus of the field F; the derivative $\dot{F}$ is estimated at F/TF, where $T_F$ is the lifetime of the ion field F and is determined below (see also [7]).

It is easy to formulate, with the aid of (4.20), a criterion for a static field [44]. In fact, the field F can be regarded as static if the atom loses coherence more rapidly than it can alter the ion field, i.e., at

$$T_\Phi(F) \ll T_F, \quad \text{or} \quad C_2FT_F \gg 1. \tag{4.21}$$

To obtain the corresponding criterion for a spectrum it is obviously necessary to replace F in (4.21) by the quantity $\Delta\omega/C_2$ that enters in the static line profile.

Reversal of the field direction leads to two important effects. First, if the reversal of $\mathbf{F}(t)$ is slow enough, the dipole moment $\mathbf{d}$ of the atom can follow the field adiabatically and maintain at all times a constant projection on the direction at $\mathbf{F}$. Such a reorientation of the atom changes the projection of **d** on the propagation direction of the light wave **k**, and since the square of this projection determines the intensity of the light radiated by the atom, a certain amplitude modulation of the radiation should set in.

The second effect of the rotation is connected with the lag of the atom relative to the field **F**(t) (the inertia of the atom). Indeed, if the axis of the rotating coordinate frame is aligned with the field **F**(t), there appears in this system, according to [45], a "magnetic" perturbation $L_{\perp}\dot{\varphi}$, where $\dot{\varphi}$ is the instantaneous angular velocity of the field rotation, equal to

$$|\dot{\varphi}(0)| = \left| \frac{\mathbf{F}(0) \times \dot{\mathbf{F}}(0)}{F^2} \right| = \left| \frac{\dot{\mathbf{F}}_{\perp}}{F} \right|. \tag{4.22}$$

This perturbation leads, obviously, to a change of the (parabolic) wave function and of the energy of the radiating atom. This manifests itself in the dipole-moment matrix elements and in the phase of the radiated wave.

All these effects occur, of course, also for paired (binary) broadening collisions (see Section 3). In the considered multiparticle case, however, in contrast to the binary one, these effects can be accounted for only by perturbation theory, since neither the dynamics of the operator nor the statistical properties of the field **F**(t) are fully known.

The perturbation-theory calculations are based on successive expansion of the atom's wave function (and hence of the line profile in powers of the time derivatives $\dot{\mathbf{F}}$ of the field, or, equivalently, in powers of the parameter $g_2^{-1} \ll 1$. For a spherically symmetrical particle distribution in velocity, the terms of first nonvanishing order are expressed through the mean squares $\langle \dot{\mathbf{F}}^2 \rangle_F$ of the field derivatives for a given F. These quantities were calculated by Chandrasekhar and von Neumann [7] and are given by the following limited expressions:

$$\langle \dot{\mathbf{F}}^2_{\parallel,\perp} \rangle_F = \frac{45}{8} (\omega_F F_0)^2 \begin{cases} \frac{1}{3}(\parallel), \quad \frac{2}{3}(\perp), \quad \beta \ll 1, \\ \frac{8\beta^3}{45}[2(\parallel), \quad 1(\perp)], \quad \beta \gg 1. \end{cases} \tag{4.23}$$

The symbols $\parallel$ and $\perp$ refer here to the components of the derivative $\dot{\mathbf{F}}$ which are parallel and perpendicular to $\mathbf{F}$, and $\omega_F = \lambda^{1/2} v_0 N^{1/3}$ is the characteristic scale of the field-variation frequency. It can be seen that in the limiting cases $\beta \gtrless 1$ the two values differ only by numerical factors of order unity.

With the aid of (4.23) we can easily determine the lifetime $T_F$ of the field:

$$T_F \sim \dot{F} < |\mathbf{F}|^2 >_F^{-1/2} \sim (N^{1/3} v_0)^{-1} \begin{cases} \beta, & \beta \ll 1, \\ \beta^{-1/2}, & \beta \gg 1; \end{cases} \tag{4.24}$$

i.e., the maximum lifetime is possessed by the intermediate fields ($\beta \sim 1$), while the weak ($\beta \ll 1$) and strong ($\beta \gg 1$) fields live a short time.

Substituting (4.24) in (4.21), we obtain the following criteria for the validity of the static approximation:

$$\Delta\omega \gg \begin{cases} v^2/C_2 \equiv \Omega_B, & \Delta\omega \gg \Delta\omega_0, \\ \sqrt{C_2 N v_0}, & \Delta\omega \ll \Delta\omega_0. \end{cases} \tag{4.25}$$

More consistent criteria for static properties are obtained by calculating the line profile for the fact that the ion field is not static, using the perturbation-theory scheme indicated above. The hydrogen-line profile is expressed in these calculations by the sum of a static (Holtsmark) profile $I^{(0)}$ and a correction profile $I^{(1)}$ that describes the effects of thermal motion [45]:

$$I(\omega) = I^{(0)}(\omega) + I^{(1)}(\omega) = \frac{1}{\Delta\omega_0}\left[\mathcal{H}\left(\frac{\Delta\omega}{\Delta\omega_0}\right) + g^{-2/3}\,\Pi\left(\frac{\Delta\omega}{\Delta\omega_0}\right)\right], \tag{4.26}$$

$$\Pi(x) \approx \begin{cases} a_1 x^{-7/6}, & x \gg 1, \\ a_2 x^{-2}, & x \ll 1. \end{cases} \tag{4.27}$$

The function $\Pi(x)$ is expressed here in terms of the functions $< \dot{\mathbf{F}}^2{}_{\parallel,\perp} >_F$ and takes into account the effects described above, i.e., those of phase modulation, amplitude modulation, and nonadiabaticity; $a_1$ and $a_2$ are numerical coefficients that depend on the line considered.

On the line wing ($x \ll 1$) the "thermal" correction $\Pi(x) \propto x^{-7/2}$, i.e., decreases more rapidly than the Holtsmark function, while the contributions of all three effects to the correction are of the same order. In this frequency region, both the line profile and the correction to it are proportional to the particle density N, i.e., are binary. Clearly, these results can be obtained also directly from the binary theory of broadening in the line wing.

At the line center ($x \ll 1$), the thermal corrections increase sharply (in proportion to $x^{-2}$) as a result, easily understood, of the nonadiabaticity of the rotation. In fact, it can be seen from (4.22), (4.26), and (4.27) that the rotation-induced nonadiabatic corrections to the wave function $\psi$ are determined by the ratio $\dot{\varphi}/d_x F$, i.e., by the quantity $|\dot{\mathbf{F}}_\perp|^2/F^2$ that increases rapidly as the line center is approached ($F \propto \Delta\omega \to 0$).

The condition for the validity of the Holtsmark static theory is, obviously, that the second term of (4.26) be small compared to the first. It can be verified that in limiting cases this is equivalent to the criterion (4.25) obtained above.* At $\Delta\omega \sim \Delta\omega_0$ (i.e., at intermediate values $F \sim F_0$ of the microfield strength) the criterion of static behavior reduces, by virtue of the order-of-magnitude relations $\mathscr{H}(x \sim 1) \sim 1$ and $\Pi(x \sim 1) \sim 1$, simply to the condition $g_2 \gg 1$ introduced in Section 3. It can be seen from (4.26), moreover, that the validity of the static approximation is better the larger the parameter $g_2$. It breaks down even further, however, for $g_2 \gg 1$ near the line center, at $\Delta\omega \lesssim \sqrt{C_2 N v_0} \ll \Delta\omega_0$. On the other hand, even for $g_2 \ll 1$ (binary region) the static approximation is valid in the far line wing $\Delta\omega \gg \Omega_B \gg \Delta\omega_0$. Thus, favoring the static approximation are large values of the density N and of the Stark constant $C_2$ (highly excited levels), large frequency shifts $\Delta\omega$ (line wings), and small $v_0$ (low temperatures).

The foregoing results pertained to line broadening by particles of one species, ions. A situation may arise, however, when allowance for additional broadening by electrons improved the validity of the static theory for ions at the line center. The physical basis for this effect is the follow-

*This criterion was obtained above on the basis of qualitative nonrigorous considerations that take only phase modulation into account. A complete analysis is contained in [45].

ing [46]. Broadening by electrons is via impact (see Section 5) and leads to a finite lifetime $\gamma^{-1}$ [$\gamma \sim N(C_2^2/v_e)\ln \rho_{max}/\rho_{min}$ is the impact electronic width of the atom at a given Stark sublevel]. If this lifetime is short compared with the characteristic time $N^{-1/3}/v_0$ of variation of the ion field, this field can be regarded as static.

The relevant estimates can be easily obtained by analyzing the structure of the thermal corrections (4.27) at the line center. As already noted, they increase in this region like $\Delta\omega^{-2}$ as $\Delta\omega \to 0$. Allowance for the additional electron impact broadening leads, roughly speaking, to replacement of the divergent quantity $\Delta\omega^{-1}$ by the finite $(\Delta\omega + i\gamma)^{-1}$. Substituting, therefore, $\gamma$ for $\Delta\omega$ in (4.26) and (4.27) at $x \ll 1$, we estimate the corrections $I^{(1)}(0)$ for the thermal motion at the line center:

$$I^{(1)}(0) \sim \frac{1}{\Delta\omega_0} g_i^{-2/3} g_e^{-2/3} \Lambda^{-2}, \qquad (4.28)$$

where we have introduced the parameters $g_{e,i} = N(C_2/v_{e,i})^3$ for the electron and ion broadening and took into account that $\gamma/\Delta\omega_0 \sim g_e^{1/3}\ln(\rho_{max}/\rho_{min}) \equiv g_e^{1/3}\Lambda$.

The quantity $I^{(1)}(0)$ should be compared with $I^{(0)}(0)$ of the static profile at the line center. Estimating $I^{(0)}(0)$ by a convolution of the Holtsmark and Lorentz (electron) profiles, we get $I^{(0)}(0) \sim (1/\Delta\omega_0)g_e^{1/3}\Lambda$. The criterion $I^{(0)} \gg I^{(1)}$ for static behavior at the line center then takes the form [46]

$$g_i^{2/3} g_e \Lambda^3 \gg 1, \quad \text{or} \quad \frac{\rho_{ef}\gamma}{v_i} \gg 1. \qquad (4.29)$$

The value of $\rho_{ef}$ is close in practice to the mean interatomic distance $N^{-1/3}$. The criterion (4.29) thus supplements at the line center the criterion (4.25).

The intensity changes produced by thermal motion of the ions in the central region of the line are described in [46]. The calculation results agree fairly well with experiment [47].

Effects of Microfield Inhomogeneities. The first moments $\langle \dot{F} \rangle_F$ of the rate of change of the ion microfield are closely related to its spatial characteristics that determine the deviations from homogeneity. Actually, since a radiating atom moving with velocity $\mathbf{v}$ "probes," so to speak, the spatial structure of the field, the rate of change of the field at a given point $\mathbf{x}$ can be expressed in terms of the moments of the inhomogeneity-tensor components:

$$\langle \dot{F} \rangle_F = \left\langle \frac{\partial F}{\partial x_k} \right\rangle_F \dot{x}_k. \tag{4.30}$$

In turn, it is just the inhomogeneity of the ion microfield which is responsible for the experimentally observed [48] asymmetry of the spectral lines of hydrogenlike systems. In fact, the symmetric splitting of the lines corresponds to allowance for only the Stark-component shifts that are linear in F, and to the use of zeroth-approximation eigenfunctions. Actually, the microfield of the ions is inhomogeneous in space, and the scale of this inhomogeneity is the average interparticle distance $R_{av}$. The ratio of the quadrupole-interaction energy to the dipole-interaction energy can be treated as a quantity of first order of smallness with respect to the parameter $\varepsilon = n^2 a_0 / R_{av}$, which is the ratio of the atom size to the field-inhomogeneity scale. Accordingly, the expansion of the energy of the electrostatic interaction between the radiator and the plasma ions can be written in the form $V = \varepsilon^2 V_2 + \varepsilon^3 V_3 + \varepsilon^4 V_4 + \ldots$, where $V_2$, $V_3$, and $V_4$ are the dipole, quadrupole, and octupole contributions. As shown in [49], for a degenerate level of a quantum-mechanical system subjected to a perturbation of this type, the first terms of the corrections to the eigenfunctions of the unperturbed Hamiltonian can be taken to be those linear in $\varepsilon$. It is necessary for this purpose that the degeneracy be lifted by the interaction $\varepsilon^2 V_2$ and the interaction $\varepsilon^3 V_3$ must have nonzero matrix elements of the transitions between previously degenerated states, the difference between which is now $\Delta E_{nm} = \varepsilon^2 (V_{nn}^{(2)} - V_{mm}^{(2)})$. This is precisely the situation when simultaneous account is taken of the dipole and quadrupole interactions between a radiating hydrogen atom and a perturbing plasma-ion field. The quadrupole interaction causes transitions between states of equal azimuthal and principal quantum numbers m and n, and have correspondingly parabolic quantum numbers $n_1$ and $n_2$ that differ by ±1. Taking into account the principal correction terms that are linear in $\varepsilon$, we can express the eigenfunctions of the hydrogen atom, in a parabolic coordinate frame, in the form

$$\Psi^{(1)}_{n_1 n_2 m}(\xi, \eta, \varphi) = \Psi^{(0)}_{n_1 n_2 m} +$$

$$+\frac{na_0}{2R}\sqrt{(n_1+1)(n-n_1-1)n_2(n-n_2)}\,\Psi^{(0)}_{n_1+1,n_2-1,m}$$

$$+-\frac{na_0}{\sqrt{2}R}\sqrt{(n-n_1)n_1(n-n_2-1)(n_2+1)}\,\Psi^{(0)}_{n_1-1,n_2+1,m}. \quad (4.31)$$

Here $\psi_{n_1n_2m}^{(0)}$ are the unperturbed eigenfunctions in the parabolic coordinates $\xi$, $\eta$, and $\varphi$. The perturbing ion is located at a distance R on the positive 0Z axis.

A consistent perturbation theory in terms of the parameter [50] yields an expression for the line profile with account taken of the asymmetry. This expression is not restricted to the "nearest neighbor" approximation, in contrast to the initial treatment of the effect [51], and takes the form

$$I(\Delta\omega)=d\beta\int_0^\infty \frac{\gamma}{\left(\Delta\omega-\frac{C_2F_0\beta}{e}\right)^2+\gamma^2}\left\{\mathcal{H}(\beta)+\frac{n^2a_0}{R}\left[a_1\Lambda(\beta)+a_2\chi(\beta)\right]\right\}. \quad (4.32)$$

Here $C_2$ is the Stark constant of the component and $\gamma$ is its impact half-width. The functions $\Lambda(\beta)$ and $\chi(\beta)$ are connected with the first moments of the function $W(F, \partial F_i/\partial x_k)$ with respect to the ion-field inhomogeneity tensor, i.e., with $(\partial F_i/\partial x_k)$. They are tabulated in [50], where they are used to obtain numerical estimates of the effect of shifts of the centroids of the lines.

Calculation of the $\mathcal{H}_\beta$ profile leads to an asymmetry wherein the "blue" maximum is more intense than the "red" one. A similar asymmetry occurs also in the case of the lines $L_\beta$, $L_\delta$, $H_\delta$ and HeII $\lambda$ = 3203 Å. The first terms of the series ($L_\alpha$, $H_\alpha$, HeII $\lambda$ = 4686 Å) are characterized by a different pattern, viz., the intensity of the red line wing increases and that of the blue one decreases. The central components of these lines, however, is slightly shifted in the blue direction. It is convenient to express the asymmetry of the far wings by using the deviations $(\Delta\lambda)_1$ and $(\Delta\lambda)_2$, connected with the equal-intensity points $\lambda_1$ and $\lambda_2$ on opposite line wings: $I_+(\lambda_1) = I_-(\lambda_2)$:

$$(\Delta\lambda)_1=\frac{1}{2}\,|\lambda_1-\lambda_2|,\qquad (\Delta\lambda)_2=\frac{1}{2}(\lambda_1+\lambda_2)-\lambda_0.$$

For large enough distances from the line center, there is a simple functional relation between $(\Delta\lambda)_2$ and $(\Delta\lambda)_1$:

$$(\Delta\lambda)_2 = C\,(\Delta\lambda_1)^{3/2},$$

with the coefficient C determined only by parameters contained in the quadrupole corrections to the frequencies, and by the Stark-component parameters [49].

The described treatment of the asymmetry of the hydrogen spectral-line profile is based on the static description of the ions and on the impact mechanism of broadening by the electrons. In this case, the contributions from oppositely charged particles are of substantially different nature, and the quadrupole corrections necessitated by them cannot be mutually canceled out. The region where the electrons might also be treated statically is usually on a rather far wing that does not influence the main qualitative effects of the asymmetry.

The principal role in the formation of the asymmetry profile is usually played by just the corrections of first order in the parameter $\varepsilon^2 = n^2 a_0/R$, which are due to the quadrupole interactions. Attempts to take into account higher-order corrections, especially for central unshifted components, are usually at variance with the assumed smallness of $\varepsilon$, since their role becomes significant only at $\varepsilon \sim 1$. It is more natural to use, in this case, elliptic coordinates for the two-center problem, i.e., to consider the excited states of the molecular ion. This approach [52, 53] predicts the presence of "satellites" on the far wings of the line $L_\alpha$, which describe in essence the excited states of an HX molecule. The question of population of such states and of their actual manifestations under experimental conditions remains unclear.

## 5. Impact Theory of Broadening by Particles and Waves

Impact broadening of atomic states by collisions with plasma particles (usually electrons) was investigated in detail in Section 3. It was stated there that in impact broadening of nonhydrogenlike atoms a distinction must be made between adiabatic and nonadiabatic broadening, which differ by the parameter (3.34). In the adiabatic theory the line profile has a Lorentz shape and a constant (impact) width $\gamma \sim Nv\sigma_B$, determined by the cross-section for optic (Weisskopf) collisions. Thus, in the case of the quadratic Stark effect the width $\gamma$ is equal to [1]

$$\gamma \simeq 11.4 N C_4^{2/3} v^{1/3}. \tag{5.1}$$

The averaged evolution parameter T of the atom is expressed, according to (3.36) and (3.37), in terms of the impact electronic broadening operator $\hat{\Phi}_{ab}$. Substituting T(t) in the general equations (1.14) for the line profile, we get

$$I_{ab}(\omega) = -\mathrm{Re} \sum_{\alpha, \alpha', \beta, \beta'} \langle\beta \mid \mathbf{d} \mid \alpha\rangle\langle\alpha' \mid \mathbf{d} \mid \beta'\rangle$$
$$\times \langle\langle\alpha\beta \mid [i\omega - i/\hbar(\widehat{H}_{0a} - \widehat{H}_{0b}) + \widehat{\Phi}_{ab}]^{-1} \mid \alpha'\beta'\rangle\rangle, \tag{5.2}$$

where the symbol $\langle\langle\alpha\beta|$ means the use of matrix elements with respect to the wave functions of the upper and lower levels of the atom (the so-called doubled atom):

$$\langle\langle\alpha\beta \mid U_a U_b \mid \alpha'\beta'\rangle\rangle = \langle\alpha \mid U_a \mid \alpha'\rangle\langle\beta \mid U_b^* \mid \beta'\rangle.$$

It can be seen from (5.2) that in the impact case the line profile is expressed in terms of the operator of a resolvent that contains the operator $\hat{\Phi}$. Since the latter operator is obtained by averaging over an aggregate of spherically symmetric collisions, it is also spherically symmetric. The resolvent operator in (5.2) can therefore be diagonalized in spherical coordinates. In the case of hydrogenlike line broadening, however, the "correct states of the atom are not spherical but parabolic wave functions with 0Z axis directed along the static electric field **F**. The operator $\Phi$ of impact electronic broadening has therefore diagonal as well as off-diagonal matrix elements. The nonzero matrix elements of the operator $\mathbf{r}\cdot\mathbf{r}$ (which enter in the operator $\Phi$) take in parabolic coordinates the form [54]

$$\langle n_1 n_2 m \mid \mathbf{r}\cdot\mathbf{r} \mid n_1 n_2 m\rangle = a_0^2 \frac{9}{4} n^2 [n^2 + (n_1 - n_2)^2 - m^2 - 1], \tag{5.3}$$

$$\langle n_1' n_2' m \mid \mathbf{r}\cdot\mathbf{r} \mid n_1 n_2 m\rangle = a_0^2 \frac{9}{4} n^2 \Big[\delta_{n_1', n_1-1}$$
$$\times \delta_{n_2', n_2+1} \sqrt{n_1(n - n_1)(n_2 + 1)(n - n_2 - 1)}$$
$$+ \delta_{n_1', n_1-1} \delta_{n_2', n_2-1} \sqrt{n_2(n - n_2)(n_1 + 1)(n - n_1 - 1)}\Big]. \tag{5.4}$$

The real part of the operator $\mathrm{Re}\,\Phi = \Gamma$ determines the impact line width, while the imaginary part $\mathrm{Im}\,\Phi = D$ determines its width. Only the diagonal matrix element $\mathrm{Re}\,(\Phi_a)_{\alpha\alpha}$ has the real meaning of a Lorentz-contour width, whereas in the general case even the general form of the contour is not Lorentzian (see below).

The determination of the impact line contour $I_{ab}(\omega)$ for a fixed ion field F calls for laborious calculations entailing diagonalization of the operator of the resolvent (5.2); this turns out to be more complicated the larger the number of line components. An analytic investigation is possible only for the simplest lines $L_\alpha$ and $L_\beta$. We consider, following [55, 56], the character of the spectrum of the $L_\alpha$ line.

The atom Hamiltonian $\hat{H}_0(F)/\hbar - i\hat{\Phi}$, which includes the interaction with the ion field F and the impact electronic broadening, is not Hermitian. We shall construct its eigenfunctions $|\psi_\alpha\rangle$ by starting from the parabolic states $|\varphi_\beta\rangle$ that diagonalize the operator $\hat{H}_0(F)$. We construct for this purpose the matrix $C_{\beta\alpha}$ that determines the transition from the basis $|\varphi_\beta\rangle$ to the basis $|\psi_\alpha\rangle$. Since the resolvent operator is not Hermitian, the obtained system of ket-vectors $|\psi_\alpha\rangle$ must be supplemented by an orthonormal system of bra-vectors $\langle\chi_\alpha|$. This system is obtained from $\langle\varphi_\beta|$ with the aid of the matrix $C_{\alpha\beta}^{-1}$, while the matrix $\hat{C}$ is not unitary: $\hat{C}^{-1} \neq \hat{C}^{+}$. By obtaining the functions $|\psi_\alpha\rangle$ and $\langle\chi_\alpha|$, we diagonalize, obviously, the resolvent operator and calculate line profile. Thus, for the simplest line $L_\alpha$ (level n = 2) the eigenvalues $E_{1,2}$ of the operator $\hat{H}_0 - i\hat{\Phi}$ are equal to

$$E_{1,2}/\hbar = \omega \pm \Omega + iw, \qquad \Omega \equiv \sqrt{(\Delta/2)^2 - \beta^2}, \tag{5.5}$$

where w and $\beta$ are, respectively, the diagonal and off-diagonal matrix elements of $\hat{\Phi}$, and $\Delta$ is the level splitting in the ion field. Equation (5.5) reveals the unique role of the off-diagonal element $\beta$: owing to the presence of $\beta$, the energy levels $E_1$ and $E_2$ are not pushed apart but, conversely, are effectively attracted. At $\beta = \Delta/2$, in particular, there is a point $E_1 = E_2$ at which the two states become degenerate or coalesce (collapse). The character of the spectrum is significantly altered in this state, since the quantity $\Omega$ becomes here purely imaginary. The contribution of the sideband (displaced) Stark components to the line profile outside the coalescence region is of the form [55]

$$I(\omega) = \frac{1}{\pi}\left[\frac{w + (\beta/\Omega)(\Delta\omega + \Omega)}{(\Delta\omega + \Omega)^2 + w^2} + \frac{w - (\beta/\Omega)(\Delta\omega - \Omega)}{(\Delta\omega - \Omega)^2 + w^2}\right]. \tag{5.6}$$

It can be seen that the line contour is not described by a simple dispersion (Lorentzian) curve. At $\Delta/2 < \beta$ (in the coalescence region) the character of the spectrum is altered; the quantity $\Omega$ enters here additively with the diagonal matrix element w. The coalescence effect leads to a certain narrowing of the line.

For a line with many components (e.g., the Balmer series), analytic diagonalization of the resolvent is difficult. The procedure of matrix inversion in (5.2) is usually numerical (see [57, 58] and also Section 8).

The line profile resulting from simultaneous static broadening by ions and impact broadening by electrons is calculated from the static distribution W(F) of the ion microfields:

$$I_{ab}(\omega)=\frac{1}{\pi}\,\mathrm{Re}\int_0^\infty dFW(F)\sum\langle\alpha\mid\mathbf{d}\mid\beta\rangle\langle\beta'\mid\mathbf{d}\mid\alpha'\rangle\langle\langle\alpha\alpha'\mid[\mathrm{i}(\Delta\omega-C_{\alpha\beta}F)-\hat{\Phi}_{ab}]^{-1}\mid\beta\beta'\rangle\rangle, \tag{5.7}$$

where $C_{\alpha\beta}$ is the difference between the Stark constants of the upper ($\alpha$) and lower ($\beta$) states between which the radiative transition is considered.

The line splitting at the most probable field strength $F_0 \sim eN^{2/3}$ greatly exceeds the impact line width [it is easy to verify that their ratio coincides with the parameter $g_2 = N(C_2/v_e)^3 \ll 1$ that determines the validity of the impact approximation]. In such fields the role of the off-diagonal matrix elements of the operator $\hat{\Phi}_{ab}$ is minor [see (5.6)] and the impact contour is well described by a dispersion (Lorentz) profile whose width $\Gamma_{\alpha\beta}$ is determined by the diagonal matrix elements of the operator $\hat{\Phi}_{ab}$. In the region $\Delta\omega \gg \Gamma_{\alpha\beta}$ we thus have

$$I_{ab}(\omega)\approx\frac{1}{\pi}\int_0^\infty dFW(F)\sum_{\alpha,\beta}\mathbf{d}_{\alpha\beta}d_{\beta\alpha}\frac{\Gamma_{\alpha\beta}}{(\Delta\omega-C_{\alpha\beta}F)^2+\Gamma_{\alpha\beta}^2}\,. \tag{5.8}$$

As noted in Section 1, line broadening is produced in a plasma by electric fields not only of individual particles, but also of collective plasma oscillations. The latter can naturally be subdivided into low-frequency (LF) and high-frequency (HF). LF noise is due, as a rule, to oscillations of the ion component of the plasma and lead to a quasistatic broadening of the atom state (see Section 2). The faster HF noise, due to electron oscillations, causes impact broadening. The validity of the impact approximation is based, in this case, on the smallness of the parameter

$$d\mathscr{E}_0\tau_c/\hbar\ll 1, \tag{5.9}$$

where $\mathcal{E}$ is the amplitude of the HF noise (d is the dipole moment of the atom); $\tau_c$ is the HF-noise characteristic correlation time and is of the order of the reciprocal half-width $\gamma_p^{-1}$ of the noise spectrum. The parameter (5.9) is the analog of the modulation index introduced above.

When condition (5.9) is satisfied, the evolution of the atomic state over times $t \gg \tau_c \sim \gamma_p^{-1}$ can be described, in accordance with the general principles of impact theory, by averaged quantities of the type of the impact-broadening parameter $\hat{\Phi}$ (cf. Section 3).

Let us consider, following [59], the character of the evolution of a hydrogen atom acted upon by the stochastic HF field of (say, Langmuir) noise:

$$E_l(t) = \sum_{j=1}^{J} E_j(t) \cos\{\Omega_j t + \varphi_j(t)\}, \tag{5.10}$$

where $\varphi_j$ and $E_j$ are the random phases and amplitudes of the HF oscillations.

In the impact approximation it can be assumed that the changes of the phase and of the amplitude take place instantaneously, and that during the time intervals $t \gg \gamma_p^{-1}$ of interest to us the number of such changes of Q is large: $Q \sim \gamma_p t \gg 1$, so that we have, for example, for the phase $\varphi_j$,

$$\varphi_j(t) = \sum_{q}^{Q(t)} \varphi_{jq}\theta(t - T_{jq})\,\theta(T_{jq+1} - t) \qquad (\varphi_{jq} = \text{const}). \tag{5.11}$$

Here $\theta(x)$ is the theta step function, $T_{jq}$ is the moment of the q-th relaxation of the phase and amplitude of the j-th wave packet; the interval $\Delta t_{jq} = T_{jq+1} - T_{jq}$ between the phase relaxations is a random quantity with a distribution density $\gamma_p = \exp[-\gamma_p \Delta T_{jq}]$.

We now find the evolution operator $\hat{T}(0, \tau)$ of the atomic level on the interval $(0, \tau)$.

The equations for the evolution-operator matrix elements are

$$\left.\begin{aligned} i\hbar \dot{\hat{T}}_{\alpha\alpha''} &= \sum_{\alpha'} e^{i\omega_{\alpha\alpha'}\tau} d_{\alpha\alpha'} E(\tau) \cos[\Omega\tau + \varphi(\tau)]\, T_{\alpha'\alpha''}, \\ T_{\alpha\alpha''}(0) &= \delta_{\alpha\alpha''}. \end{aligned}\right\} \tag{5.12}$$

We use second-order perturbation theory to solve (5.12). It is easy to verify that the ensuing double summation over the variables q and j leaves only terms with q = q' and j = j', yielding*

$$[\hat{T}_l(\tau, 0) - 1]^{(2)}_{\alpha\alpha} = -\frac{1}{12\hbar^2} \sum_{\alpha', j, q} |d_{\alpha\alpha'}|^2 E^2_{jq} \left[ \frac{i\,\Delta T_{jq}}{\omega_{\alpha\alpha'} - \Omega_j} \right.$$

$$\left. + \frac{1 - \exp\{i(\omega_{\alpha\alpha'} - \Omega_j)\Delta T_{jq}\}}{(\omega_{\alpha\alpha'} - \Omega_j)^2} + \frac{i\,\Delta T_{jq}}{\omega_{\alpha\alpha'} + \Omega_j} + \frac{1 - \exp\{i(\omega_{\alpha\alpha'} + \Omega_j)\Delta T_{jq}\}}{(\omega_{\alpha\alpha'} + \Omega_j)^2} \right]. \tag{5.13}$$

Assuming for simplicity the distribution $P(\Omega_j)$ of the frequencies $\Omega_j$ to be narrow enough, $P(\Omega_j) \propto \delta(\Omega - \Omega_j)$, averaging over the $\Delta T_{jq}$ distribution, and putting $E_0^2 = \sum_{j=1}^{J} E_j^2$, we get from (5.13)

$$[\hat{T}_l(\tau, 0) - 1]^{(2)}_{\alpha\alpha} = \left\{ -\frac{E_0^2}{12\hbar^2} \sum_{\alpha'} |d_{\alpha\alpha'}|^2 \right.$$

$$\left. \times \left[ \frac{1}{\gamma_p - i(\omega_{\alpha\alpha'} - \Omega)} + \frac{1}{\gamma_p + i(\omega_{\alpha\alpha'} + \Omega)} \right] \right\} \tau. \tag{5.14}$$

It can be seen from (5.14) that, just as the usual impact theory, the increment of the evolution operator on the interval $\tau \gg \tau_c$ is found to be proportional to $\tau$, so that

$$[T_l(\tau, 0)]_{\alpha\alpha} = \exp[\Phi^c_{\alpha\alpha}(F)\tau], \tag{5.15}$$

where $\Phi_{\alpha\alpha}{}^c(F) = \Gamma_\alpha{}^c(F) + iD_\alpha{}^c(F)$ is the operator of impact broadening by collective HF oscillations of the plasma. The real and imaginary parts of this operator determine the width $\Gamma_\alpha{}^c$ and the shift $D_\alpha{}^c$ of the line:

$$\Gamma^c_\alpha(F) = \frac{E_0^2 \gamma_p}{12\hbar^2} \left[ \frac{2|d_{\alpha\alpha}|^2}{\gamma_p^2 + \Omega_p^2} + (|d_{\alpha\alpha-1}|^2 + |d_{\alpha\alpha+1}|^2) \right.$$

$$\left. \times \left( \frac{1}{\gamma_p^2 + (\omega_F - \Omega_p)^2} + \frac{1}{\gamma_p^2 + (\omega_F + \Omega_p)^2} \right) \right], \tag{5.16}$$

*The HF-noise distribution is assumed isotropic, so that $(\mathbf{d}_{\alpha\alpha'} E_{jq}{}^2 (\mathbf{d}_{\alpha'\alpha} E_{jq}) = |\mathbf{d}_{\alpha\alpha'}|^2 E_{jq}{}^2/3$. Here and elsewhere, we write only the diagonal matrix elements of the operator $\hat{T}(0, \tau)$, since they play the principal role far from the line center.

$$D_\alpha^c(F) = \frac{F_0^2}{12\hbar^2}\left(|\mathbf{d}_{\alpha\alpha-1}|^2 - |\mathbf{d}_{\alpha\alpha+1}|^2\right)\left[\frac{\omega_p - \Omega_p}{\gamma_p^2 + (\omega_p - \Omega_p)^2} + \frac{\omega_p + \Omega_p}{\gamma_p^2 + (\omega_p + \Omega_p)^2}\right], \quad (5.17)$$

$$\mathbf{d}_{\alpha\alpha}^2 = e^2 a_0^2 \frac{9}{4} n^2 (n_1 - n_2)^2, \qquad |\mathbf{d}_{\alpha\alpha-1}|^2 - |\mathbf{d}_{\alpha\alpha+1}|^2 = \mathbf{d}_{\alpha\alpha}^2/(n_1 - n_2),$$

$$|\mathbf{d}_{\alpha\alpha-1}|^2 + |\mathbf{d}_{\alpha\alpha+1}|^2 = \frac{\mathbf{d}_{\alpha\alpha}^2\left[n^2 - (n_1 - n_2) - m^2 - 1\right]}{2(n_1 - n_2)^2}.$$

It can be seen from (5.14) and (5.15) that the criterion of the validity of the impact theory is, as above, the possibility of choosing an interval $\Delta\tau$, that is, on the one hand large ($\Delta\tau \gg \tau_c$) and on the other small compared with the characteristic time of variation $\tau_{ef} \sim |\Gamma_\alpha{}^c|^{-1}$ of the evolution operator itself. Compatibility of these conditions leads to the criterion (5.9).

The action of the HF noise on an atom in a plasma thus leads to an increase of the impact width of the line. For the central components of Lyman lines the "collective" width $\Gamma_\alpha{}^c$ is of the order

$$\Gamma_\alpha^c \approx \frac{2\gamma_p\, \varepsilon_\alpha^2}{\Omega_p^2}, \qquad \varepsilon_\alpha^2 = \frac{\left(|\mathbf{d}_{\alpha\alpha-1}|^2 + |\mathbf{d}_{\alpha\alpha+1}|^2\right) E_0^2}{12\hbar^2}. \quad (5.18)$$

The central-component profile is Lorentzian with a width $\Gamma$ equal to the sum of the collision $\Gamma^e$ (electronic) and collective $\Gamma^c$ widths.

Dips on Hydrogen-Line Profiles. A distinctive property of the collective width $\Gamma^c$ is its resonant character. Indeed, when the sideband-component Stark-splitting frequency becomes resonant with the HF noise frequency $\Omega_p$, the line width increases sharply. Therefore, according to (5.8), the atoms located in static fields $F^* = \hbar\Omega_p / |\mathbf{d}_{\alpha\alpha} - \mathbf{d}_{\alpha-1\,\alpha-1}|$ emit a wider (compared to other atoms) Lorentz profile, which is therefore less intense in its central part. The envelope of the profile has thus an intensity dip near the frequency $\omega = -\mathbf{d}_{\alpha\alpha}F^*/\hbar \equiv \omega_1$. The frequencies corresponding to the appearance of intensity dips are equal to, with allowance for the broadening of both the upper ($\alpha$) and lower ($\beta$) levels [59],

$$\left.\begin{aligned} \omega_1 &= \left[(n_1 - n_2)_\alpha - \frac{n_b}{n_a}(n_1 - n_2)_\beta\right]\Omega_p, \\ \omega_2 &= \left[\frac{n_a}{n_b}(n_1 - n_2)_\alpha - (n_1 - n_2)_\beta\right]\Omega_p = \frac{n_a}{n_b}\,\omega_1. \end{aligned}\right\} \quad (5.19)$$

The described picture of the dips was obtained in the impact approximation corresponding to the condition (5.9). Actually, cases of strong turbulent fields are frequently encountered, with the inequality sign in the condition (5.9) reversed. It can then be seen from (5.12) that the characteristic variation time of the evolution operator turns out to be of the order of $|\mathbf{d}_{\alpha\alpha'}\mathbf{E}_0|^{-1} \ll \tau_c \sim \gamma_p^{-1}$. Therefore, in contrast to the impact approximation, the time dependences of the phase and amplitude of the HF noise can be disregarded here. The character of the evolution of the atom is easily clarified by using a resonance approximation that meets the condition

$$|\omega_{\alpha\alpha'} - \Omega_p| \ll \varepsilon_\alpha. \tag{5.20}$$

To this end, we note that the system (5.12) of the equations relating the quantities $T_{\alpha\alpha'}$ includes a relatively independent block of equations of the type

$$i\hbar\dot{T}_{\alpha\alpha} = \frac{1}{2}\mathbf{E}\,(\mathbf{d}_{\alpha\alpha-1}T_{\alpha-1\alpha}e^{-i\varphi} + \mathbf{d}_{\alpha\alpha+1}T_{\alpha+1\alpha}e^{+i\varphi}), \tag{5.21}$$

$$i\hbar\dot{T}_{\alpha-1\alpha} = \frac{1}{2}\mathbf{E}\,(\mathbf{d}_{\alpha-1\alpha}T_{\alpha\alpha}e^{+i\varphi} + \mathbf{d}_{\alpha-1\,\alpha-1}T_{\alpha-1\alpha}e^{-i\varphi}), \tag{5.22}$$

$$i\hbar\dot{T}_{\alpha+1\alpha} = \frac{1}{2}\mathbf{E}\,(\mathbf{d}_{\alpha+1\alpha}T_{\alpha\alpha}e^{-i\varphi} + \mathbf{d}_{\alpha+1\,\alpha+2}T_{\alpha+2\alpha}e^{+i\varphi}), \tag{5.23}$$

where the subscripts $\alpha + k$ denote states separated by the energy splitting $k(3/2)n(\hbar/me)F$.

Substituting (5.22) and (5.23) in the differentiated equation (5.21) and averaging over the random phases $\varphi$, we get

$$-\hbar^2\ddot{T}_{\alpha\alpha} = \frac{1}{4}[(\mathbf{d}_{\alpha\alpha-1}\mathbf{E})(\mathbf{d}_{\alpha-1\alpha}\mathbf{E}) + (\mathbf{d}_{\alpha\alpha+1}\mathbf{E})(\mathbf{d}_{\alpha+1\alpha}\mathbf{E})]\,T_{\alpha\alpha}, \tag{5.24}$$

from which it follows that the solution oscillates at a frequency

$$\varepsilon_{n_1n_2m} = \frac{E_0}{\hbar\sqrt{12}}\left[\left|\mathbf{d}^{n_1n_2m}_{n_1'n_2'm-1}\right|^2 + \left|\mathbf{d}^{n_1n_2m}_{n_1''n_2''m+1}\right|^2\right]^{1/2}$$
$$= (\sqrt{6}/8\pi)(ea_0E_0/\hbar)\,[n^2 - (n_1 - n_2)^2 - m^2 - 1]^{1/2}. \tag{5.25}$$

These oscillations lead to an additional splitting of each Stark component into two sublevels separated in frequency by $2\varepsilon_\alpha$. What is essentially realized here is a quasi-energy spectrum of a two-level system acted upon by a resonant monochromatic field [60].

The presence of the additional splitting leads, obviously, also to formation of dips on the hydrogen-line profiles under the resonance conditions (5.20). Since the formation of such dips is due, as above, to resonance between the HF-field frequency and the frequency of the Stark splitting in LF fields, the dip locations are determined by the previous equations (5.19). The half-width $\Delta\omega_{1/2}$ of these dips depends on the splitting:

$$\Delta\omega_{1/2} = \omega_\nu \mathcal{E}_\nu / \Omega_p, \qquad \nu \equiv \alpha, \beta\,; \tag{5.26}$$

i.e., it is proportional to the HF-noise amplitude $E_0$.

It is thus easy to determine the amplitude $E_0$ of the Langmuir noise (in kV/cm) from the measured dip width $\Delta\lambda_{1/2}$ for a line of wavelength $\lambda_0$:

$$E_0 = \frac{7.6\cdot 10^9\,[\Delta\lambda_{1/2},\ \text{Å}]}{\left[n_\nu^2 - (n_1 - n_2)_\nu - m_\nu^2 - 1\right]^{1/2} \left| n_\alpha (n_1 - n_2)_\alpha - n_\beta (n_1 - n_2)_\beta \right| [\lambda_\theta\ \text{Å}]}\,. \tag{5.27}$$

The dip positions calculated from Eqs. (5.19) agree well with the dips experimentally observed on turbulent-plasma hydrogen-line profiles [61, 62] (see also Section 8).

## 6. Transition from the Impact to the Static Broadening Mechanism

The relation between the impact and static mechanisms of broadening was considered in detail in Section 3, where the character of the time evolution of the atom was analyzed. Since the line contour is obtained, according to (1.13), by simply taking the inverse Fourier transform of the evolution operator, the question of the relation between the static and impact regions in the line spectrum is solved on the basis of a simple "uncertainty relation" in the Fourier analysis: $\Delta\omega\Delta\tau \sim 1$. Thus, the static approximation corresponds to short times $\tau \ll \tau_c$, or to large frequency shifts $\Delta\omega \gg \tau_c^{-1}$, i.e., to the line wings. The impact approximation, on the contrary, is valid for long times $\tau \gg \tau_c$ and, accordingly, to small frequency shifts, i.e., near the line center $\Delta\omega \ll \tau_c^{-1}$.

We wish to call attention here to the fundamental importance of this question for a correct construction of the convolution of the static-ion and impact-electron broadening. In fact, if we use for the construction of the total hydrogen-line contour a pure Lorentzian contour for the electron broadening, it is easy to perceive that in the far wing (i.e., as $\Delta\omega \to \infty$) this contour breaks up into two terms:

$$I(\omega) \underset{\Delta\omega\to\infty}{\to} \frac{NC_2^{3/2}}{\Delta\omega^{5/2}} + \frac{1}{\pi}\frac{\gamma}{\Delta\omega^2}. \qquad (6.1)$$

The first term in (6.1) corresponds to binary static broadening by ions, and the second to impact broadening by electrons. Comparing both terms, we verify that the second begins to prevail over the first at $\Delta\omega \gtrsim v_e^2/C_2 = \tau_c^{-1}$. But this is just the region where the electrons cease to cause broadening by impact. A correct construction of the contour in the asymptotic region therefore requires allowance for the replacement of the impact broadening mechanism by the static one for electrons in the far wing of the line.

It is customary to describe the transition to static broadening by the so-called relative number R of statically broadening particles:

$$R(\Delta\omega) = I(\omega)/I_{CT}(\omega) = I(\omega)\,\Delta\omega^{5/2}/NC_2^{3/2}. \qquad (6.2)$$

Only the ions are static at R = 1, and both electrons and ions at R = 2.

It is particularly easy to track the transition from the impact to the static limit in the binary adiabatic model of the broadening. Indeed, taking the Fourier transform of the correlation function (3.19) and integrating once by parts, we get

$$I(\omega) = \frac{1}{\pi}\,\mathrm{Re}\left[-\frac{1}{i\Delta\omega} + \frac{N}{i\Delta\omega}\int_0^\infty d\tau e^{-i\Delta\omega\tau} V'(\tau)\, e^{-NV(\tau)}\right]. \qquad (6.3)$$

The characteristic scale of the change of the functions $V(\tau)$ and $V'(\tau)$, as already noted, is the quantity $\Omega^{-1}$. We subject (6.3) to an identity transformation by adding to and subtracting from $V'(\tau)$ its value $V'(\infty)$ [see (3.27)]. One of the resultant integrals reduces then to $I(\omega)$, while in the second we can put $e^{-NV(\tau)} \approx 1$. Indeed, this integral includes in its integrand the difference $V'(\tau) - V'(\infty)$, which vanishes at $\tau \gtrsim \Omega^{-1}$, but it is just in this region that $NV(\tau)$ is small; $N|V(\tau \lesssim \Omega^{-1})| \sim N|V'(\infty)|\Omega^{-1} = g \ll 1$, which coincides precisely with the parameter g that determines the validity of the binary approximation. Further simple transformations yield [28, 63]

$$I(\omega)=\frac{1}{\pi}\frac{\gamma(\omega)}{\Delta\omega^2+\gamma^2(0)}, \tag{6.4}$$

$$\gamma(\omega)=N\,\mathrm{Re}\int_0^\infty d\tau e^{i\Delta\omega\tau}V''(\tau)=\pi N v\int_0^\infty d\rho\rho\left|\int_{-\infty}^{\infty}d t\varkappa(t)\exp\left[i\,\Delta\omega t-i\int_0^t d\tau\varkappa(\tau)\right]\right|^2. \tag{6.5}$$

It follows from (6.4) and (6.5) that in the binary approximation $g \ll 1$ the line contour has a Lorentzian structure, except that the width of the contour is *not* constant but is itself dependent on $\Delta\omega$. The scale of the change of this "variable" width is the Weisskopf frequency $\Omega$, as follows from obvious considerations connected with the "uncertainty relation" of the Fourier analysis of the collision volume $V(\tau)$. It is therefore convenient to transform to a dimensionless frequency $x = \Delta\omega/\Omega$, with the dimensionless contour $I(x)$ given by $I(x)dx = I(\omega)d\omega$. It then follows from (6.5) that

$$I(x)=\frac{1}{\pi}\frac{g\gamma(x)}{x^2+g^2\gamma^2(x)}. \tag{6.6}$$

At $x \ll 1$ the variable width $\gamma(x)$ attains a constant value $\gamma(0)$ that coincides with the impact adiabatic line width.

At $x \gg 1$, the change of $\gamma(x)$ is such that $I(x)$ is certain to reach its static limit. For the case of the linear Stark effect (hydrogen lines), the function $\gamma(x)$ has an explicitly analytic form [12]:

$$\gamma(x)=\pi\int_0^\infty\frac{dy}{y}k_{1/y}^2(2xy),\quad k_\nu(z)=\frac{W_{\frac{\nu}{2},\frac{1}{2}}(2z)}{\Gamma\left(1+\frac{\nu}{2}\right)}, \tag{6.7}$$

where $W_{\lambda,\nu}(z)$ is a confluent hypergeometric function (the Whittaker function [64]), and $\Gamma(z)$ is the gamma function.

The limiting values of $\gamma(x)$ are

$$\gamma(x)\approx\begin{cases}-\pi^3[\ln x+C]\equiv\gamma_{\mathrm{imp}}, & x\ll 1,\\ 2\pi^2/\sqrt{x}\equiv\gamma_{\mathrm{st}}(x), & x\gg 1.\end{cases} \tag{6.8}$$

Substitution of (6.8) in (6.6) ensures that an impact Lorentz formula with constant width $\gamma_{imp}$ is obtained* at $x \ll 1$ and that the result of the binary static theory (4.5) is obtained at $x \gg 1$. Equation (6.6) ensures thus a smooth transition from the impact to the static broadening limit. This transition occurs at $x \sim 1$, when the term $g^2\gamma^2(x) \ll 1$ in the denominator of (6.6) can already be disregarded. With the aid of the function $\gamma(x)$, the parameter R in (6.7) takes, obviously, the form

$$R(\Delta\omega) = 1 + \frac{\gamma(\Delta\omega/\Omega)}{\gamma_{st}(\Delta\omega/\Omega)}, \quad \gamma_{st}(z) = \frac{2\pi^2}{\sqrt{z}}. \tag{6.9}$$

Equation (6.9), with allowance for (6.8), describes the monotonic increase of the fraction of the statistically broadening particles on moving toward the line wing. The use of the variable width $\gamma(\omega)$ obviously eliminates the incorrect asymptotic form (6.1) of the total line contour.

The foregoing results were based on the adiabatic model. However, as shown in Section 3, the evolution of the hydrogen states in the exact nonadiabatic theory is similar in many respects to the adiabatic theory. This is due to the specific features of the hydrogen degeneracy. The transition from the impact to the static broadening in the exact theory therefore meets only with complications of a technical nature, connected with allowance for the nonadiabatic coupling of a large number of degenerate states. Thus, the use of the exact wave functions (3.54) and (3.55) yields for the $L_\alpha$ line contour the expression

$$I(\omega) = \frac{9}{\pi} N \frac{(\hbar/m)^2}{v} \left[\gamma\left(\frac{3\Delta\omega\hbar}{mv^2}\right) + \gamma\left(-\frac{3\Delta\omega\hbar}{mv^2}\right)\right] \Delta\omega^{-2}, \tag{6.10}$$

where $\gamma(x)$ is the variable line width and can, just as (6.7), be expressed analytically:

*The logarithmic divergence of $\gamma(x)$ as $x \to 0$ ($\Delta\omega \to 0$) is actually immaterial, since no account was taken above of the Debye screening; when this screening is allowed for, $\Delta\omega$ in (6.8) must be replaced by the plasma frequency $\omega_p$ at $\Delta\omega < \omega_p$ (see [1, 2]). Note, however, that at $\Delta\omega > \omega_p$ the upper cutoff length in (6.8) is the so-called Lewis radius $\rho_L \sim v/\Delta\omega$, the appearance of which is due to the long-range character of the Coulomb interaction (3.7); see Section 3 above.

$$\gamma(x)=\frac{\pi^3}{2}\int_0^\infty \frac{dt}{t(1+t^2)}\left[k^2_{\sqrt{1+t^{-2}}-1}(xt)+k^2_{\sqrt{1+t^{-2}}+1}(xt)+2t^2k_1^2(xt)\right]. \quad (6.11)$$

The limiting values of $\gamma(x)$ are given by

$$\gamma(x)\approx\begin{cases}-4\pi\ln x\equiv\gamma_{\text{imp}}, & x\ll 1,\\ 2\pi^2/\sqrt{x}=\gamma_{\text{st}}, & x\gg 1.\end{cases} \quad (6.12)$$

The largest difference (by a factor $\pi^2/4$) from the results (6.8) of the adiabatic model are observed in the impact region $x \ll 1$, whereas the static limits ($x \gg 1$), where the influence of the nonadiabaticity effects is negligible, are equal.

In the intermediate region $x \sim 1$ the difference between the two functions does not exceed 20%. This is of great practical importance for the description of contours of complex lines, since the results of the adiabatic theory are universal, while those of the complete nonadiabatic theory require separate calculations for each line. On the whole, the adiabatic theory of broadening describes well the contours of the hydrogen lines at frequencies of the transition region [28]; see also Section 8.

The quantum approach to the calculation of line contours, based on the general method first proposed by Born (see [65], Chap. 22, Section 9) for the calculation of the bremsstrahlung spectra, was further developed in [66-70]. The gist of the method reduces to consideration of emission of a photon not by a single atom but by the entire interacting system "atom + broadening particle" (the particle, for the sake of argument, can be an electron). In this general approach the line contour can be expressed in terms of the cross-section for photon emission when the broadening electron collides with an atom.

## 7. Comparison of Theory with Experiment. Conclusions

We report here some experimental data that, on the one hand, permit assessment of the accuracy of modern theory of broadening and, on the other, illustrate a number of basic premises of the physics of line broadening in a plasma.

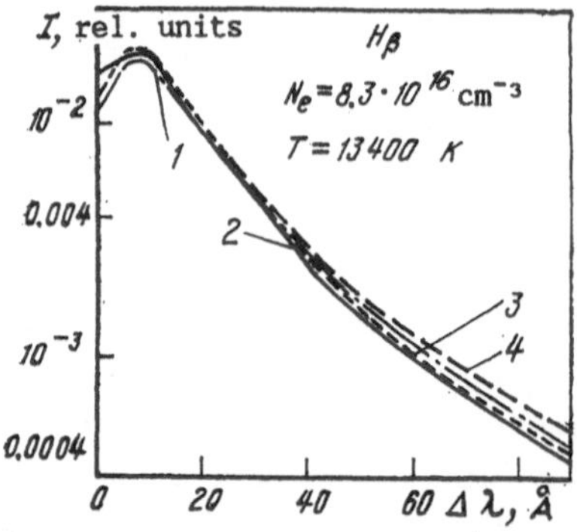


Fig. 5. Hβ line contour: 1) experiment ("red" wing of line); 2) "blue" wing; 3) calculations [57]; 4) calculations [58].

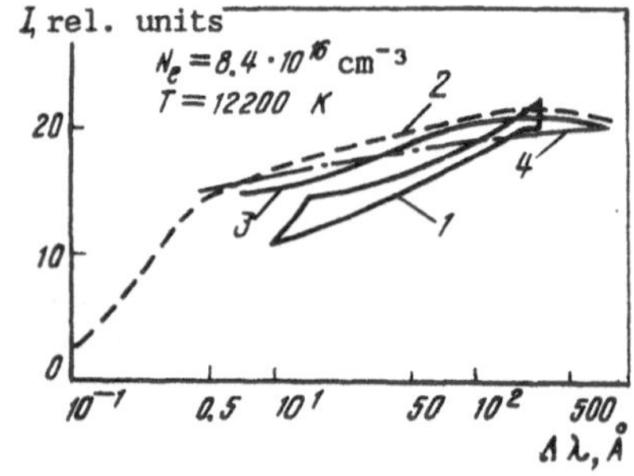


Fig. 6. Lα line contour in a far wing: 1) experiment [71]; 2) calculations [58]; 3) calculations [34]; 4) adiabatic model [28].

Figure 5 shows the contour of the Hβ hydrogen line, measured by Wiese et al. [47] and calculated theoretically [57, 58]. It can be seen that, on the whole, the calculations agree well with experiment, except in the central part of the line contour. This divergence is due apparently to thermal motion of the ions, not accounted for in the calculations. The line half-width is determined mainly by the static (Holtsmark) broadening produced by the plasma ions. The role of the electrons is manifested in the fact that the intensity dip at the line center is not too large (in the case of pure static broadening the intensity at the center is at any rate zero). It can also be seen that the line contour is somewhat asymmetric.

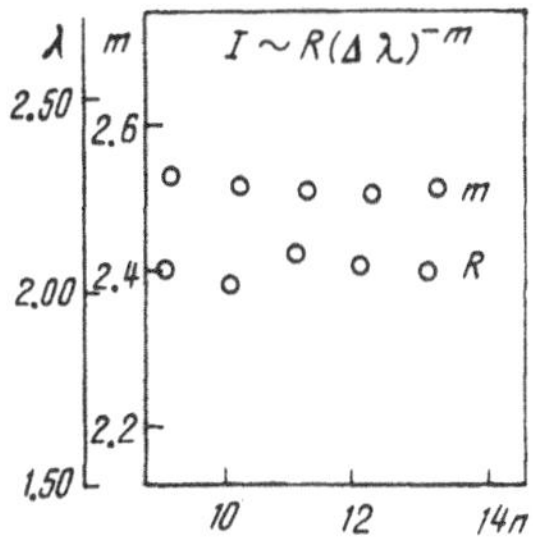


Fig. 7. Relative number R of statically broadening particles and the exponent m, which determine the law governing the intensity decrease in the line wing for higher terms of the Balmer series [75].

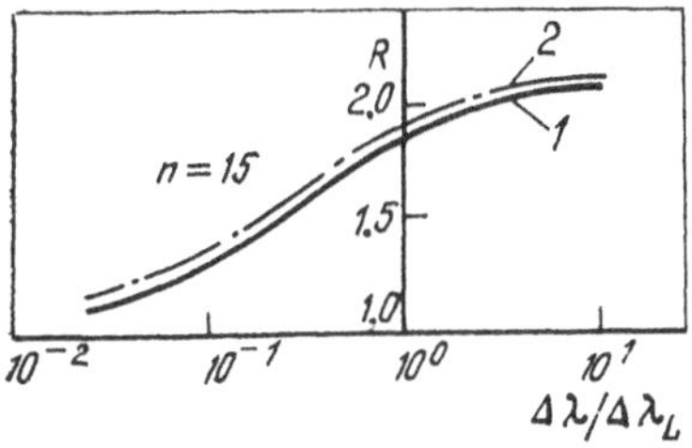


Fig. 8. Number $R(\Delta\lambda)$ for $H_{15}$ line: 1) experiment [75]; 2) calculation [28].

The electrons influence not only the central part of the line, but also the far wings not shown in Fig. 5. These wings correspond to static broadening by ions as well as by electrons. Static broadening by electrons should manifest itself, according to the general results of the theory, at $\Delta\omega \gtrsim v_e^2/C_2 \equiv \Omega_{Be}$. This region can be observed either in the far wings of the line (large $\Delta\omega$) or for highly excited levels with large enough Stark constant $C_2 \sim n^2$ (small $\Omega_{Be}$). Both possibilities were implemented in experiment. The far wings of the $L_\alpha$ line were observed by Boldt and Cooper [71], Elton and Griem [72], and Preston [73]. Highly excited Balmer-series lines ($H_8$-$H_{15}$) were observed by Schlüter and Avic [74, 75] and by Himmel [76]. The results of both

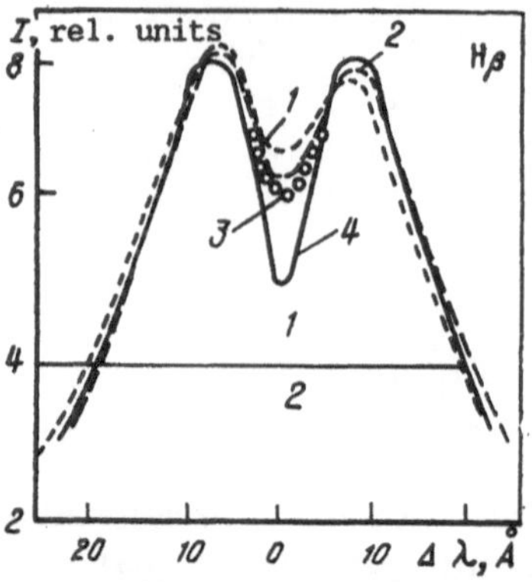


Fig. 9. Effect of "reduced mass" for the $H_\beta$ line [47]: 1) $\mu = 0.5$; 2) $\mu = 1$; 3) $\mu = 2$; 4) theory [57].

types of experiment are conveniently represented in the form of a parameter $R(\Delta\lambda)$ that determines the relative number of statically broadening particles.

Figure 6 shows the results [71, 72] for the $L_\alpha$ line wing together with the theoretical calculations [28, 34]. It can be seen that the farther along the line wing the greater the fraction of electrons that cause static broadening. At $\Delta\lambda \sim 50$ Å, the influence of almost all particles is described by the static theory ($R \approx 2$).

In Schlüter's experiments [75], the borderline $\Delta\lambda_e$ for the transition to the static broadening for electrons is well described by the equation

$$\Delta\lambda_e = \frac{0.62 \cdot 10^{12} v_e^2 \lambda}{n(n-1)} . \tag{7.1}$$

According to the static theory of broadening, the intensity distribution in the line wing should take the form

$$I(\Delta\lambda) \sim R(\Delta\lambda)^{-m},$$

where $R \approx 2$, and the exponent is $m = 5/2$.

Figure 7 shows the observed values of R and m as functions of the principal quantum number n of the upper level. These values are close to the theoretical predictions.

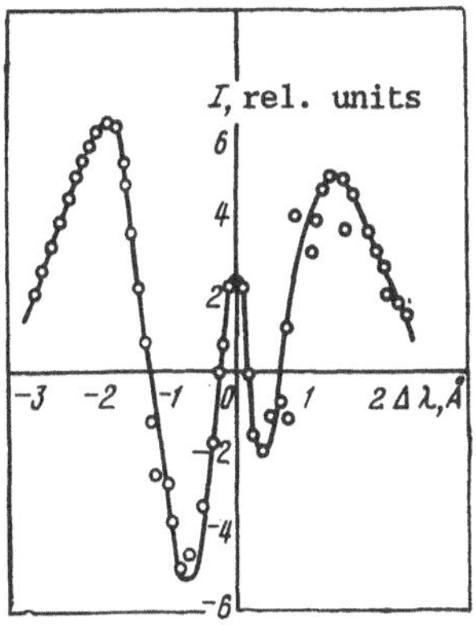


Fig. 10. Experimentally observed [22] difference contour of the $H_\alpha$ line.

Figure 8 demonstrates the behavior of the parameter $R(\Delta\lambda/\Delta\lambda_e)$ for the $H_{15}$ line, both measured and theoretically calculated. The agreement between the two values is quite satisfactory.

Recent experiments (primarily by Wiese et al. [47]) revealed effects connected with the thermal motion of plasma ions. A dependence of the depth of the dip at the center of the $H_\beta$ line on the reduced mass $\mu$ of the "radiating atom + perturbing ion" pair was detected (Fig. 9). The broadening of hydrogen (H) and deuterium (D) by $H^+$, $D^+$, and $Ar^+$ ions ($\mu_{HH} = 0.5$, $\mu_{HAr} \approx 1$, $\mu_{DAr} \approx 1.9$) was investigated. The dependence of the dip depth on $\mu$ and the line shape near the center are satisfactorily described by the thermal-correction theory developed in Section 4; see [46]. For a line with central components ($H_\alpha$, $H_\gamma$) there was also observed a reduced mass effect, with decrease of $\mu$ (increase of the thermal-motion intensity), viz., the maximum line intensity also decreases, while the half-width increases [47]. A detailed explanation of this effect, based on allowance for the thermal-motion effects described in Section 4, can be found in Griem's paper [77].

We proceed to discuss experiments in a turbulent plasma. In the case of low-frequency ion-sound oscillations, the measured spectral-line contours in a plasma with a high noise level agree well with the static theory of broadening, in which a convolution of Rayleigh and Holtsmark distribu-

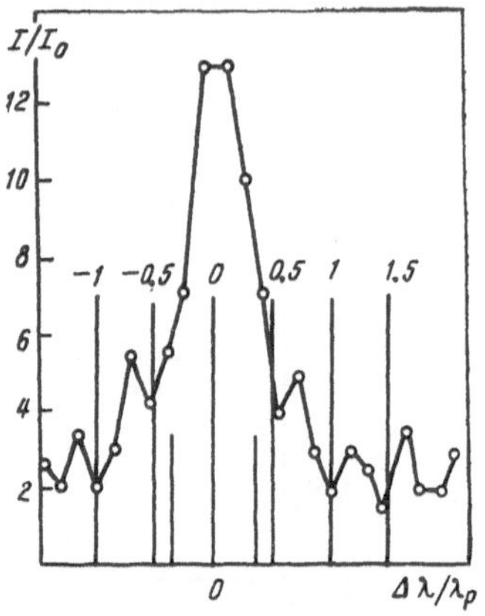


Fig. 11. Formation of dips on the contour of the $D_\alpha$ line in a plasma with Langmuir turbulence [62].

tions is used; see Sections 2 and 4. An anomalous (compared with the pure Holtsmark) increase of the line width in turbulent-oscillation fields was observed [62, 78], as well as the power-law decrease of the intensity in the line wings [79], a decrease typical of such a convolution [80].

A series of experiments [22] has revealed also anisotropy of the LF noise distribution. The measurements were based on the use of the polarization procedure described in Section 4. The difference contour of the line (obtained by subtracting the observed contours with different polarizations) agrees well with the theoretical predictions (Fig. 10).

In the case of high-frequency (Langmuir) turbulence, besides the effect of additional broadening of lines with strong central components [61, 62], characteristic "reliefs" were observed in experiment on the main "body" of the line [61, 62]. The origin of these reliefs, which lead to the appearance of numerous intensity maxima and minima,* is attributed by the theory, according to Section 5, to resonances between the HF-field frequency and the frequencies of the splitting of the Stark components in LF fields. Fig-

*With increasing noise intensity, the line contour can become so jagged that it is difficult even to determine its half-width [62].

ure 11 shows the experimental $D_\alpha$ profile obtained in compression of a Z pinch [62] past the second singularity of the current. In accord with the expected growth of the turbulent-noise intensity on compression of a Z pinch, the line contour becomes more and more jagged. The vertical lines in Fig. 11 designate the dip positions calculated theoretically from Eqs. (5.18); the numbers above these lines are the distances of the dips from the line center in units of $\lambda_p = \omega_p \lambda_0{}^2/2\pi c$ ($\omega_p$ is the plasma frequency). It can be seen that the theoretically calculated locations of the dips agree well with the experimental ones. A similar agreement was also noted for other lines [62].

## REFERENCES


1. I. I. Sobelman, Introduction to the Theory of Atomic Spectra, Pergamon, Oxford (1973).
2. H. R. Griem, Spectral Line Broadening in Plasmas, Academic Press, New York (1978).
3. L. A. Vainshtein, I. I. Sobel'man, and E. A. Yukov, Excitation of Atoms and Broadening of Spectral Lines [in Russian], Nauka, Moscow (1979).
4. Yu. L. Klimontovich, Kinetic Theory of Electromagnetic Processes [in Russian], Nauka, Moscow (1980).
5. NBS Special Publication 336 (1972) [Suppl. I (1973), Suppl. II (1975)].
6. J. Holtsmark, Ann. Phys., 58, 577 (1919).
7. S. Chandrasekhar, Rev. Mod. Phys., 15, 1 (1943).
8. C. G. Hopper, Jr., Phys. Rev., 165, 215 (1968).
9. L. P. Kudrin, Statistical Physics of Plasma [in Russian], Atomizdat, Moscow (1973).
10. G. Ecker, Theory of Fully Ionized Plasmas, Academic Press, New York (1971).
11. V. I. Kogan and A. D. Selidovkin, Beitr. Plasmaphysik, 9, 199 (1969).
12. V. S. Lisitsa, Usp. Fiz. Nauk, 122, 449 (1977).
13. V. I. Kogan, in: Plasma Physics and the Problem of Controlled Thermonuclear Reactions, Vol. 4, M. A. Leontovich (ed.) [in Russian], Izd. Akad. Nauk SSSR, (1958).
14. J. D. Hey, J. Quant. Spectrosc. Radiat. Transfer, 16, 947 (1976).
15. D. Bohm and D. Pines, Phys. Rev., 82, 625 (1951); 92, 609 (1953); 92, 626 (1953).
16. G. Ecker and K. Fischer, Z. Naturforsch., 26a, 1360 (1971).

17. G. Ecker and K. Müller, Z. Phys., 153, 317 (1958).
18. G. V. Sholin, Dokl. Akad. Nauk SSSR, 195, 589 (1970).
19. E. A. Oks and G. V. Sholin, Zh. Tekh. Fiz., 46, 254 (1976).
20. G. V. Sholin and E. A. Oks, Dokl. Akad. Nauk SSSR, 209, 1318 (1973).
21. E. K. Zavoiskii, Yu. G. Kalinin, E. A. Skoryupin, et al., Pis'ma Zh. Eksp. Teor. Fiz., 13, 19 (1971).
22. M. V. Babykin, A. N. Zhuzhunashvili, E. A. Oks, et al., Zh. Eksp. Teor. Fiz., 65, 175 (1973).
23. S. M. Rytov, Introduction to Statistical Radiophysics [in Russian], Nauka, Moscow (1966).
24. S. G. Rautian, G. I. Smirnov, and A. M. Shalagin, Nonlinear Resonances in the Spectra of Atoms and Molecules [in Russian], Nauka, Novosibirsk (1979).
25. M. I. Podgoretskii and A. V. Stepanov, Zh. Eksp. Teor. Fiz., 40, 561 (1961).
26. D. D. Burgess, in: Lectures at the 11th International Conference on Phenomena in Ionized Gases, Ed. Yanev, Belgrade (1975).
27. P. W. Anderson, Phys. Rev., 86, 809 (1952).
28. V. I. Kogan and V. S. Lisitsa, J. Quant. Spectrosc. Radiat. Transfer, 12, 881 (1972).
29. H. R. Griem, M. Baranger, A. C. Kolb, and G. Oertel, Phys. Rev., 125, 177 (1962).
30. L. D. Landau and E. M. Lifshitz, Quantum Mechanics, Nonrelativistic Theory, Pergamon, Oxford (1975).
31. Yu. N. Demkov, B. S. Monozon, and V. N. Ostrovskii, Zh. Eksp. Teor. Fiz., 57, 1431 (1969).
32. Yu. N. Demkov, V. N. Ostrovskii, and E. A. Solov'ev, Zh. Eksp. Teor. Fiz., 55, 126 (1974).
33. V. B. Berestetskii, E. M. Lifshitz, and L. P. Pitaevskii, Quantum Electronics, Pergamon, Oxford (1971).
34. V. S. Lisitsa and G. V. Sholin, Zh. Eksp. Teor. Fiz., 61, 9 (1971).
35. R. L. Green, J. Coper, and E. W. Smith, J. Quant. Spectrosc. Radiat. Transfer, 15, 1025 (1975).
36. V. I. Kogan, Dokl. Akad. Nauk SSSR, 135, 1374 (1960).
37. M. Lewis, Phys. Rev., 121, 501 (1961).
38. A. B. Underhill and J. Waddell, NBS Circ. No. 603 (1959).
39. H. Griem, Astrophys. J., 132, 883 (1960).
40. E. K. Zavoiskii, Yu. G. Kalinin, V. A. Skoryupin, et al., Pis'ma Zh. Eksp. Teor. Fiz., 13, 19 (1971).
41. G. V. Sholin and E. A. Oks, Dokl. Akad. Nauk SSSR, 200, 1318 (1973).

42. W. Pauli, Papers on Quantum Theory [Russian translation], Nauka, Moscow (1975), p. 99.
43. T. Holstein, Phys. Rev., 79, 744 (1950).
44. H. K. Wimmel, J. Quant. Spectrosc. Radiat. Transfer, 1, 1 (1961).
45. G. V. Sholin, V. S. Lisitsa, and V. I. Kogan, Zh. Eksp. Teor. Fiz., 59, 1390 (1970).
46. A. V. Demura, V. S. Lisitsa, and G. V. Sholin, Zh. Eksp. Teor. Fiz., 73, 400 (1977).
47. W. L. Wiese, D. E. Kelleher, and V. Helbig, Phys. Rev., A11, 1854 (1975).
48. W. L. Wiese, D. E. Kelleher, and D. R. Paquette, Phys. Rev., 6, 1132 (1972).
49. G. V. Sholin, Opt. Spektrosk., 26, 289 (1969).
50. A. V. Demura and G. V. Sholin, J. Quant. Spectrosc. Radiat. Transfer, 15, 881 (1975).
51. L. P. Kudrin and G. V. Sholin, Dokl. Akad. Nauk SSSR, 197, 342 (1962).
52. J. C. Stewart et al., Astrophys. J., 179, 983 (1973).
53. Rang Le Quang and D. Voslamber, J. Phys., B8, 331 (1975).
54. G. V. Sholin, A. V. Demura, and V. S. Lisitsa, Zh. Eksp. Teor. Fiz., 64, 2097 (1973).
55. M. L. Strekalov and A. I. Burshtein, Zh. Eksp. Teor. Fiz., 61, 101 (1971).
56. H. Pfennig, Z. Naturforsch., 26a, 1071 (1971).
57. P. Kepple and H. R. Griem, Phys. Rev., 123, 317 (1968).
58. C. R. Vidal, J. Cooper, and E. W. Smith, Astrophys. J., 25, 37 (1973).
59. E. A. Oks and G. V. Sholin, Zh. Eksp. Teor. Fiz., 68, 975 (1975).
60. Ya. B. Zel'dovich, Usp. Fiz. Nauk, 110, 139 (1973).
61. A. I. Zhuzhunashvili and E. A. Oks, Zh. Eksp. Teor. Fiz., 73, 2142 (1977).
62. E. A. Oks and V. A. Rantsev-Kartinov, Zh. Eksp. Teor. Fiz., 79, 99 (1980).
63. V. V. Yakimets, Zh. Eksp. Teor. Fiz., 51, 1469 (1966).
64. A. Erdelyi (ed.), Higher Transcendental Functions, McGraw-Hill, New York (1953).
65. N. F. Mott and H. S. W. Massey, The Theory of Atomic Collisions, Clarendon Press, Oxford (1965).
66. A. Jablonski, Phys. Rev., 68, 78 (1945).
67. V. S. Lisitsa, Acta Phys. Pol., A55, 87 (1978).
68. F. F. Baryshnikov and V. S. Lisitsa, Zh. Eksp. Teor. Fiz., 72, 1797 (1977).

69. N. Tran Minh, N. Feautrier, and H. van Regemorter, J. Quant. Spectrosc. Radiat. Transfer, 16, 849 (1976).
70. F. F. Baryshnikov and V. S. Lisitsa, Zh. Eksp. Teor. Fiz., 80, 926 (1981).
71. G. Boldt and W. B. Cooper, Z. Naturforsch., 19a, 968 (1964).
72. R. C. Elton and H. R. Griem, Phys. Rev., 135, 1550 (1964).
73. R. Preston, J. Phys., B10, 523 (1977).
74. H. Schlüter and C. Avic, Astrophys. J., 144, 785 (1966).
75. H. Schlüter, J. Quant. Spectrosc. Radiat. Transfer, 18, 140 (1969).
76. G. Himmel, J. Quant. Spectrosc. Radiat. Transfer, 16, 529 (1976).
77. H. R. Griem, Phys. Rev., A20, 1114 (1980).
78. S. P. Zagorodnikov, Yu. E. Smolkin, E. A. Striganova, and G. V. Sholin, Dokl. Akad. Nauk SSSR, 195, 861 (1970).
79. H. R. Griem and H. J. Kunze, Phys. Rev. Lett., 14, 1261 (1959).
80. G. V. Sholin, Proceedings of the 15th International Conference on Phenomena in Ionized Gases, Invited Papers, Minsk, USSR, July 14-18, 1981, pp. 341-357.

# ELECTRON CYCLOTRON PLASMA HEATING IN TOKAMAKS

**A. D. Piliya and V. I. Fedorov**

## Introduction

It is universally accepted at present that additional plasma heating is needed to ignite a thermonuclear reaction in a tokamak. Highly promising for this purpose are assumed to be high-frequency methods based on feeding energy to the plasma in the form of electromagnetic waves. High-frequency (HF) heating is possible in various frequency bands, each with its own advantages and shortcomings [1-6].

We attempt in this paper a systematic elucidation of the theoretical problems encountered in research into plasma heating in tokamaks at the highest possible frequencies, namely, close to the electron-cyclotron frequency. For modern and future-generation tokamaks, this means millimeter wavelengths.

From the theoretical standpoint, the picture of excitation, propagation, and absorption of waves in electron-cyclotron plasma heating in tokamaks can be described in general outline within the framework of linear electrodynamics of a weakly inhomogeneous medium with spatial dispersion. From among the nonlinear effects, an important role can be played under the heating condition only by the distortion of the electron-distribution function in velocity. At the same time, the complicated and difficultly controllable parametric-turbulence methods will apparently play no significant role.

This circumstance raises hopes that the linear and quasilinear methods developed to date for the calculation of the wave-absorption efficiency, methods that take into account the real character of the plasma and magnetic-field inhomogeneities in tokamaks, yield reliable results. What makes these calculations unusual is that geometric-optics methods are used to describe wave phenomena in inhomogeneous media with strong spatial and temporal dispersion and with an essentially non-Hermitian dielectric tensor. An assessment of the validity of geometric optics under such conditions is one of the principal topics of this article.

## 1. Electromagnetic Waves in the Region of Electron-Cyclotron Resonance Frequencies

Before we proceed to a discussion of the influence of the real character of the inhomogeneity on wave propagation, we recall briefly the main data on wave properties in the homogeneous-plasma model.

We are interested in the frequency band close to the electron-cyclotron frequency $\omega_B(|\omega - \omega_B| \lesssim \omega_B)$. We disregard initially, however, the immediate vicinity of the cyclotron resonance, putting $|\omega - \omega_B| \gg k_\parallel v_T$, where $k_\parallel$ is the projection of the wave vector on the direction of the magnetic field $\mathbf{B}$, and $v_T = \sqrt{2T_e/m_e}$ is the thermal velocity of the electrons. If we assume, in addition, that $k_\perp v_T/\omega_B \ll 1$, where $k_\perp$ is the wave-vector component perpendicular to $\mathbf{B}$, we can neglect spatial dispersion and describe the waves in the cold-plasma approximation. In a Cartesian frame with z axis directed along $\mathbf{B}$, the dielectric tensor $\varepsilon_{\alpha\beta}$ takes the form

$$\begin{gathered}\varepsilon_{xx} = \varepsilon_{yy} = \varepsilon, \quad \varepsilon_{xy} = -\varepsilon_{yx} = ig, \quad \varepsilon_{zz} = \eta, \\ \varepsilon_{xz} = \varepsilon_{zx} = \varepsilon_{yz} = \varepsilon_{zy} = 0,\end{gathered} \tag{1}$$

where, neglecting the ion contribution,

$$\varepsilon = 1 - q/(1-u), \quad g = q\sqrt{u}/(1-u), \quad \eta = 1 - q, \tag{2}$$

$q = \omega_{pe}^2/\omega^2$, $u = \omega_B^2/\omega^2$, and $\omega_{pe}$ is the electron-plasma frequency. Besides the Cartesian components $E_\alpha$ of the electric field of the wave (where $\alpha$ takes the values x, y, and z), it is convenient to introduce "rotating" field components $E_1 = E_x - iE_y$ and $E_2 = E_x + iE_y$ with respective rotation directions toward the electrons and ions. In contrast to the Cartesian components, the "rotating" components will be designated by the Latin letter i or k (i, k = 1, 2, 3, with $E_3 \equiv E_z$). By analogy with $E_i$, we can introduce rotating components $D_i$ of the induction vector and the tensor $\varepsilon_{ik}$ defined by the relation $D_i = \varepsilon_{ik} E_k$.

The tensor $\varepsilon_{ik}$ is diagonal, with components

$$\begin{gathered}\varepsilon_{11} \equiv \varepsilon - g = 1 - \frac{q}{1-\sqrt{u}}, \\ \varepsilon_{22} \equiv \varepsilon + g = 1 - \frac{q}{1+\sqrt{u}}, \qquad \varepsilon_{33} \equiv \eta = 1 - q.\end{gathered} \tag{3}$$

It is known that the basic properties of the waves are determined by a dispersion equation, which takes in the cold-plasma approximation the form [7]

$$An^4 + Bn^2 + C = 0, \tag{4}$$

where $n = kc/\omega$ is the refractive index of the wave, $A = \varepsilon \sin^2\theta + \eta\cos^2\theta$, $B = -\varepsilon_{11}\varepsilon_{22}\sin^2\theta - \varepsilon\eta(1 + \cos^2\theta)$, $C = \eta\varepsilon_{11}\varepsilon_{22}$, $\theta$ is the angle between **k** and **B**.

Two solutions of the dispersion equation $n_{1,2}^2(\theta)$ describe two oscillation modes of a cold anisotropic plasma. These modes were named, in accord with the value of $n^2$ at $\theta = \pi/2$; viz., the wave with $n^2(\pi/2) = (\varepsilon^2 - g^2)/\varepsilon$ is called extraordinary (mode x) and that with $n^2(\pi/2) = \eta$ ordinary (mode 0).

The coefficients of the dispersion equation, and hence its solutions $n(\theta)$, depend on the plasma parameters and the frequency $\omega$ only in terms of the dimensionless quantities u and q. It is therefore convenient to interpret the solutions of the dispersion equation on the (u, q) plane (Fig. 1). The figure shows also the lines that demarcate regions with the topological structures of the $n_{0,x}(\theta)$ constant in the interior but capable of changing on the boundaries.* All these lines are special for the dispersion-equation coefficients. The coefficient C vanishes on the lines $\varepsilon_{11} = 0$ and $\varepsilon_{22} = 0$. The refractive index of one of the waves [the extraordinary one,

*The (u, q) plane is usually used to interpret much less convenient phase-velocity plots, i.e., $1/n(\theta)$ (called the Klemov–Melely–Allis diagram).

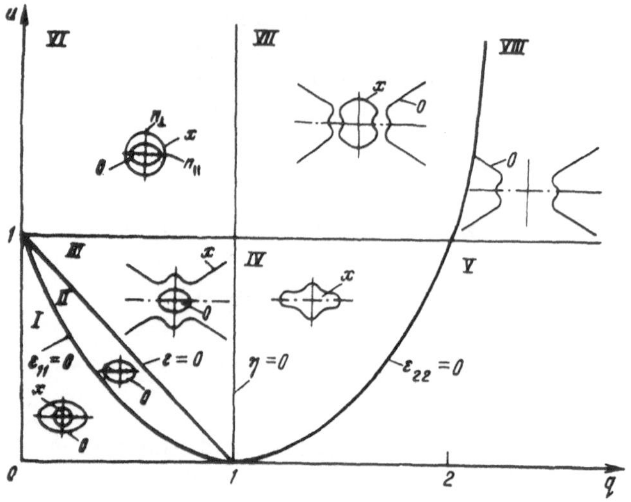


Fig. 1. Cutoff and resonance lines on the (u, q) plane.

as follows from the expression given for $n(\pi/2)$] vanishes, therefore, on this line regardless of the angle $\theta$, so that the $n_x(\theta)$ curve for this wave contracts to a point. This line is accordingly called the cutoff point of the extraordinary wave. Similarly, the critical-density line $q = 1$, on which $\eta = 0$, is the cutoff line for the ordinary wave. An exception here, however, is the angle $\theta = 0$, when $\eta$ turns out to be a common factor in (4), so that $n^2(0) \neq 0$ at $\eta = 0$. The plot of $n_0(\theta)$ of the ordinary wave degenerates thus into a segment of the z axis on the $q = 1$ line.

The line $\varepsilon = 0$, called the line of (upper, in this case) hybrid resonance, separates the region III in which $\varepsilon$ and $\eta$ have opposite signs, so that the coefficient A of the dispersion equation vanishes at a certain resonant value $\theta_r$ of the angle $\theta$:

$$\theta_r = \arctan\sqrt{\frac{-\eta}{\varepsilon}} = \arctan\sqrt{\frac{(1-u)(1-q)}{u+q-1}}.$$

The same situation obtains in regions VI and VII. (The second branch of the $\varepsilon = 0$ curve, the lower hybrid resonance line that demarcates these regions at large u, is outside the range of the considered parameters and is not shown in Fig. 1.) Clearly, as $\theta \to \theta_r$, the refractive index of one of the waves tends to infinity.* In region III this takes place for the extraordinary wave, and in this case $n_x^2(\theta) > 0$ in the angle interval $\theta_r \leq \theta \leq \pi/2$ and $n_x^2(\theta) < 0$ outside this interval. In regions VII and VIII the infinite refractive index occurs for the ordinary wave, which can propagate $[n_0{}^2(\theta) > 0]$ at $0 \leq \theta \leq \theta_r$.

Note that the ordinary wave in regions VII and VIII in the angle region close to $\theta_r$ (where $n_0{}^2 \gg 1$) is usually called oblique Langmuir wave (in the non-Soviet literature it is called the Trivelpiece–Gould mode), while at small angles it is called the whistler mode.

We now consider directly the vicinity of the cyclotron resonance, but disregard, for the time being, as before, the spatial dispersion. It can be seen from (2) and (3) that the "resonant" component $\varepsilon_{11}$ of the tensor $\varepsilon_{ik}$ has a singularity as $u \to 1$. Since $\varepsilon_{11}$ enters linearly in all three coefficients of the dispersion equation (4), it takes at $|u - 1| \ll 1$ the approximate form

$$\varepsilon_{11}\Lambda_0 = 0, \tag{5}$$

where

$$\Lambda_0 = \frac{1}{2}\sin^2\theta n^4 - \left[\varepsilon_{22}\sin^2\theta + \frac{1}{2}\eta(1+\cos^2\theta)\right]n^2 + \eta\varepsilon_{22}.$$

At $\theta \neq 0$ the refractive indices of both waves remain thus finite at cyclotron resonance, the cold-plasma approximation. At $\theta = 0$ the refractive index of one of the waves becomes infinite at $u = 1$; this singularity of $n(\theta)$ is a continuation of the oblique hybrid resonance (since $\theta_r = 0$ at $u = 1$). We can conclude, therefore, that this singularity takes place for the extraordinary wave at $q < 1$, and for the whistler mode at $q > 1$.

Qualitative plots of $n(\theta)$ for all regions are shown in Fig. 1.

Besides the refractive index, a most important property of plane waves is their polarization. In the general case, both waves are elliptically polarized. Interesting features appear in the vicinity of cyclotron and oblique hybrid reso-

---

*This phenomenon is called oblique hybrid resonance (at $\theta_r \neq \pi/2$).

nances. We examine them, using the wave equations [7]

$$D_{\alpha\beta}E_\beta = 0, \qquad D_{\alpha\beta} = \delta_{\alpha\beta}k^2 - k_\alpha k_\beta - \frac{\omega^2}{c^2}\varepsilon_{\alpha\beta}, \tag{6}$$

after determining from them the following field-component ratios:

$$\frac{E_y}{E_x} = \frac{ig}{n^2 - \varepsilon}, \qquad \frac{E_z}{E_x} = \frac{n_\perp n_\parallel}{n_\perp^2 - \eta}, \qquad \frac{E_1}{E_2} = \frac{n^2 - \varepsilon_{22}}{n^2 - \varepsilon_{11}}, \tag{7}$$

where $n_\parallel = n\cos\theta$, $n_\perp = n\sin\theta$, and $k_y$ is assumed equal to zero. In oblique hybrid resonance (i.e., for $\theta \to \theta_r$ and $u \neq 1$) we have $n^2 \to \infty$ and obtain

$$E_y/E_x \to 0, \quad E_z/E_x \to k_z/k_x, \tag{8}$$

so that the wave polarization becomes longitudinal and the field **E** can be regarded as potential ($\mathbf{E} = -\nabla\varphi$). This situation obtains, however, only far from the cyclotron resonance, when $\theta_r \neq 0$. As $u \to 1$ we have $\theta_r \to 0$, and we find from (6) and (7) that $E_2/E_1 \to 0$ for this wave, which has a singular refractive index. The presence of the component $E_1$ with which the electrons interact resonantly is in fact the cause of the refractive-index singularity. Conversely, at $\theta \neq 0$ and $u \to 1$ we have $E_1/E_2 \to 0$ for both waves, so that both waves can have in a plane perpendicular to the external magnetic field an identical "nonresonant" polarization and differ by the relative value of the longitudinal field component $E_z$. At $\theta = \pi/2$, in particular, $E_z$ is the only nonzero ordinary-wave field component.

Problem. Investigate in detail the change of the shapes of the $n(\theta)$ plots on going through the critical density line at $u > 1$ at small values of $\theta$.

## 2. Electron-Cyclotron Resonance in a Homogeneous Plasma with Allowance for the Thermal Motion of the Electrons

In the cold-plasma approximation and at oblique propagation ($\theta \neq 0$) the refractive indices of both waves have thus no singularities in cyclotron resonance. The component $\varepsilon_{11}$ of the tensor $\varepsilon_{ik}$, however, becomes infinite here. Allowance for the spatial dispersion, naturally, removes this singularity. In contrast to a cold plasma, the screening of the resonant field component $E_1$ becomes incomplete, so that the wave is damped. Analysis [5, 8] shows that under the conditions of plasma-heating experiments this damping, as any influence of thermal motion of the particles on the wave, turns out to be weak and the "cold" approximation yields a reasonable estimate of the refractive index $n \approx 1$. We then have $(k_\perp v_T/\omega_B) \sim (k_\parallel v_T/\omega) \sim \beta$, where $\beta = v_T/c \ll 1$, and the known expressions for $\varepsilon_{ik}$ can be written as series in powers of this small parameter:

$$\left.\begin{aligned} &\varepsilon_{11} \equiv \frac{1}{2}(\varepsilon_{xx} + \varepsilon_{yy}) + i\varepsilon_{xy} = 2i\sigma + 1 - 4i\alpha, \\ &\varepsilon_{22} \equiv \frac{1}{2}(\varepsilon_{xx} + \varepsilon_{yy}) - i\varepsilon_{xy} = 1 - \frac{q}{1+\sqrt{u}}, \quad \varepsilon_{33} = 1 - q + \varepsilon_{33}', \\ &\varepsilon_{12} = \varepsilon_{21} = i\alpha, \quad \varepsilon_{31} = \frac{1}{2}\varepsilon_{13} = \varepsilon_{xz}, \quad \varepsilon_{32} = \varepsilon_{23} = 0, \end{aligned}\right\} \quad (9)$$

where

$$i\sigma = -\frac{qZ(\zeta)}{2\zeta\beta n\cos\theta} \sim \beta^{-1}, \quad \alpha = (\beta n \sin\theta)^2\sigma, \quad \varepsilon_{13} = q\tan\theta[1 - Z(\zeta)],$$

$$\varepsilon_{33}' = \beta n\zeta\sin\theta\varepsilon_{13}, \quad \zeta = \frac{\omega - \omega_B}{k_\parallel v_T} = \frac{\omega - \omega_B}{\omega\beta n\cos\theta},$$

$$Z(\zeta) = X(\zeta) - iY(\zeta), \quad X(\zeta) = 2\zeta e^{-\zeta^2}\int_0^\zeta e^{t^2}\,dt, \quad Y(\zeta) = \sqrt{\pi}\,\zeta e^{-\zeta^2}.$$

In the vicinity of the cyclotron resonance the dispersion equation can also be written as an expansion in $\beta$ [8]. It is convenient here to use the following procedure [9]. The wave equation (6) expressed in terms of the components $E_i$ is rewritten as

$$n^2E_i - \frac{1}{2}n_\perp^2(E_1 + E_2) - \frac{1}{2}n_\parallel n_\perp E_3 - D_i = 0, \quad i = 1, 2, \quad (10)$$

$$n_\parallel n_\perp(E_1 + E_2) + n_\perp^2 E_3 - D_3 = 0. \quad (11)$$

From the first of these expressions (with i = 1) we determine with the desired accuracy the resonant field component $E_1$:

$$E_1 = \frac{i}{2\sigma}\left[\frac{1}{2}n_\perp^2E_2 + \left(\frac{1}{2}n_\parallel n_\perp + \varepsilon_{31}\right)E_3\right] \sim 0(\beta). \quad (12)$$

Substituting this expression in the remaining two expressions and subtracting the determinant of the resultant system of two equations for $E_2$ and $E_3$, we find that the principal term of this determinant coincides with $\Lambda_0$ of Eq. (5). The term of first order in $\beta$ in the resultant dispersion equation can be simplified by putting $\Lambda_0 = 0$, an expression valid in the zeroth order in $\beta$. As a result, we obtain, accurate to quantities of order of $\beta$ [9],*

$$\Lambda_0 + \frac{1}{i\sigma}\Lambda_1 + \frac{\omega - \omega_B}{\omega}\Lambda_2 = 0, \tag{13}$$

where

$$\Lambda_1 = \frac{n_\perp^4}{4}\frac{(n^2 + 2q - 1)^2}{n_\perp^2 + q - 1}, \qquad \Lambda_2 = n_\perp^4 \frac{\left(n^2 + \frac{3}{2}q - 1\right)}{n_\perp^2 + q - 1},$$

and at the accuracy considered one should put $\omega = \omega_B$ in the expressions for $\Lambda_0$, $\Lambda_1$, and $\Lambda_2$.

Thus, allowance for spatial dispersion leads at $\beta \ll 1$ and $|(\omega - \omega_B)/\omega| \ll 1$ to the appearance of small complex corrections of order $\beta$ in the dispersion equation. These corrections describe, in particular, the wave damping, which is found naturally to be weak (at distances on the order of the wavelength).

The dispersion equation (13) can be solved by successive approximations, putting

$$n = n^{(0)} + n^{(1)} + i\varkappa,$$

where $n^{(0)}$ is the refractive index of the cold plasma [the solution of Eq. (5)], and $n^{(1)}$ and $\varkappa$ are small corrections of order $\beta$. For the absorption coefficient we thus obtain [8]

$$\varkappa = \Phi(\theta, q) f(\zeta)\beta, \tag{14}$$

where

$$f(\zeta) = \frac{\pi\zeta^2 e^{-\zeta^2}}{|Z(\zeta)|^2}; \qquad \Phi = \Lambda_1 \frac{\partial\Lambda_0}{\partial n}\frac{2n\cos\theta}{q\sqrt{\pi}}\bigg|_{\Lambda_0=0}. \tag{15}$$

*Strictly speaking, the assumption that $|\sigma| \gg 1$ is also used in the derivation of (13). Since, however, in tokamaks as a rule $q \sim 1$, the requirement $\beta \ll 1$ at $n_\parallel \neq 0$ is sufficient.

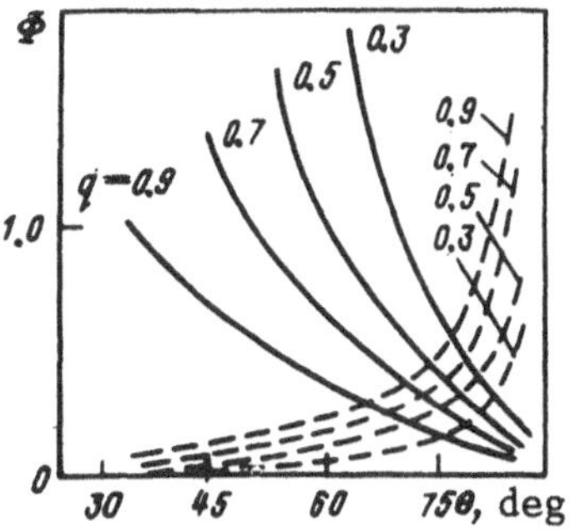

Fig. 2. Angular dependences of the absorption coefficient for the extraordinary (dashed) and ordinary waves.

The quantity $\varkappa(\theta)$ is not convenient for use in HF-heating problems, and therefore will not be analyzed in detail. To gain an idea, Fig. 2 shows plots of the function $\Phi(\theta, q)$ for several values of q. It can be seen that the damping, which is of the same order in the small parameter $\beta$ for both wave types, differs greatly numerically and has a different angular dependence.

The extraordinary-wave damping increases strongly as $\theta \to 0$, but this case cannot be realized in the tokamak geometry and will not be considered.

The cyclotron-absorption line contour (Fig. 3) is symmetric about $\omega_B$; its width $\Delta\omega \sim \beta\omega_B$ is determined by the Doppler broadening of the cyclotron resonance.

Expressions (9)-(15) cease to hold in a narrow interval of angles $\theta$ close to $\pi/2$, $|(\pi/2) - \theta| \leq \beta$, when the Doppler shift becomes so small that it encounters competition from the relativistic resonance broadening due to the dependence of the electron's cyclotron frequency on its energy. The role of the relativistic effects can be discerned already in the calculation of the terms of zeroth order in the parameter $k_\perp v_\perp/\omega_B$ in the components of the tensor $\varepsilon_{\alpha\beta}(\alpha, \beta = x, y)$. Obviously, these components can be expressed in terms of the integral [13]

$$I = -\mathrm{i}\left\langle \int\limits_0^\infty \exp\left[\mathrm{i}\left(\omega - \omega_B - k_\parallel v_z\right)t\right] dt \right\rangle, \qquad (16)$$

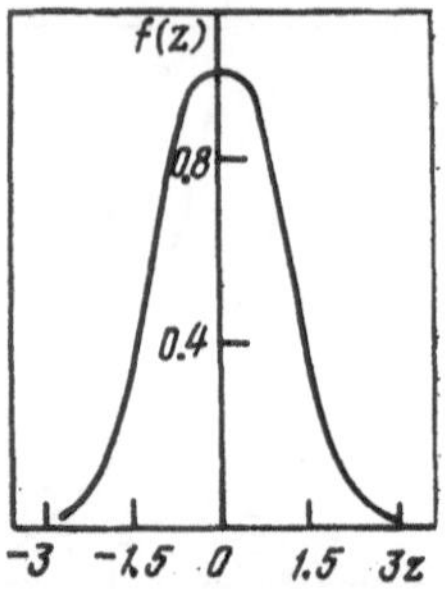

Fig. 3. Cyclotron-absorption line shape.

where the averaging is over the particle velocities whose distribution function can be regarded as Maxwellian in the weak relativism approximation. If the relativistic dependence of $\omega_B$ on $v^2$ is ignored, then

$$I = \frac{1}{\omega - \omega_B} Z\left(\frac{\omega - \omega_B}{k_{\|} v_T}\right).$$

In the weakly relativistic approximation, as $\omega_B \to \omega_B(1 - v^2/2c^2)$, we obtain, instead,

$$I = \frac{2}{\beta^2 \omega_B} \mathcal{F}_p\left(\frac{\delta}{\beta^2}, \frac{n_{\|}}{\beta}\right), \tag{17}$$

where $\delta = (\omega - \omega_B)/\omega_B$, $p = 3/2$,

$$\mathcal{F}_p = -i \int_0^\infty \exp\left\{\frac{1}{\beta^2}\left[2\, i\delta\tau - \frac{n_{\|}^2 \tau^2}{(1 - i\tau)}\right]\right\}(1 - i\tau)^{-p}\, d\tau. \tag{18}$$

The function $\mathcal{F}_p$ differs greatly in its analytic properties from the function Z contained in the nonrelativistic expression for $\varepsilon_{\alpha\beta}$. First, it depends on two independent variables $\delta/\beta^2$ and $n_{\|}/\beta$, rather than on one. Next, rewriting the argument of the exponential in (18) in the form

$$i\left(2\delta - n_{\|}^2\right)\beta^{-2}\tau + \frac{i\, n_{\|}^2}{\beta^2}\frac{i\tau}{1 - i\tau}$$

and turning in obvious manner the integration contour in the complex $\tau$ plane, we easily verify that for real $\omega$ and $n_{\|}$ the function $\mathcal{F}_p \equiv \mathcal{F}_p(\zeta, n_{\|}^2/\beta^2)$ is real at $\zeta \equiv (1/\beta^2)(2\delta - n_{\|}^2) > 0$ and is complex at $\zeta < 0$. Thus, this function, and with it the $\varepsilon_{\alpha\beta}$ tensor components regarded as functions of the variable $\zeta$, have a branch point at $\zeta = 0$ (i.e., on the line $\omega - \omega_B = 2\omega_B n_\parallel^2$ in the $\omega$, $n_\parallel$ plane); this branch point vanishes in the nonrelativistic limit. The fact that $\mathscr{F}_p(\zeta, n_\parallel^2/\beta^2)$ is real at $\zeta > 0$ means absolute absence of cyclotron damping under these conditions. The origin of the relativistic-plasma singularity becomes obvious if we write the condition for resonance between a particle of velocity v and a wave $\omega - \omega_B - \mathbf{kv} = 0$ in the form

$$n_\parallel \frac{v \cos \alpha}{c} - \frac{v^2}{2c^2} = \delta,$$

where $\alpha$ is the angle between $\mathbf{v}$ and the z axis. The maximum value of the left-hand side of this equation, reached at $\alpha = 0$, $v/c = n$, is equal to $(1/2)n_\parallel^2$, so that at $2\delta > n_\parallel^2$ cyclotron resonance between the wave and the particle is impossible at any velocity. When relativism is taken into account, the cyclotron-absorption line shape is thus asymmetric about the cyclotron frequency $\omega_B$. In particular, at $n_\parallel = 0$ absorption takes place only if $\omega < \omega_B$.

The transition to the nonrelativistic limit is effected in (17) by simultaneously satisfying the two conditions $n_\parallel^2 \gg \beta^2$ and $n_\parallel^2 \gg 2\delta$, when the main contribution to the integral is made by the region of small $\tau$. The first of them was mentioned above, and the second is that the frequency should not be close to the absorption-line boundary $\delta = n_\parallel^2/2$, the very presence of which is a purely relativistic effect. The condition $2\delta \ll n_\parallel^2$ means, somewhat unexpectedly, that even at the "nonrelativistic" wave-propagation angles $|\theta - \pi/2| > \beta$ the usual expressions for $\varepsilon_{\alpha\beta}$ are, strictly speaking, invalid far from resonance. It must be borne in mind, however, that this occurs in the frequency region where the spatial dispersion is already of low significance and, in particular, the damping is exponentially small.

A consistent allowance for the relativistic corrections, at the accuracy required for the calculation, leads to a very cumbersome dispersion. We shall not present it here (see [10]), since its explicit form is not needed to obtain the quantity of greatest interest in heating problems, viz., the total damping in the cyclotron layer. We present only, for reference, the expressions for the $\varepsilon_{\alpha\beta}$ components in the simplest case $n_\parallel = 0$ for the vicinity of resonance with a harmonic of the cyclotron frequency $|\omega - S\omega_B| \ll \omega_B$, $S = 1, 2, 3, \ldots$ [10]:

$$\varepsilon_{\alpha\beta}=\varepsilon_{\alpha\beta}^{(c)}+\varepsilon_{\alpha\beta}^{(w)}. \tag{19}$$

Here $\varepsilon_{\alpha\beta}{}^{(c)}$ is the contribution made to $\varepsilon_{\alpha\beta}$ by the nonresonant terms, calculated for a cold plasma,

$$\left.\begin{array}{l}\varepsilon_{xx}^{(c)}=\varepsilon_{yy}^{(c)}=\varepsilon,\quad \varepsilon_{zz}^{(c)}=\eta,\quad \varepsilon_{xy}^{(c)}=-\varepsilon_{yx}^{(c)}=ig \ \text{ for } S>1,\\ 1-\varepsilon_{xx}^{(c)}=1-\varepsilon_{yy}^{(c)}=i\varepsilon_{xy}^{(c)}=-i\varepsilon_{yx}^{(c)}=\dfrac{q}{2(1+\sqrt{u})},\quad \varepsilon_{zz}^{(c)}=\eta \ \text{ for } S=1.\end{array}\right\} \tag{20}$$

The dispersive increment to $\varepsilon_{\alpha\beta}{}^{(w)}$ in the lowest order of $\lambda = k_\perp{}^2T_e/m_e\omega_B{}^2$ is given by the equations

$$\begin{gathered}\varepsilon_{xx}^{(w)}=\varepsilon_{yy}^{(w)}=i\varepsilon_{xy}^{(w)}=-i\varepsilon_{yx}^{(w)}=-\frac{qS^2}{\beta^2 2^{s-1}S!}\lambda^{s-1}F_{s+\frac{3}{2}}(\zeta),\\ \varepsilon_{zz}^{(w)}=-\frac{q\lambda^s}{\beta^2 2^{s-1}S!}F_{s+\frac{5}{2}}(\zeta),\end{gathered} \tag{21}$$

where $F_{s+1/2}(\zeta) \cong \mathscr{F}_{s+1/2}(\zeta, 0)$, $\zeta = 2\delta/\beta^2$. The function $F_p$ is connected with the plasma dispersion function by the relation

$$F_p(\zeta)=\sum_{l=0}^{p-3/2}(-\zeta)^l\frac{\Gamma(p-1-l)}{\Gamma(p)}-\frac{\sqrt{\pi}}{\Gamma(p)}(-\zeta)^{p-\frac{3}{2}}Z(i\zeta^{1/2}). \tag{22}$$

The remaining components of the tensor ($\varepsilon_{xz}$, $\varepsilon_{zx}$, $\varepsilon_{yz}$, $\varepsilon_{zy}$) at $n_\parallel = 0$ are equal to zero.

Problem. Calculate the damping of the extraordinary wave for perpendicular propagation, in the lowest order in $\beta$ in the S-th harmonic of the cyclotron frequency ($S \geq 1$).

Solution. The dispersion equation for the ordinary wave takes at $n_\parallel = 0$ the form

$$n^2=\varepsilon_{zz}.$$

Using (21), (22), and the properties of $Z(\zeta)$, we get

$$\varkappa=\frac{4\pi s^2 q}{\Gamma\left(s+\frac{3}{2}\right)2^s S!}\frac{\lambda_0^{s+1}}{\beta^2}\left|\frac{2\delta}{\beta^2}\right|^{s+\frac{1}{2}}e^{-\left|\frac{2\delta}{\beta^2}\right|},$$

where $\lambda_0 = (1-q)S^2\beta^2$, $\delta < 0$. The result is valid in the case $S = 1$ only for a low-density plasma.

## 3. Poynting Vector and Energy Absorption in Cyclotron Resonance

It can be seen from (9) that the anti-Hermitian part of the tensor $\varepsilon_{ik}$ is in the vicinity of the cyclotron resonance (at $|\zeta| \leq 1$) of the same order as the Hermitian part. The weakness of the damping is explained in this case by a peculiarity of the wave polarization, viz., smallness of the resonant component of the field $E_1$. Under these conditions caution must be exercised when using concepts such as energy density, wave-energy flux, or group velocity, since the derivation of the wave-energy transport equation that relates this quantity is essentially based on the assumed smallness of the anti-Hermitian part of the dielectric tensor $\varepsilon_{\alpha\beta}$ (see, e.g., [7]). To elucidate the result of dispensing with this last assumption (but, of course, still assuming weak wave damping), we consider the Poynting theorem in a form that follows directly from the Maxwell equations:

$$\frac{1}{8\pi}\frac{\partial}{\partial t}(\tilde{B}^2+\tilde{E}^2)+\frac{c}{4\pi}\operatorname{div}[\tilde{\mathbf{E}}\tilde{\mathbf{B}}]+\tilde{\mathbf{j}}\cdot\tilde{\mathbf{E}}=0, \qquad (23)$$

where $\tilde{\mathbf{E}}$ and $\tilde{\mathbf{B}}$ are the physical (real) values of the electric and magnetic fields of the wave; $\tilde{\mathbf{j}}$ is the density of the current produced by the wave. In a medium with spatial dispersion, the connection between the field and the current is nonlocal:

$$\tilde{j}_\alpha(\mathbf{r}, t)=\int_{-\infty}^{t} dt' \int d\mathbf{r}' \sigma_{\alpha\beta}(\mathbf{r}-\mathbf{r}', t-t')\tilde{E}_\beta(\mathbf{r}', t'), \qquad (24)$$

and the tensor $\varepsilon_{\alpha\beta}(\omega, \mathbf{k})$ is connected with the kernel $\sigma_{\alpha\beta}\times(\mathbf{r}-\mathbf{r}', t-t')$ by the relation

$$\varepsilon_{\alpha\beta}(\omega, \mathbf{k})=\delta_{\alpha\beta}+\frac{4\pi i}{\omega}\sigma_{\alpha\beta}(\omega, \mathbf{k}),$$

$$\sigma_{\alpha\beta}(\omega, k)=\int_0^\infty d\tau \int d\mathbf{R}\sigma_{\alpha\beta}(\tau, \mathbf{R})\exp(i\omega\tau-i\mathbf{kR}).$$

We now consider the problem of damping of a quasimonochromatic wave of the form

$$\tilde{\mathbf{E}}(\mathbf{r}, t)=\operatorname{Re}\{\mathbf{E}(\mathbf{r}, t)\exp[i\mathbf{kr}-i\omega t]\},$$

assuming $\omega$ and $\mathbf{k}$ to be real. For weak damping, the amplitude $\mathbf{E}(\mathbf{r}, t)$ can be regarded as a slowly varying function and represented in the integral of (24) by an expansion in (t − t') and (r − r'). After integration by parts, Eq. (24) then takes the form [7]

$$j_\alpha = \mathrm{Re}\exp[i(\mathbf{kr}-\omega t)]\left\{\sigma_{\alpha\beta}E_\beta - i\frac{\partial\sigma_{\alpha\beta}}{\partial\mathbf{k}}\frac{\partial E_\beta}{\partial\mathbf{r}} + i\frac{\partial\sigma_{\alpha\beta}}{\partial\omega}\frac{\partial E_\beta}{dt} + \ldots\right\}, \quad (25)$$

where the terms with higher derivative of the smooth function $\mathbf{E}(\mathbf{r}, t)$ with respect to coordinate and time were discarded. Substituting this expression in (23), we obtain after averaging over the time

$$\frac{\partial W}{\partial t} + \operatorname{div}\mathbf{S} + Q = Q', \quad (26)$$

where the quantities

$$W = \frac{\partial(\omega\varepsilon'_{\alpha\beta})}{\partial\omega}\frac{E^*_\alpha E_\beta}{16\pi} + \frac{\mathbf{BB}^+}{16\pi}, \quad (26a)$$

$$\mathbf{S} = \frac{c}{8\pi}[\mathbf{EB}^*] - \omega\frac{\partial\varepsilon'_{\alpha\beta}}{\partial\mathbf{k}}\frac{E^*_\alpha E_\beta}{16\pi}, \quad (26b)$$

$$Q = \frac{\omega}{8\pi}\varepsilon''_{\alpha\beta}E^*_\alpha E_\beta \quad (26c)$$

are usually interpreted as the energy density, the wave-energy flux density, and the energy dissipated in the plasma, respectively; $\varepsilon_{\alpha\beta}'$ and $i\varepsilon_{\alpha\beta}''$ are, respectively, the Hermition and anti-Hermitian parts of the dielectric tensor $\varepsilon_{\alpha\beta}$,

$$2iQ' = \frac{1}{8\pi}\frac{\partial(\omega\varepsilon''_{\alpha\beta})}{\partial\omega}\left(E^*_\alpha\frac{\partial E_\beta}{\partial t} - E_\beta\frac{\partial E^*_\alpha}{\partial t}\right) - \frac{\omega}{8\pi}\frac{\partial\varepsilon''_{\alpha\beta}}{\partial\mathbf{k}}\left(E^*_\alpha\frac{\partial\mathbf{E}_\beta}{\partial\mathbf{r}} - \frac{\partial E^*_\alpha}{\partial\mathbf{r}}E_\beta\right). \quad (26d)$$

At $|\varepsilon_{\alpha\beta}''| \ll |\varepsilon_{\alpha\beta}'|$, the quantity Q' should obviously be neglected, and we get Poynting's theorem in standard form for media with spatial dispersion. If we now attempt to use Eq. (26) in the vicinity of cyclotron resonance, the following difficulty arises. The components of the tensor $\varepsilon_{\alpha\beta}$ depend in this case on the frequency, in particular, also via the argument $\zeta = (\omega - \omega_B)/\omega n_{\parallel}\beta$ of the dispersion function $Z(\zeta)$. The real and imaginary parts of this function and their derivatives with respect to the argument are of the order of unity at $\zeta \sim 1$; at the same time, $(\partial\zeta/\partial\omega) = (1/\omega_B n_{\parallel}\beta)$, so that $(\partial^n\varepsilon_{\alpha\beta}/\partial\omega^n) \sim [n/(\beta\omega_B n_{\parallel})^n]$. The damping rate in the vicinity of the cyclotron resonance can be roughly estimated at $\gamma \sim \beta\omega_B$. The small parameter of the time derivatives vanishes thus in the expansion (25), and there are no grounds now for retaining only the first terms of the expansion. Accordingly, the usual expressions for the energy density of a quasimonochromatic wave and for its group velocity $v_g = \partial\omega/\partial\mathbf{k}$ also become meaningless. This is indeed the explanation why formal calculations can yield for the group velocity values higher than that of light [11, 12].

Consideration of the damping of quasimonochromatic plasma oscillations with time as $\omega \to \omega_B$ encounters thus fundamental difficulties. In research on plasma heating, however, greatest interest attaches to propagation and spatial damping of a wave whose frequency is set beforehand by an external generator. In this stationary formulation the problem is noticeably simpler, for when $Z(\zeta)$ is differentiated with respect to $k_{\parallel}$ no large factors appear at $k_{\parallel} \neq 0$. Nonetheless, Eq. (26) can still not be used directly as yet to calculate $\mathbf{E}_0(\mathbf{r})$, since the ratio of the different field components of the quasimonochromatic wave (more accurately, of a beam of waves of fixed frequency) has not yet been established. Of course, no problem arises at all for a homogeneous plasma. In essence, all the necessary information can be obtained directly from Eqs. (6) by using the Fourier expansion. Examination of this problem, however, is of definite methodological interest from the standpoint of generalizing Eq. (26) to include a weakly inhomogeneous medium. We consider, therefore, in greater detail the stationary energy transport of a beam of weakly damped waves in a plasma in which the anti-Hermitian part of the tensor $\varepsilon_{\alpha\beta}$ is not small.

Equations (6) for a wave with a smoothly varying amplitude can be represented in analogy with (25) in the form

$$D_{\alpha\beta}E_\beta - i\frac{\partial D_{\alpha\beta}}{\partial \mathbf{k}}\frac{\partial E_\beta}{\partial \mathbf{r}} = 0.$$

It is natural to represent the wave amplitude $\mathbf{E} \equiv \mathbf{E}(\mathbf{r})$ by an expansion

$$\mathbf{E} = \mathbf{E}^{(0)} + \mathbf{E}^{(1)} + \ldots,$$

where it is assumed that $|E^{(n+1)}| \sim (1/kL)|E^{(n)}| \ll |E^{(n)}|$, and L is the characteristic scale of the amplitude change. Usually, $iD_{\alpha\beta}''$ (the anti-Hermitian part of the tensor $D_{\alpha\beta}$)

is regarded in the case of weak damping as a quantity of first order in smallness, so that in the zeroth approximation we get for the field amplitude the system of equations [7]

$$D'_{\alpha\beta}E^{(0)}_{\beta} = 0,$$

where $D_{\alpha\beta}'$ is the Hermitian part of the tensor $D_{\alpha\beta}$.

In our case, however, we must proceed differently at $|D_{\alpha\beta}''| \sim |D_{\alpha\beta}'|$. It is convenient to introduce a new tensor $\tilde{D}_{\alpha\beta} = D_{\alpha\beta} + \delta D_{\alpha\beta}$ with $\delta D_{\alpha\beta} \sim O(1/kL)$ such that the dispersion equation

$$\det|\tilde{D}_{\alpha\beta}| = 0 \tag{27}$$

yields solutions with real **k** close in magnitude to the real part of the corresponding solutions of the exact dispersion equation $\det|D_{\alpha\beta}| = 0$. In this case, we can choose for the field amplitude, in the zeroth approximation, the following system of equations:

$$D_{\alpha\beta}E^{(0)}_{\beta} = 0. \tag{28}$$

In first order in 1/kL we obtain

$$\tilde{D}_{\alpha\beta}E^{(1)}_{\beta} - \delta D_{\alpha\beta}E^{(0)}_{\beta} - i\frac{\partial D_{\alpha\beta}}{\partial \mathbf{k}}\frac{\partial E^{(0)}_{\beta}}{\partial \mathbf{r}} = 0. \tag{29}$$

Obviously, $\delta D_{\alpha\beta} = (\partial D_{\alpha\beta}/\partial \mathbf{k})[\mathbf{k}(\omega) - \mathbf{k}]$, where $\mathbf{k}(\omega)$ is the solution of the exact dispersion equation $\det|D_{\alpha\beta}| = 0$, and **k** is the solution of Eq. (27). The leeway in the choice of $\delta D_{\alpha\beta}$ is thus due to the possibility of including part of the phase of the wave in the complex amplitude $\mathbf{E}_r^{(0)}$.

Multiplying Eq. (29) by $E_\alpha^{(0)*}$ and separating next the real and imaginary parts, we obtain after simple transformations, using the system (28), the following two equations:

$$\operatorname{div}\mathbf{S}_0 + Q_0 = Q_1 + Q_0'; \tag{30}$$

$$\frac{1}{2}\frac{\partial D'_{\alpha\beta}}{\partial \mathbf{k}}\left(E^{(0)*}_{\alpha}\frac{\partial E^{(0)}_{\beta}}{\partial \mathbf{r}} - \frac{\partial E^{(0)*}_{\alpha}}{\partial \mathbf{r}}E^{(0)}_{\beta}\right) - iE^{(0)*}_{\alpha}\delta D'_{\alpha\beta}E^{(0)}_{\beta}$$
$$= -\frac{i}{2}\frac{\partial D''_{\alpha\beta}}{\partial \mathbf{k}}\frac{\partial}{\partial \mathbf{r}}\left(E^{(0)*}_{\alpha}E^{(0)}_{\beta}\right) + Q_1', \tag{31}$$

where

$$2S_0 = \frac{c^2}{8\pi\omega} E_\alpha^{(0)*} E_\beta^{(0)} \frac{\partial D'_{\alpha\beta}}{\partial \mathbf{k}}, \quad Q_0 = \frac{\omega}{8\pi} \varepsilon''_{\alpha\beta} E_\alpha^{(0)*} E_\beta^{(0)},$$

$$-Q_1 = \frac{\omega}{8\pi} \varepsilon''_{\alpha\beta} (E_\alpha^{(0)*} E_\beta^{(1)} + E_\alpha^{(1)*} E_\beta^{(0)}),$$

$$Q'_0 = +\omega \frac{\partial \varepsilon''_{\alpha\beta}}{\partial \mathbf{k}} \left( E_\alpha^{(0)*} \frac{\partial E_\beta^{(0)}}{\partial \mathbf{r}} - \frac{\partial E^{(0)*}}{\partial \mathbf{r}} E_\beta^{(0)} \right) \frac{i}{16\pi}$$

$$Q'_1 = + D''_{\alpha\beta} (E_\alpha^{(0)*} E_\beta^{(1)} - E_\alpha^{(1)*} E_\beta^{(0)}).$$

It is readily seen that Eq. (30) is a stationary variant of Eq. (26), in which the wave amplitude $\mathbf{E}(\mathbf{r})$ is represented by a sum of two terms of zeroth and first order in the parameter 1/kL. The density of the energy dissipated in the plasma is in this case also represented by a sum of two terms, $Q_0$ and $Q_1$. Although we assume $|E_1^{(1)}| \ll |E_1^{(0)}|$, however, there are, generally speaking, no grounds for neglecting $Q_1$ compared with $Q_0$ if $|\varepsilon_{\alpha\beta}''| \sim |\varepsilon_{\alpha\beta}'|$.

Equation (31) determines the complex phase of the amplitude $\mathbf{E}^{(0)}(\mathbf{r})$ or, in other words, the first-order correction to the wave vector.

We now use Eqs. (9) to estimate the quantities $Q_1$, $Q_0'$, and $Q_1'$ in the vicinity of the cyclotron resonance within the framework of the approximations used in Section 2, viz., weakness of the transverse spatial dispersion and weakness of the relativism, $\beta \ll 1$. It can be seen from (28) that $E_1^0 \sim O(\beta)$, since $\varepsilon_{||} \sim 1/\beta$.

As already noted, smallness of $E_1$ ensures precisely weakness of wave absorption (smallness of $Q_0$). It can be easily seen from (29) that the quantity

$$|E_1^{(1)}| \sim \beta \left( |E_z^{(1)}| + |E_2^{(1)}| + \frac{1}{kL} |E^{(0)}| \right)$$

is similarly anomalously small, so that $Q_1' \sim (1/kL)Q_0 \ll Q_0$. The term with $Q_1$ can therefore be left out of (30). We note next that

$$|Q'_0| \sim \left| \omega \frac{\partial \varepsilon''_{\alpha\beta}}{\partial \mathbf{k}} \frac{\partial E^{(0)*} E_\beta^{(0)}}{\partial \mathbf{r}} \right| \sim \left| \omega \frac{\partial \varepsilon'_{\alpha\beta}}{\partial \mathbf{k}} \frac{\partial E^{(0)*} E^{(0)}}{\partial \mathbf{r}} \right| \beta \approx \beta \operatorname{div} [\mathbf{E}^{(0)} \mathbf{B}^{(0)*}].$$

It is clear, therefore, that in the lowest order in $\beta$ the energy flux in the wave is determined entirely by the electromagnetic-energy flux, and the energy transport by the plasma particles can be neglected. The result is the Poynting theorem in the following customary form:

$$\operatorname{div} \mathbf{S}_0 + Q_0 = 0, \tag{32}$$

where $\mathbf{S}_0 = (c/8\pi)[\mathbf{E}^{(0)}\mathbf{B}^{(0)*}]$ is the energy flux in a wave and coincides in the lowest order in $\beta$ with the local value of the energy flux in the cold-plasma approximation. The expression for $Q_0$ can then be written in the form

$$Q_0 = \frac{\omega}{8\pi}\left\{\frac{\varepsilon_{11}''}{2}\,|E_1^{(0)}|^2 + \frac{1}{2}\varepsilon_{xz}''\operatorname{Re}(E_1^{(0)}E_z^{(0)}) + \varepsilon_{33}''\,|E_z^{(0)}|^2\right\},$$

where, just as in (12),

$$E_1^{(0)} = \frac{-1}{\varepsilon_{11}}\left[\frac{n_\perp^2}{2}E_2^{(0)} + (n_\parallel n_\perp + 2\varepsilon_{xz})E_z^{(0)}\right],$$

and the x axis is directed along $\mathbf{k}_\perp$.

Since the cold-plasma approximation describes correctly the components $E_2^{(0)}$ and $E_z^{(0)}$ in the zeroth order in $\beta$, and since $E_1^{(0)} \sim \beta$, we can formulate a simple rule for the calculation of the damping of a beam of waves. First, in the cold-plasma approximation, we calculate the distribution of the wave-beam field. The field is, naturally, localized in space around the beam, whose direction coincides with that of the group velocity calculated in the cold-plasma approximation.

This is followed by calculation of the wave damping along the beam. It is possible here either to use Eq. (32), in which we must substitute $E_2^{(0)}$ and $E_z^{(0)}$ from the cold-plasma approximation, or to integrate along the beam the wave-vector imaginary part determined to within a value of the order of $\beta$ from the dispersion equation (13). Equation (31) determines then the change of the phase of the complex amplitude $\mathbf{E}^{(0)}(\mathbf{r})$ along the beam. To the accuracy employed, we can set its right-hand side equal to zero.

It will be shown below that the formulated calculation procedure can be used practically unchanged, within the geometric-optics approximation, in the case of a nonhydrogen plasma.

To conclude this section, we discuss the limits of applicability of our results. In the analysis of the wave beams we used essentially Eq. (32), which is applicable because $\varepsilon_{\alpha\beta}$ can be differentiated with respect to k for the entire spectrum of the wave vector of the given beam. It turns out that this requirement imposes certain restrictions on the value of $k_{\parallel}$ and on the characteristic scale ($L_{\parallel}$) of the inhomogeneity of the amplitude of the considered wave along the constant magnetic field.

The point is that $\varepsilon_{\alpha\beta}'$ is known to be a nonanalytic function of $k_{\parallel}$ (it contains $|k_{\parallel}|$), so that $\partial\varepsilon_{\alpha\beta}'/\partial k_{\parallel}$ becomes meaningless at $k_{\parallel} = 0$. It is necessary, therefore, that the Fourier components with small values of $|k_{\parallel}|$ be present in the wave spectrum with small weights. If this requirement is not met, the spatial localization of the current produced in the plasma by the wave beam will differ substantially from the electric-field localization.

Consequently, validity of the approximation (25), which corresponds to allowance for the weak nonlocality with the aid of the spatial derivatives of the amplitude, calls for satisfaction of the following conditions (see the problem at the end of this section):

$$k_{\parallel} L \gg 1, \qquad k_{\parallel}^2 L_{\parallel}^2 \gg \left(\frac{\omega - \omega_B}{k_{\parallel} v_{\mathrm{T}}}\right)^2. \tag{33}$$

Problem. Neglecting the transverse spatial dispersion, calculate the current produced in a plasma by a "resonant" component of an electric field, in the form

$$\widetilde{E}_1(\mathbf{r}_{\perp}, z) = \sqrt{\frac{\alpha}{\pi}}\, E(\mathbf{r}_{\perp}) \int\limits_{-\infty}^{\infty} \exp\left[\mathrm{i}\, kz - \alpha (k - k_z)^2\right] dk.$$

Solution. Using the following representation of the plasma-dispersion function

$$\frac{1}{\omega - \omega_B} Z\left(\frac{\omega - \omega_B}{k_z v_{\mathrm{T}}}\right) = -\,\mathrm{i} \int\limits_0^{\infty} \exp\left[\mathrm{i}\,(\omega - \omega_B)\, t - \frac{1}{4} k^2 v_{\mathrm{T}}^2 t^2\right] dt,$$

we obtain, after integrating with respect to k:

$$-4\pi\tilde{j}_1 = qE(\mathbf{r}_\perp)\int_0^\infty \exp\left[i(\omega-\omega_B)t + \frac{\left(\alpha k_z - \frac{iz}{2}\right)^2}{\alpha + \frac{v_T^2 t^2}{4}} - \alpha k_z^2\right]\frac{dt}{\sqrt{1+\frac{v_T^2 t^2}{4\alpha}}}.$$

Consider the region $|z| \ll 2\alpha k_z$. The main contribution to the integral is made by the initial part of the integration path in the vicinity of one of the saddle points. When conditions (33) with $L_\parallel^2 = 4\alpha$ are met, the contribution of the saddle point at $z^2/4\alpha \lesssim 1$ can be neglected accurate to values of order $\exp[-\alpha k_z^2]$, and we obtain $j_1 = \sigma_{11}(k_z)\tilde{E}_1$ at $\omega \neq \omega_B$. This expression corresponds to the principal term in Eq. (25), so that the connection between the current and the field inside the packet is local. The second condition of (44) is needed here so that, in the entire interval of t which determines $Z[(\omega - \omega_B)/k_z v_T]$ we can assume that $v_T^2 t^2/4\alpha \ll 1$.

It is easy to verify that at $|Z| \gg \alpha k_z$ the main contribution to the initial integral will be made by the saddle point. In this case, $j_1 \sim e^{-\frac{\sqrt{3}}{2}\left[\frac{z^2(\omega-\omega_B)^2}{4v_T^2}\right]^{1/3}}$ decreases with increasing z much more slowly than the field $E_1 \sim e^{-z^2/4\alpha}$, and Eq. (25) no longer holds. At $\alpha k_z^2 \gg 1$, however, the current in the region $z^2 > \alpha^2 k_z^2$ is exponentially small and can be neglected.

## 4. Accessibility of Resonances in a Tokamak

It was shown in Section 2 that cyclotron damping of waves is substantial only in a narrow frequency interval $|(\omega - \omega_B)/\omega| \lesssim \beta$. When the plasma in a tokamak is heated, the frequency $\omega$ is fixed, while $\omega_B$ is a function of the distance to the symmetry axis of the torus. The energy is therefore absorbed in a narrow layer having a thickness of order $\beta R_0 \ll a$ around the cyclotron-resonance surface $\omega = \omega_B(\mathbf{r})$ (here $R_0$ and $a$ are, respectively, the major and minor radii of the tokamak). To attain high plasma-heating efficiency it is desirable that the region where the wave energy is absorbed be in the central part of the pinch. Particular importance attaches then to the problem of accessibility of the resonance region to waves excited on the periphery of the

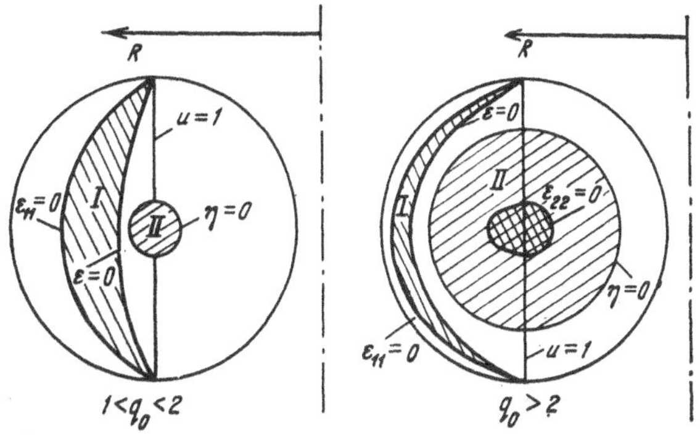


Fig. 4. The (u, q) plane in a tokamak: I) opacity region of the x mode; II) opacity region of the 0 mode.

plasma pinch. The most general idea of the accessibility can be gained by simply portraying the arrangement of the regions, described in Section 1, of the parameters q, and in a small section of the plasma (Fig. 4). Clearly, the cyclotron-resonance region is inaccessible to the extraordinary wave launched from the outer side of the torus. Inaccessible to a wave launched from the inner side (i.e., from the side of the strong magnetic field) is at any rate the plasma region where the density is double the critical value ($q > 2$). For the ordinary wave, the region $q > 1$ is inaccessible. This region is transparent to the whistler mode – that of an oblique Langmuir wave. For a tokamak with typical parameters, however, this mode cannot be excited by an antenna located outside the plasma, in view of the presence of an extensive opacity region at low densities.

The described accessibility picture is, of course, very crude, and a reasonable second step is consideration of wave propagation in the equatorial plane of a tokamak in the model where the medium is planarly layered. We neglect for the time being the thermal motion of the electrons and the poloidal component of the magnetic field. We assume that the plasma, whose parameters depend on one Cartesian coordinate x, occupies the region $-a \leq x \leq a$, the q(x) profile has a maximum at the center of this interval, and u(x) decreases monotonically with increasing x. Assuming the plasma to be weakly Inhomogeneous ($\omega a/c \gg 1$), we solve the wave equation in the quasiclassical approximation, putting

$$\mathbf{E}(\mathbf{r}, t) = \mathbf{E}(x) \exp\left\{i \int^{x} k_x(x')\, dx' + i k_z z - i\omega t\right\}, \qquad (34)$$

where $\mathbf{E}(x)$ and $k_x(x)$ are slowly varying functions, while $k_z$ is a constant (the z axis is directed along the magnetic field). Bearing in mind wave propagation in the equatorial plane, we have put $k_y = 0$. The dimensionless components $n_\perp = k_x c/\omega$ and $n_\parallel = k_z c/\omega$ of the wave vector are connected at any point of the plasma by an equation that coincides in form with the dispersion equation (4) of a cold plasma, with coefficients that are determined by the local values of the parameters $q(x)$ and $u(x)$. In this equation, however, it is now necessary to regard the quantity $n_\perp = n \sin\theta$ as given, and the quantity $n_\parallel = n \sin\theta$ as unknown. It is accordingly convenient to rewrite Eq. (4) in the form

$$\tilde{A} n_\perp^4 + \tilde{B} n_\perp^2 + \tilde{C} = 0, \tag{35}$$

where $\tilde{A} = \varepsilon$, $\tilde{B} = (\varepsilon + \eta) n_\parallel^2 - \varepsilon_{11}\varepsilon_{22} - \varepsilon\eta$, $\tilde{C} = \eta(\varepsilon_{11} - n_{11}^2)(\varepsilon_{22} - n_\parallel^2)$. Wave propagation is described in the considered planar model by the dispersion curves $n_\perp(x, n_\parallel)$ at $n_\parallel = \text{const}$. The character of these curves can be easily understood qualitatively with the aid of the plots of the function $n(\theta)$, since $n_\perp$ is determined by the intersection of the straight line $n_\parallel = \text{const}$ with the plot of $n(\theta)$. We note right away that, as can be seen from the shapes of the $n(\theta)$ curves (Fig. 1), the two roots $n_{\perp 1/2}^2$ of Eq. (35) can pertain to different modes as well as to one and the same wave. Cutoff of waves with fixed $n_\parallel$ (i.e., vanishing of $n_\perp^2$) is possible on the lines $\eta = 0$, $\varepsilon_{11} = n_\parallel^2$, and $\varepsilon_{22} = n_\parallel^2$, i.e., at values of q given by the relations

$$q = 1, \quad q = (1 + \sqrt{u})(1 - n_\parallel^2), \\ q = (1 - \sqrt{u})(1 - n_\parallel^2). \tag{36}$$

With increasing $n_\parallel^2$, the wave-propagation region is thus generally decreased. The demarcation lines $\varepsilon_{11} = 0$, $\varepsilon_{22} = 0$, $q = 1$ in Fig. 1, determined from the condition $n_\parallel^2 + n_\perp^2 = 0$, coincide with the cutoff lines $n^2 = 0$ only in the case of perpendicular propagation.

Figures 5-7 show typical $n_\perp^2(x, n_\parallel)$ curves for $q(x) = q_0(1 - x^2/a^2)$ and $u = 1$ at the point $x = 0$. These curves were calculated with allowance for the fact that in the axisymmetric configuration of the tokamak we have $n_\parallel(x) = n_{\parallel 0} \times [1 - (a + x)/R_0]$, where $n_{\parallel 0}$ is the initial value of $n_\parallel$ specified on the inner edge of the plasma. The toroidal dependence of $n_\parallel$ on x can, of course, be taken into account also in Eqs. (35).

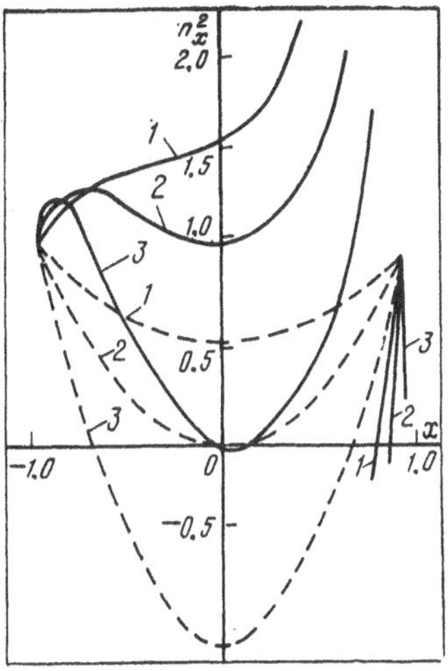


Fig. 5. Dispersion curves $n_\perp^2(x)$ at $n_\parallel^2 = 0, 1$: dashed curves – ordinary wave; solid curves – extraordinary wave. The curves correspond to: 1) $q_0 = 0.5$; 2) $q_0 = 1.0$; 3) $q_0 = 2.0$.

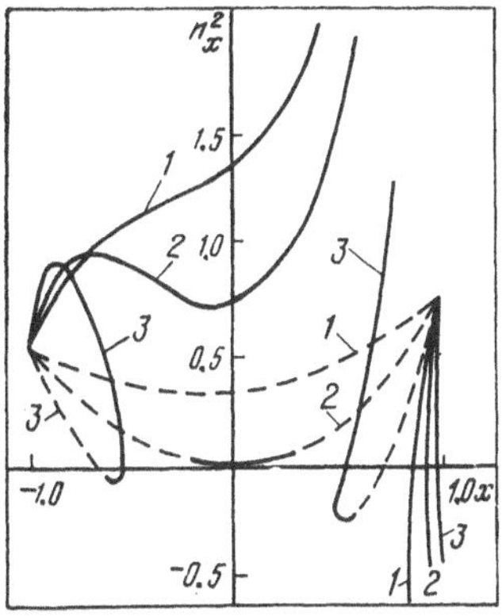


Fig. 6. The same as Fig. 5 for $n_\parallel^2 = 7$.

Figures 5-7 show clearly how the accessibility conditions deteriorate with increase of $q_0$ and $n_{\parallel 0}$. Figure 7 shows the situation mentioned above, where two branches of

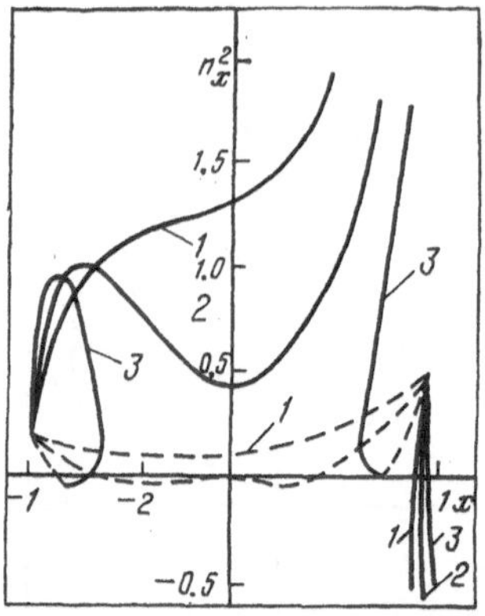


Fig. 7. The same as Fig. 5, for $n_{\parallel}^2 = 0.9$.

the $n_{\perp}^2$ plot pertain to one (extraordinary) mode. The group velocities of these waves along the x axis, at equal sign of $n_{\perp}$, are oppositely directed. At the point where the values of $n_{\perp}$ for these waves coincide, they are linearly transformed [15-18]. It can be seen from Fig. 7 that it is precisely the position of the transformation point and not of the cutoff point which determines in this case the maximum value of q accessible to the wave. The condition for accessibility of cyclotron resonance for a wave propagating from the left (from the side of the strong magnetic field) is that the transformation point, defined by the condition

$$\tilde{B}^2 - 4\tilde{A}\tilde{C} = 0, \tag{37}$$

must not be in the region $u(x) > 1$. Substituting $u = 1$ in (37), we determine the maximum value $q = q_{max}$ corresponding to the maximum plasma density at the point at which the resonance still remains accessible:

$$q_{\max} = \frac{(1 + n_{\parallel}^2)^2}{4n_{\parallel}^2}, \qquad \frac{1}{3} \leqslant n_{\parallel}^2 \leqslant 1,$$

where $n_{\parallel}(x)$ is taken at the cyclotron-resonance point. At $n_{\parallel}^2 < {}^1/_3$ for the extraordinary wave, and also in the case of the ordinary wave, the value of $q_{max}$ is determined by the cutoff, and it can be found from Eqs. (47) with $u = 1$. The conditions obtained in this manner for accessibility of a resonance from the direction of the strong magnetic field, when the resonance layer is located at the density maximum, or to the left of it, are summarized in Fig. 8.

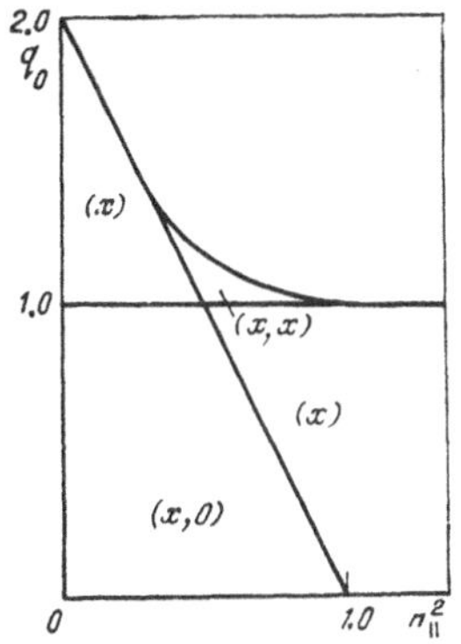


Fig. 8. Cyclotron-resonance accessibility in a tokamak. The regions where waves can exist at $\omega = \omega_B$ are marked by the symbols 0 and x in parentheses.

For waves launched from the outer side of the torus, the cyclotron-resonance accessibility conditions are noticeably more stringent. Thus, the extraordinary wave cannot reach resonance at all. The ordinary wave is cut off at $q_0 \gtrsim 1$ if $n_{\|}^2 < {}^1/_2$ and at $q_0 \geq 2(1 - n_{\|}^2)$ for values $n_{\|}^2 > {}^1/_2$. At $n_{\|}^2 \cong {}^1/_2$ and $q > 1$, the wave can, in principle, be transformed on the surface $q = 1$ into an extraordinary wave propagating first into the region of the denser plasma ($1 - q < {}^9/_8$), and then turning around toward the hybrid resonance [15-18].

In the case $n_y^2 \neq 0$, the only accessible part of the plasma is where $n_x^2 = n_\perp^2 - n_y^2 > 0$. Using the $n_\perp^2(x, n_{\|})$ curves shown in Figs. 5-7, it is easy to track qualitatively the change of the accessibility picture at $n_y^2 \neq 0$. It is clear that the accessibility of the resonance deteriorates noticeably with increasing $n_y^2$.

## 5. Cyclotron Damping of Waves in the Model of One-Dimensional Inhomogeneity

We consider now effects connected with thermal motion of electrons, paying principal attention, naturally, to wave damping. It is clear from the results of Section 2 that the damping is concentrated in a narrow layer $[\omega - \omega_B(x)]^2 \ll k_{\parallel}^2 v_T^2$, within which one can neglect the changes of the plasma density and of its temperature, while the nonuniformity of the magnetic field need be taken into account only in the argument of the dispersion function $Z[(\omega - \omega_B)/k_{\parallel} v_T]$ in Eq. (13). To calculate the absorption coefficient of the wave (the optical thickness of the plasma)

$$\Gamma = 2 \int_{-\infty}^{\infty} k_x'' dx ,$$

it is necessary in fact to calculate the integral [see (9)]

$$\int_{-\infty}^{\infty} f(\zeta)\, dx = L n_{\parallel} \beta \int_{-\infty}^{\infty} \mathrm{Im}\, [\zeta Z^{-1}(\zeta)]\, d\zeta,$$

where it is taken into account that $\zeta = (\omega - \omega_B)/k_{\parallel} v_T = x/L n_{\parallel} \beta$, where L is the characteristic size of the magnetic-field nonuniformity. In tokamaks, obviously, $L = R_0$. To calculate the integral it is convenient to shift the integration contour into the complex region $\mathrm{Im}\,(\omega - \omega_B) > 0$, so as to satisfy everywhere on the integration contour the condition $|\omega - \omega_B| \gg k_{\parallel} v$ and to be able to use the asymptotic form of the function $Z_1(\zeta) = 1 + (2\zeta^2)^{-1} + \ldots.$

Expanding the integrand in powers of $1/\zeta$ and taking into account the known relation $\mathrm{Im}\,(1/\zeta) = -\pi\delta(\zeta)$, which is equivalent to Landau's circuiting rule, and using also (5), we obtain after calculating $\partial\Lambda_0/\partial k_x$ (see [9]), a final expression for $\Gamma$:

$$\Gamma = \frac{\pi\beta^2\omega L}{4cn_x} \frac{q n_\perp^4 (2 - q - n_\perp^2)(2n_\perp^2 - 3 + 3q)^2}{[2(1-q) - n_\perp^2]^2 [(n_\perp^2 - 2 + 2q)^2 + q(1-q)]}, \tag{38}$$

$$n_\perp^2 = n_x^2 + n_y^2.$$

The cyclotron damping of the waves can be calculated also by a somewhat different method, using the dispersion equation in the weak-spatial-dispersion approximation, which can be written in the vicinity of the cyclotron resonance in the form

$$\Lambda_0 + \frac{2\beta^2}{(u-1)q} \Lambda_1 + \frac{(u-1) n_\perp^4}{4} = 0, \tag{39}$$

where $\Lambda_0$ and $\Lambda_1$ are defined by (13). Solutions of this equation at $|\beta^2/(1-u)| \ll 1$ can be easily obtained by perturbation theory, and take the form $n_x = n_{x0} + \beta^2\Lambda_2(1-u)^{-1}$, where $\Lambda_2=2\Lambda_1\left(q\frac{d\Lambda_0}{dn_x}\right)^{-1}\Big|_{n_x=n_{x0}}$ : $n_{x0}$ is the solution of the equation $\Lambda_0 = 0$. Putting $1 - u = 2x/L$ we get

$$\int k_x dx = \frac{\omega}{c} n_{x_0} x + \frac{\omega L}{2c}\beta^2\Lambda_2 \ln(u-1) + \text{const.} \tag{40}$$

A quasiclassical solution of (34) in the form $E \sim \exp(i\int k_x dx)$ is correct, naturally, not only on the real axis but also in the complex x plane. It can be uninterruptedly continued through those complex-plane sectors in which this solution decreases exponentially as $|x| \to \infty$. In our case, depending on the sign of $n_{x0}$, this is the upper or lower half-plane. On the other hand, the asymptotic representation (39) of the dispersion equation is valid only if $\mathrm{Im}(\omega - \omega_B) \geq 0$. We find thus that we can continue a solution of the form (34) into the complex plane only for one sign of $n_{x0}$, corresponding to a wave propagating away from the strong magnetic field. When the point $u - 1 = 0$ in the complex plane is circuited, the argument of the logarithm in (40) changes by $\pi$, and this leads to the appearance of an imaginary term in the expression for the quasiclassical integral

$$\frac{1}{2}\Gamma = \mathrm{Im}\int k_x dx = \pi\frac{\omega L}{2c}\beta^2\Lambda_2. \tag{41}$$

It is easy to verify that this quantity coincides with the wave absorption coefficient (38).

Two important conclusions can be drawn from the foregoing analysis. First, Eqs. (39)-(41) were derived with no restrictions whatever on $n_\parallel^2$ Accordingly, Eqs. (38) and (41) should describe correctly the absorption coefficient at $n_\parallel^2 = 0$. This fact is most surprising, inasmuch as at $n_\parallel^2 = 0$ we have for the ordinary wave $n_\perp^2 \to 1 - q$ and the last two terms in (13) increase without limit. The method of successive approximations in terms of the small parameter $1/\sigma$, used in the derivation of (13), turns out to be unsuitable. On going to the complex plane, however, these terms cancel each other, so that expansion in powers of $\beta^2/(1-u)$ is perfectly possible.

The second conclusion is that the cyclotron-absorption coefficient can be rigorously calculated only for a wave propagating away from the strong magnetic field. Of course, we can continue through the $\mathrm{Im}(\omega - \omega_B) > 0$ half-plane also the increasing quasiclassical solution and obtain the correct expression for the amplitude of the transmitted plane (the "absorption" coefficient is unchanged thereby). It is clear, however, that in this case the Stokes phenomenon can give rise to solutions corresponding to a reflected wave, so that the attenuation of the wave on passing through the resonant layer can be due not only to absorption of its energy but also to reflections [19-21].

Let us analyze now in detail expression (38) for the absorption coefficient $\Gamma$. We take it into account first of all that Eq. (5) can be transformed into

$$n_{\parallel}^2 = (n_{\perp}^2 - 1 + q)(n_{\perp}^2 - 2 + q)(2 - 2q - n_{\perp}^2)^{-1}. \qquad (42)$$

Thus, in the case $n_x = n_{\perp}$ we obtain a parametric representation of the absorption coefficient $\Gamma$ in terms of $n_{\perp}^2$. Specifying $n_{\perp}^2$, we determine the absorption coefficient $\Gamma$ and simultaneously from Eq. (42) the value of $n^2_{\parallel}$ corresponding to this absorption. The range of variation of $n_{\perp}^2$ can be easily determined from the form of the $n_{\parallel}^2 = n_{\parallel}^2(n_{\perp}^2)$ curves.

It can be seen from (42) that $n_{\parallel}^2 \to 0$ as $n_{\perp}^2 \to 2 - q$ and $n_{\perp}^2 \to 1 - q$. The first case corresponds to the extraordinary wave. As seen from (38), the absorption coefficient $\Gamma$ of the extraordinary wave then vanishes at the accuracy considered. The absorption of the ordinary wave at $n_{\parallel}^2 = 0$ remains finite and equal to

$$\Gamma = \frac{\pi\beta^2\omega L}{4c} q\sqrt{1-q}.$$

As $n_{\perp}^2 \to 2(1 - q)$ we have also $\Gamma \to \infty$. It can be seen from (42) that this case corresponds to $n_{\parallel}^2 \to \infty$. As already noted, under tokamak conditions a wave with $n_{\perp}^2 > 1$ cannot be launched from the vacuum into the interior of the plasma, and we shall therefore not consider this case.

At $[n_{\perp}^2 - 2(1 - q)]^2 = (q - 1)q$, the wave transformation described in Section 4 takes place precisely in the vicinity of the cyclotron resonance. Naturally, the geometric-optics approximation is then violated. This is indeed the

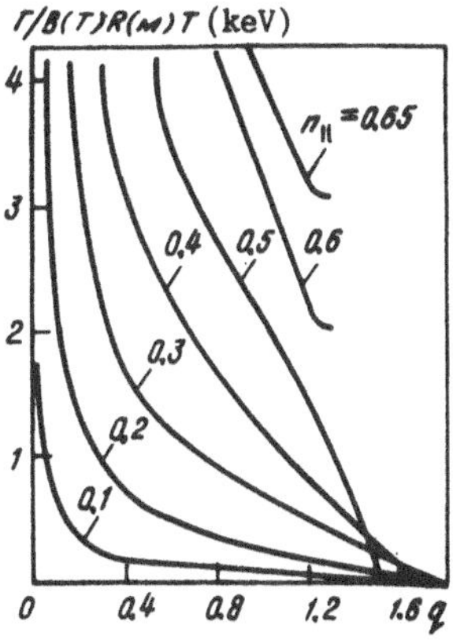


Fig. 9. Dependence of the optical thickness Γ of the cyclotron layer on q for the extraordinary wave.

reason why Γ becomes infinite at this point. In the interval $1 < q < 4/3$ at $n_{\|}^2 \geq 2q - 1 - 2\sqrt{q(q-1)}$ the cyclotron layer is inaccessible to waves ($n^2$ is complex in this case). At $0 < n_{\perp}^2 < 2(1-q) + \sqrt{q(q-1)}$ and in the interval $4/3 > q > 1$, Γ becomes formally negative. This is due to reversal of the sign of the wave's group velocity. In this band we can put $n_x = -n_{\perp} < 0$ for the circuiting considered above.

A more complete idea of the dependence of the absorption coefficient Γ on the plasma density (on the perimeter $q|_{u=1}$) and on the value of $n_{\|}^2|_{u=1}$ can be obtained from Figs. 9 and 10, which show the numerical calculation results.

Worthy of special attention is the absorption of the extraordinary wave at $n_{\|}^2 = 0$. It has already been noted that, accurate to terms of order $\beta^2$, there is no absorption in this case. We can use for this calculation relativistic equations for $\varepsilon_{\alpha\beta}$ such as (19)-(21), but calculated accurate to $\beta^4$. It is convenient to use for the calculation an asymptotic representation of the functions $F_p$ and encircle the resonance point $\omega = \omega_B$ in the complex plane in the region $|\omega - \omega_B| \gg \omega\beta^2$. As a result we have [22]

$$\Gamma = (2-q)^{3/2}\left(-1 + \frac{q}{4} + \frac{5}{4q}\right)\frac{\pi R_0 \omega}{c}\beta^4.$$

The results presented in this section are valid only for a sufficiently dense plasma $q \gg \beta$.

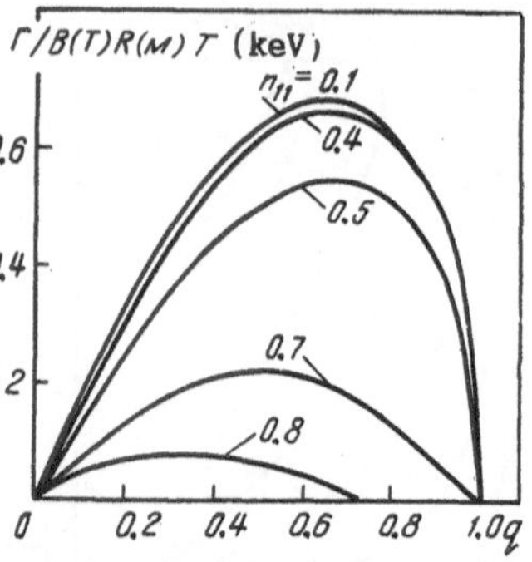


Fig. 10. Dependence of the optical thickness Γ of the cyclotron layer on q for the ordinary wave.

## 6. Reflection of Waves from a Resonance Region

It has already been noted that in the vicinity of cyclotron resonance we can have not only absorption but also reflection of waves [19-21]. We consider this phenomenon using as an example an ordinary wave propagating perpendicular to the magnetic field (at $n_\parallel = 0$), neglecting for simplicity the changes of q in the resonant layer. We use a model equation obtained from the dispersion equation $n^2 = \varepsilon_{zz}$ by making the substitution $k_x \to -i(d/dx)$:

$$(1+\alpha(x))\frac{d^2E}{dx^2}+\frac{\omega^2}{c^2}(1-q)E=0, \tag{43}$$

where

$$E=\int E_z dx \text{ (see [20])}, \quad \alpha=\frac{q}{2}F_{7/2}\left(\frac{\omega-\omega_B}{\omega\beta^2}\right).$$

The function $F_{7/2}$ is defined in Eq. (22). It is convenient to transform in (43) to the new variables

$$\psi=\sqrt{n}\,E, \quad \xi=\frac{\omega}{c}\int_0^x n(x')\,dx', \quad n^2(x)=\frac{1-q}{1+\alpha(x)},$$

the "quasiclassical" refractive index, and the origin is chosen at the resonance point. In the new variables, we get

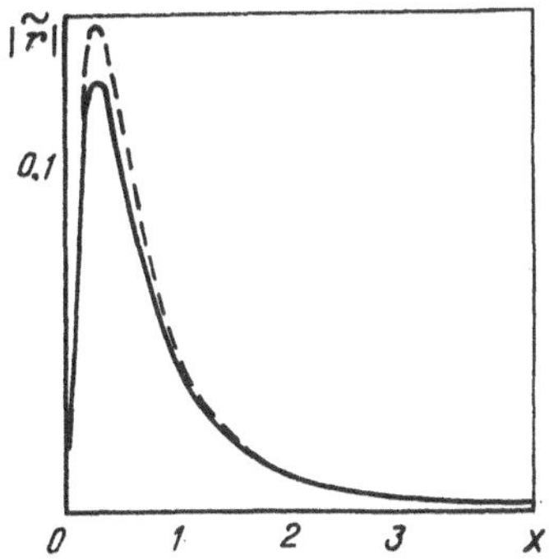


Fig. 11. Coefficient of ordinary-wave reflection from resonance layer vs. χ: dashed – calculated using Eq. (44), solid – numerical calculation of Eq. (43) [20].

$$\frac{d^2\psi}{d\xi^2}+\psi=V\psi,$$

where $(\omega^2/c^2)V = n''/2n^3 - 3n'^2/4n^4$, and the prime denotes here differentiation with respect to the coordinate x. This equation can be solved at $\alpha \ll 1$ (and at arbitrary $|\alpha|L$) by perturbation theory, using as the zeroth approximation $\psi = e^{-i\xi}$ (i.e., a quasiclassical wave incident from the direction of the weaker magnetic field). In first-order approximation, as $x \to \infty$, we then get $\psi = e^{-i\xi} + \tilde{r}e^{i\xi}$, where the reflection coefficient is defined by the integral

$$\tilde{r}=-\frac{i\omega}{2c}\int_{-\infty}^{\infty} Ve^{-2i\xi}n d\xi.$$

In the calculation of $\tilde{r}$ we must integrate the first term of the integrand by parts and retain only terms linear in $\alpha$. It is convenient next to modify the integration contour, drawing it from minus infinity around the point x = 0 to minus infinity. The integral is then determined by that term of $F_{7/2}$ which contains a branch point. As a result we get

$$\tilde{r}=(1-i\chi)^{-7/2}\Gamma, \qquad (44)$$

where $\Gamma$ is the power-absorption coefficient defined in (38), $\chi = \sqrt{(1-q)}\,\beta^2(\omega L/c)$, and the magnetic field is assumed to be a linear function $\omega_B = \omega(1 - x/L)$.

It can be seen from the above equation that the modulus of the reflection coefficient is a maximum at $\chi = \sqrt{2/5}$, when the width of the absorption region is commensurate with the wavelength. Figure 11 shows plots of the modulus of the reflection coefficient, calculated from Eq. (44) and by using numerical integration of Eq. (43) [20]. It can be seen that Eq. (44) approximates fairly well the reflection coefficient $\tilde{r}$ even at $\tilde{q} = 0.5$, the requirement $q \ll 1$ for the applicability of the calculation notwithstanding.

Problem. Using perturbation theory, calculate the reflections by the harmonics of the cyclotron frequency.

Solution. One must start out in the calculation from Maxwell equations in which Eqs. (19)-(21) are used for the dielectric tensor, and $n_\perp^2$ can be replaced in the corrections to the tensor by the corresponding solution of the dispersion equation for a cold plasma. The remaining calculations are similar to those considered above. If the variation of the magnetic field is neglected everywhere except in the argument of the function $F_p$, we obtain an expression for the (amplitude) reflection coefficient and for the (power) absorption coefficient $\Gamma$:

$$\Gamma_1 = \frac{\pi q R_0}{\omega c}\cdot\frac{(s\sqrt{1-q})^{2s-1}}{2^{2s}s!}\beta^{2s}, \qquad \tilde{r}_1 = \Gamma_1 (1 - i\chi)^{-(s+3/2)},$$

$$\Gamma_2 = \frac{\pi\omega_p^2 R\left(\beta^2 s^2 N_2^2\right)^{s-1}}{\omega c 2^{2s-1}}\,\frac{\left(\omega^2+\omega\omega_B-\omega_p^2\right)^2(\omega-\omega_B)^2}{N^2\left(\omega^2-\omega_B^2-\omega_p^2\right)^2\omega^2},$$

$$\tilde{r}_2 = \Gamma(1 - i\chi_2)^{-s-3/2}, \quad \chi_0 = \sqrt{1-q}\,\frac{\omega R}{c}\,\beta^2, \quad s > 2, \quad \frac{\omega_p^2}{\omega^2} \ll 1.$$

The subscripts 1 and 2 pertain here to the ordinary and extraordinary waves, respectively; $N_2$ is the refractive index for the extraordinary wave in the cold-plasma approximation.

## 7. Linear Transformation of Waves Near the Upper Hybrid Resonance

As mentioned above, one more plasma region in which spatial dispersion is significant is the vicinity of the hybrid resonance $\varepsilon = 0$. According to (35), we have for the extraordinary wave in a cold plasma as $\varepsilon \to 0$

$$n_{\perp}^2 = -\tilde{B}/\tilde{A} = -(\eta n_{\parallel}^2 + g^2)/\varepsilon, \tag{45}$$

so that $n_{\perp}^2 \to \infty$ as $\varepsilon \to 0$.

The ensuing singularity is removed by taking spatial dispersion into account; it suffices [4] to use the weak spatial-dispersion approximation, i.e., regard $\lambda = k_{\perp}^2 v_T/\omega_B$ as small and retain only the lowest-order terms in the expansion of the tensor $\varepsilon_{\alpha}$ in powers of $\lambda$. It is actually necessary in this case to take the thermal correction into account only in the component $\varepsilon_{xx}$, whose "cold" part is anomalously small in the vicinity of the hybrid resonance. In this approximation we have [4]: $\varepsilon_{xx} = \varepsilon - \gamma n_{\perp}^2$, $\gamma = 1.5q\beta^2(1 - u)^{-1}(1 - 4u)^{-1}$, while the remaining components of the tensor $\varepsilon_{\alpha\beta}$ are described by Eqs. (1) and (2). When this expression is substituted in the dispersion equation (35), the order of the latter increases and a new oscillation branch appears. For solutions with large refractive index $n_{\perp}^2 \gg 1$ we now obtain in place of (45)

$$n_{\perp}^2 = \frac{\varepsilon \pm \sqrt{\varepsilon^2 + 4\gamma\tilde{B}}}{2\gamma}. \tag{46}$$

When the plasma is heated at the fundamental harmonic of the cyclotron frequency, we have $\gamma < 0$ and Eq. (46) describes two branches of oscillations that propagate at $\varepsilon < 0$. The lower branch goes over far from the hybrid resonance into the extraordinary wave (45); for the upper branch we have asymptotically $n_{\perp} \to \varepsilon/\gamma$. This is the limiting (as $k_{\perp} \to 0$) expression for the Bernstein-mode refractive index

$$k_{\alpha}k_{\beta}\varepsilon_{\alpha\beta} = 0. \tag{47}$$

The dispersion equation (35) with the thermal correction describes thus the linear transformation of the extraordinary wave into a Bernstein mode. The corresponding analysis [15, 23] shows that, in a weakly inhomogeneous plasma, assuming accessibility of the hybrid resonance, the transformation is practically complete: the extraordinary wave incident on the resonant layer is reflected from the latter in the form of a Bernstein mode. The wave propagating from the weak magnetic field is likewise transformed at the upper hybrid resonance point, owing to the below-barrier penetration through the opacity region. Under tokamak conditions, however, this effect is negligibly small. With increasing distance from the hybrid resonance, the refractive index of the Bernstein wave increases rapidly, and this wave is completely damped, on account of the Čerenkov radiation, in the plasma region between the upper-hybrid and cyclotron-resonance surfaces. Since the transformation efficiency is independent of $T_e$, it turns out that the extraordinary wave launched from the side of the strong magnetic field is totally absorbed in the plasma, regardless of $T_e$. At low temperatures the energy is absorbed via transformation and damping of the Bernstein mode, and as $T_e$ increases this process begins to encounter competition from the direct damping of the electromagnetic wave in the cyclotron layer, a process that becomes dominant at high temperatures.

## 8. Eikonal Equation in the Vicinity of the Cyclotron Resonance and Ray Trajectories of Waves

The geometric-optics approximation in its classical formulation was developed primarily for differential equations. Wave equations for a medium with spatial depression are, as already noted, integrodifferential, so that a suitable generalization of this approximation is necessary. In cases when the anti-Hermitian part of the dielectric tensor is small, this generalization is conceptually rather trivial (see, e.g., [24]), albeit somewhat elaborate. In the immediate vicinity of electron-cyclotron resonance, however, the dielectric tensor is essentially anti-Hermitian [see (9)], so that the picture becomes quite complicated. We derive geometric-optics equations suitable for the vicinity of cyclotron resonance by starting from the Maxwell equation and from the linearized kinetic equation

$$\mathrm{rot}\,\mathbf{E} = -\frac{1}{c}\frac{\partial \mathbf{B}}{\partial t}, \quad \mathrm{rot}\,\mathbf{B} = \frac{1}{c}\frac{\partial \mathbf{E}}{\partial t} + \frac{4\pi}{c}\mathbf{j}, \tag{48}$$

$$\left.\begin{aligned} &\frac{\partial f}{\partial t} + v\frac{\partial f}{\partial \mathbf{r}} + \frac{e}{mc}[\mathbf{v}\mathbf{B}_0]\frac{\partial f}{\partial \mathbf{v}} = -\frac{e}{m}\mathbf{E}\frac{\partial F}{\partial \mathbf{v}},\\ &\mathbf{j} = e\int \mathbf{v} f d\mathbf{v}. \end{aligned}\right\} \tag{49}$$

Here $\mathbf{B}_0$ is the external magnetic field; F is the unperturbed electron distribution function in velocity. We assume F to be an isotropic function in velocity space, $F \equiv F(v^2, \mathbf{r})$. By this token we neglect a large number of effects due to the inhomogeneity of the plasma, particularly those connected with electron drift. It can be assumed that these effects are inessential for the electromagnetic waves of interest to us, whose phase velocities are of the order of that of light.

To solve the kinetic equation it is convenient to introduce a local coordinate frame in which one of the Cartesian axes (e.g., $u_z'$, where $u_\alpha$ is a velocity component along the new axes) is directed along the external magnetic field. Since we are dealing with a simple rotation of the coordinate frame, we can write

$$u_\alpha = s_{\alpha\beta} v_\beta, \quad u^2 = v^2, \quad v_\alpha = s_{\alpha\beta}^{-1} u_\beta,$$

where $s_{\alpha\beta}$ are the cosines of the angles between the old and new coordinate frames. Obviously, $s_{\alpha\beta}^{-1} = s_{\beta\alpha}$.

In the new variables, the kinetic equation takes the form

$$\frac{\partial f}{\partial t} + \mathbf{u}\overset{\leftrightarrow}{s}^{-1}\left(\frac{\partial f}{\partial \mathbf{r}} + \frac{\partial f}{\partial u_\alpha}\frac{\partial s_{\alpha\beta}}{\partial \mathbf{r}} s_{\beta\gamma}^{-1} u_\gamma\right) - \omega_B \frac{\partial f}{\partial \varphi} = \frac{e}{T}(\mathbf{E}\overset{\leftrightarrow}{s}^{-1}u) F, \quad (50)$$

where $\omega_B = \omega_B(\mathbf{r})$; it was taken into account that $\partial F/\partial v = -(mv/T)F$, where T is the electron temperature; $\varphi$ is an angle, in velocity space, in a plane perpendicular to the magnetic field; $u_\alpha$ is a velocity component in the "Cartesian" coordinate frame.

In the geometric-optics approximation, it is natural to seek for solutions of the system (48)-(50) in the form

$$A = \sum_n A^{(n)}(\mathbf{r}) \exp\{i\psi(\mathbf{r}) - i\omega t\},$$

where A stands for any of the quantities $\mathbf{E}$, $\mathbf{B}$, and f. The amplitudes A(r), which vary smoothly in space and are represented by a formal geometric-optics series, are then obtained from the equations

$$\left.\begin{aligned} &[\mathbf{k}\mathbf{E}^{(0)}] - \frac{\omega}{c}\mathbf{B}^{(0)} = 0, \\ &[\mathbf{k}\mathbf{B}^{(0)}] + \frac{\omega}{c}\mathbf{E}^{(0)} + \frac{4\pi i}{c} j^{(0)} = 0, \qquad j^{(0)} = e\overset{\leftrightarrow}{s}^{-1}\int \mathbf{u} f du, \\ &i(\omega - \mathbf{k}\mathbf{u}) f^{(0)} + \omega_B \frac{\partial f^{(0)}}{\partial \varphi} + \frac{e}{T}(\mathbf{E}^{(0)}\mathbf{u}) F(u^2) = 0; \end{aligned}\right\} \quad (51)$$

$$\left.\begin{aligned} &[\mathbf{kE}^{(n)}] - \frac{\omega}{c}\mathbf{B}^{(n)} = \mathrm{i}\,\mathrm{rot}\,\mathbf{E}^{(n-1)},\\ &[\mathbf{kB}^{(n)}] + \frac{\omega}{c}\mathbf{E}^{(n)} + \frac{4\pi\mathrm{i}}{c}\mathbf{j}^{(n)} = \mathrm{i}\,\mathrm{rot}\,\mathbf{B}^{(n-1)},\\ &\mathbf{j}^{(n)} = e\overset{\leftrightarrow}{s}{}^{-1}\int \mathbf{u} f^{(n)}\,d\mathbf{u}, \qquad n = 1, 2, 3, \end{aligned}\right\} \tag{52}$$

where $\mathbf{k} = \nabla\psi$ is the wave vector in the geometric-optics approximation, $\mathbf{k}\mathbf{u} = k_\alpha s_{\beta\alpha} u_\beta$.

It is readily seen that the system (51) for zeroth-approximation quantities coincides formally with the corresponding equations of the homogeneous-plasma equations (see, e.g., [7]), so that we can write

$$\left.\begin{aligned} &\mathbf{j}^{(0)} = \overset{\leftrightarrow}{\sigma}{}^{(\mathrm{r})}(\omega, \mathbf{k})\,\mathbf{E}^{(0)},\\ &\sigma_{\alpha\beta}^{(\mathrm{r})}(\omega, \mathbf{k}) = s_{\alpha'\alpha} s_{\beta'\beta}\sigma_{\alpha'\beta'}(\omega, \overset{\leftrightarrow}{s}\mathbf{k}), \end{aligned}\right\} \tag{53}$$

where $\sigma_{\alpha'\beta'}(\omega, \overset{\leftrightarrow}{s\mathbf{k}})$ is the conductivity tensor in a coordinate frame whose z axis is directed along the magnetic field. It coincides in form with the corresponding homogeneous plasma expression in which the electron temperature and the cyclotron and plasma frequencies must now be regarded as functions of the coordinates as parameters. To emphasize that this tensor depends on the wave-vector components along the magnetic field and along directions perpendicular to it (and not on the components along the axes of the initial coordinate frame), the dependence on $s\mathbf{k}$ is indicated in the argument of $\sigma_{\alpha'\beta'}$. Naturally, the connection between the conductivity tensor $\sigma_{\alpha\beta}{}^{(\mathrm{r})}$ in the initial coordinate frame and $\sigma_{\alpha'\beta'}$ is given in the geometric-optics approximation, as can be seen from (53), by the usual rules for tensor transformations in rotations of the coordinate frame.

The first-approximation expression for the current density $\mathbf{j}^{(1)}$ is obtained from the system (52) and can be written in the form

$$\mathbf{j}^{(1)} = \overset{\leftrightarrow}{\sigma}{}^{(\mathrm{r})}\mathbf{E}^{(1)} + \overset{\leftrightarrow}{e}\overset{\leftrightarrow}{s}{}^{-1}\int \mathbf{u} h\,d\mathbf{u}, \tag{54}$$

where the function h is defined by the equation

$$\mathrm{i}(\omega - \mathbf{ku})h + \omega_B\frac{\partial h}{\partial\varphi} = u_{\alpha'} s_{\beta\alpha}\left(\frac{\partial f^{(0)}}{\partial x_{\alpha'}} + \frac{\partial f^{(0)}}{\partial u_\alpha} s_{\beta'\beta}\frac{\partial s_{\alpha\beta'}}{\partial x_{\alpha'}} u_{\beta'}\right).$$

The geometric-optics approximation is applicable, naturally, only to weakly damped waves. Equations (48)-(53), however, provide a useful method for investigating wave propagation in a smoothly inhomogeneous plasma under less stringent requirements, viz., under conditions when the Hermitian part of the conductivity tensor $\sigma_{\alpha\beta}' = {}^1/_2(\sigma_{\alpha\beta} + \sigma_{\alpha\beta}^*)$ is substantially smaller than the non-Hermitian part $i\,\sigma_{\alpha\beta}'' = {}^1/_2(\sigma_{\alpha\beta} - \sigma_{\alpha\beta}^*)$. In this case, it is convenient to neglect in the zeroth approximation of geometric optics the wave attenuation [i.e., neglect the terms $\sigma_{\alpha\beta}'$ in Eqs. (51)] and include the corresponding term in the first-approximation equations. As a result we get

$$\left.\begin{aligned} &D_{\alpha\beta}(\mathbf{k},\omega)E_\beta^{(0)} = \Big(k^2\delta_{\alpha\beta} - k_\alpha k_\beta - \frac{\omega^2}{c^2}\varepsilon_{\alpha\beta}'\ E_\beta^{(0)} = 0,\\ &[\mathbf{kE}^{(1)}] - \frac{\omega}{c}\mathbf{B}^{(1)} = i\,\mathrm{rot}\,\mathbf{E}^{(0)},\quad [\mathbf{kB}^{(1)}] + \frac{\omega}{c}\mathbf{E}^{(1)}\\ &+\frac{4\pi i}{c}j^{(1)} = i\,\mathrm{rot}\,\mathbf{B}^{(0)} - \frac{4\pi i}{c}\overset{\leftrightarrow}{\sigma}{}^{(r)}\,\mathbf{E}^{(0)}, \end{aligned}\right\} \tag{55}$$

where $\varepsilon_{\alpha\beta} = \delta_{\alpha\beta} + (4\pi i/\omega)\,\sigma_{\alpha\beta}^{(r)}$ is the plasma electric tensor in the initial coordinate frame. The condition for solvability of the system (55) yields, as for a homogeneous plasma, the "dispersion" equation

$$D = \det|D_{\alpha\beta}| = 0. \tag{56}$$

In an inhomogeneous plasma, Eq. (56) can be used directly to find the coordinate dependence of the wave vector $\mathbf{k}(\mathbf{r})$ only in the case of one-dimensional inhomogeneity, when the eikonal $\psi(\mathbf{r})$ can be represented in the form $\psi(\mathbf{r}) = k_x dx + k_y y + k_z z$ (the x axis is directed along the inhomogeneity, while $k_y$ and $k_z$ are constants). In the general case it is convenient to calculate the ray trajectories and track the variation of $\mathbf{k}$ along these trajectories by starting from the system [24, 25]

$$\left.\begin{aligned} &\frac{d\mathbf{r}}{dt} \equiv v_g = \frac{\partial\omega}{\partial\mathbf{k}} = -\frac{\partial D}{\partial\mathbf{k}}\left[\frac{\partial D}{\partial\omega}\right]^{-1},\\ &\frac{d\mathbf{k}}{\partial t} = \frac{\partial D}{\partial\mathbf{r}}\left[\frac{\partial D}{\partial\omega}\right]^{-1}, \end{aligned}\right\}$$

which is convenient for the treatment of problems dealing with the evolution of wave packets. In investigations of stationary propagation of waves it is frequently convenient to use these equations in a different form:

$$\left.\begin{aligned}\frac{d\mathbf{r}}{\partial s} &= -\frac{\partial D}{\partial \mathbf{k}}\left|\frac{\partial D}{\partial \mathbf{k}}\right|^{-1} \operatorname{sign}\frac{\partial D}{\partial \omega},\\ \frac{\partial \mathbf{k}}{\partial s} &= \frac{\partial D}{\partial \mathbf{r}}\left|\frac{\partial D}{\partial \mathbf{k}}\right|^{-1} \operatorname{sign}\frac{\partial D}{\partial \omega},\end{aligned}\right\} \tag{57}$$

where s is the arc length along the ray trajectory.

From the mathematical point of view, Eqs. (57) are the equations of the characteristics of the dispersion equation (56), which is a linear partial differential equation of first order relative to the function $\psi(\mathbf{r})$. These equations remain in force also in the case when the tensor $\varepsilon_{\alpha\beta}$ is strongly non-Hermitian, but their physical meaning as rays along which the energy propagates is lost in this case. To analyze this problem we use the small parameter $\beta = v_T/c$ introduced in Section 2. In analogy with the case of a homogeneous plasma, one can expect here the smallness of $\beta$ to permit under certain conditions [under certain restrictions on the inhomogeneity scales of the plasma and on the field $\mathbf{E}^{(0)}(\mathbf{r})$] to identify, in the lowest order in $\beta$, the total energy flux with the electromagnetic-energy flux, and calculate the wave damping either from Eq. (30) or by integrating along the ray the wave-vector imaginary part calculated from (13). The ray trajectories and the variation of $\mathbf{k}$ along the trajectories are most convenient to calculate here by using Eq. (56) with $D(\omega, \mathbf{k})$. This corresponds to the cold-plasma approximation although, of course, $D(\omega, \mathbf{k})$ can be chosen as before to be any function of $\tilde{D}(\omega, \mathbf{k})$ which differs little from the exact expression, by an amount of the order of $\beta$, and ensures real solutions for $\mathbf{k}(\omega, \mathbf{r})$. It is clear from Section 3 [see Eq. (31)] that this choice will change the phase shift of the complex amplitude $E^{(0)}(\mathbf{r})$, but not the wave damping.

In other words, this leeway in the choice of $\tilde{D}(\omega, \mathbf{k})$ is connected with a certain leeway in the subdivision of the entire phase of the field into parts due to the eikonal $\psi$ and to the complex phase of the amplitude.

We consider now the conditions under which this neglect of the second term in (54) is legitimate. Just as in [13], we neglect the curvature of the magnetic-field force lines and the shear (i.e., we put $s_{\alpha\beta} = \delta_{\alpha\beta}$). Clearly, under tokamak conditions this neglect is justified, since $R_0 \gg a$ and the poloidal field is usually small. It can be shown [26] that, in order of magnitude,

$$\frac{4\pi e}{\omega}\int hv\,dv = \frac{4\pi}{\omega}\frac{\partial}{\partial \mathbf{r}}\frac{\partial \overleftrightarrow{\sigma}}{\partial \mathbf{k}}\mathbf{E}.$$

We thus must stipulate

$$\operatorname{div} s_0 \sim \beta k s_0 \gg \left| \frac{4\pi}{\omega}\frac{\partial^2 \sigma_{\alpha\beta}}{\partial \mathbf{k}\partial \mathbf{r}} E_\alpha^* E_\beta \right|. \tag{58}$$

The components $\varepsilon_{\alpha\beta}$ and $\sigma_{\alpha\beta}$ depend on the spatial coordinates, in particular, also via the argument of the dispersion function Z. Differentiation with respect to $k_{\parallel}$, as already noted in Section 3, does not change the order of magnitude of the considered terms with respect to the parameter $\beta$, but differentiation with respect to the coordinate, just as with respect to frequency, leads to the appearance of the large factor $\beta^{-1}$. It turns out, as a result, that the condition (58) leads to the inequalities

$$n_{\parallel}\beta k_{\parallel} L \gg 1, \qquad \beta^2 (n_{\parallel}^2 + \beta^2) k_{\perp}^2 L^2 \gg 1, \tag{59}$$

where $L_{\parallel}$ and $L_{\perp}$ are characteristic scales of the inhomogeneity along and across the magnetic field.

These conditions are easily met in experiments on large tokamaks.

## 9. Applicability of Geometric Optics in the Vicinity of Cyclotron Resonance

When discussing the conditions of applicability of geometric optics in the ray treatment to problems of cyclotron absorption of waves under real conditions and in real installations, three groups of problems are encountered. First are the usual geometric-optics problems connected with the intersection of the "quasiclassical" solutions with the degeneracy of the wave polarizations. Mention must next be made of the question of wave diffraction, which becomes particularly acute in the case of non-one-dimensional inhomogeneity. Finally, in the vicinity of cyclotron resonance we encounter a new problem due to spatial dispersion, i.e., to the nonlocality of the connection between the electric field of the wave and the current in the plasma. The point is that, as follows from [9, 14, 27, 28], the inhomogeneity of the magnetic field alters the character of the resonant interaction of the electrons with the wave, for owing to the motion of the electron along the inhomogeneous magnetic field, a phase mismatch takes place between its cyclotron rotation and the wave field, so that the electrons are no longer at resonance with the wave. Becoming nonresonant, the electron "forgets" the phase of the field, and this results in coherently modulated electron beams. Owing to the thermal scatter in velocity, these beams begin to interfere far from the resonance region, but the "phase memory" can persist under certain conditions over considerable distances, and can lead to phenomena such as spatial echo, induced transparency of the barriers, and others [29]. Moreover, under certain conditions the change of the character of the resonant interaction of the electrons with the wave influences substantially the field structure in the resonance zone [9], so that approximate allowance for the nonlocality by using derivatives [representation of the current in the form (25)] becomes, in general, incorrect. This was seen already in the example, considered in Section 3, of a quasimonochromatic wave with Gaussian distribution of the electric field [see the Problem in Section 3]. Although many specific results on this subject have been obtained by now within the framework of various plasma-inhomogeneity models, the question of the role of nonlocal effects within the framework of the geometric-optics approximation has been very little studied. We shall attempt to set forth on this subject only a few considerations that are, naturally, of preliminary character.

At first glance, the expansion in Section 9 in the small parameter $1/kL$ seems to corroborate the geometric-optics approach (at any rate if $\beta \ll 1$). This, however, is not so, as can be verified by calculating the amplitude of the distribution function f without resorting to expansion in the small geometric-optics parameter. We confine ourselves for simplicity to a model with a nonuniform magnetic field having straight force lines, neglect the inhomogeneity of the plasma density and temperature, and take into account the resonant component of the field $\mathbf{E} = {}^1/{}_2 E_1(\mathbf{e}_x + i\mathbf{e}_y)$. Substituting $\tilde{f} = f \exp[i(\psi - \omega t)]$, $\mathbf{E} = \mathbf{E} \exp[i(\psi - \omega t)]$ in Eq. (50), and letting $k_\perp v_T/\omega_B \to 0$, we obtain

$$i(\omega - k_z v_z) f - \omega_B \frac{\partial f}{\partial \varphi} - v_z \frac{\partial f}{\partial v_z} = \frac{e}{2T} E_1 v_\perp F e^{i\varphi}.$$

The solution of this equation is

$$f = -\frac{e v_\perp F}{2T v_z} \int\limits_{\mp\infty}^{z} E_1 \exp\left[ i \int\limits_{z'}^{z} \frac{\omega - \omega_B(z) - k_z v_z}{v_z} dz'' \right] dz',$$

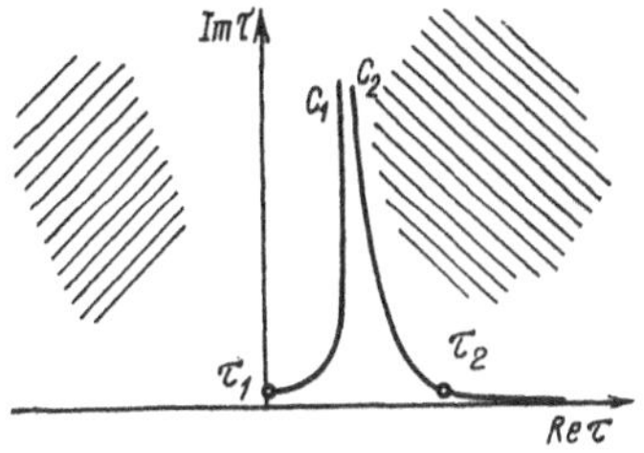


Fig. 12. Contour of integration in (61).

where the sign of the lower integration limit is the sign of $v_z$, being negative at $v_z > 0$ and positive at $v_z < 0$.

We transform in the integrals to the variables $z' = \xi + z$, $z'' = \xi' + z$ and represent the continuous functions $E_1(z + \xi)$, $\omega_B(z + \xi')$, and $k_z(z + \xi')$ as expansions in powers of $\xi$ and $\xi'$, respectively. Integration over the transverse velocities yields for the component $j_1 = -e\int(v_x - iv_y)f d\mathbf{v}$ the expression

$$4\pi j_1 = \frac{\omega_p^2 E_1}{\sqrt{\pi}} \int_0^\infty d\tau \exp\left[i(\omega - \omega_B)\tau\right] \int_{-\infty}^{\infty} \exp\left[-\frac{v_z^2}{v_T^2} - i\tau v_z\left(k_z - \frac{\omega_B'\tau}{2}\right) + iv_z^2\tau^2\left(\frac{k_z'}{2} - \frac{\omega_B''\tau}{6}\right)\right]\frac{dv_z}{v_T}, \tag{60}$$

where we have introduced a new integration variable $\tau = \xi/v_z$,

$$\omega_B' = d\omega_B(z)/dz, \quad k_z = k_z(z), \quad k_z' = dkz(z)/dz;$$

for simplicity, the derivatives of the electric fields were left out of the pre-exponential factor.

Far from the extrema of the magnetic field, in the case of weak inhomogeneity, we can expand the exponential under the integral sign in (60) in powers of $(k_z''/2 - \omega_B''\tau/6)$ and integrate over the velocity. The leading term of the expansion leads to the expression

$$j_1 = \frac{\omega_p^2 E_1}{4\pi} \int_0^\infty d\tau \exp\left[i(\omega - \omega_B)\tau - \frac{k_z^2 v_T^2 \tau^2}{4}\left(1 - \frac{\omega_B'\tau}{2k_z}\right)^2\right]. \tag{61}$$

We consider first a wave propagating toward the increasing magnetic field ($\omega_B'/k_z > 0$). It is convenient then to represent the integral in (61) by a sum of two integrals over the contours $C_1$ and $C_2$ (Fig. 12). The main contribution to the integral over $C_1$ is made by the start of the path and by the saddle point $\tau_1 = 2i(\omega - \omega_B)/k_z^2 v_T^2$. If condition (59) is satisfied on the entire $C_1$ segment that determines the main contribution to the integral, the terms with $\omega_B'$ in the argument of the exponential are small and can be neglected. The remaining integral coincides with one of the representations of the dispersion function $Z[(\omega - \omega_B)/k_z v_T]$.

The integral over the contour $C_2$ is determined by the saddle point $\tau_2 = 2k_z/\omega_B' + 2i(\omega - \omega_B)/k_z^2 v_T^2$. Simple calculations lead to an equation for the total current:

$$\tilde{j}_1 = \sigma_{11}(k_z) E_1(z) \exp[i k_z(0) z + i\psi(0)] + 2\sigma_{11}'(k_z) E_1(z) \exp[-ik_z(0) z].$$

We have taken into account here the fact that in the vicinity of the resonance we can put $(\omega - \omega_B)/\omega_B' = -z$, $\psi(z) = \psi(0) + k_z(0)z$. We find thus that when the wave is incident from the direction of the weak magnetic field, a current with an oppositely directed wave vector is generated in the resonant layer $[(\omega-\omega_B)/k_z v_T]^2 \sim 1$ and excites, naturally, a reflected wave. In a collisionless plasma and at a real frequency, this current does not tend to zero as $L_\parallel \to \infty$. To obtain the corresponding limiting transition in the case of a homogeneous plasma, we must put, in accordance with the Landau rule, $\mathrm{Im}\,\omega > 0$, so that $\mathrm{Re}[ik_z(\omega - \omega_B)/\omega_B'] \to -\infty$ as $\omega_B' \to 0$.

Consider now a wave propagating toward the decreasing magnetic field. At $\omega_B'/k_z < 0$, the integrand in (61) decreases monotonically after passing through the first saddle point $\tau_1$, so that the integral reduces at $\omega_B/k_z^2 v_T L_\parallel \ll 1$ to the dispersion function, and we obtain

$$j_1 = \sigma_{11} E_1.$$

We can conclude, therefore, that the expansion, described in Section 8, in powers of the geometric-optics small parameter $1/kL$, is applicable, strictly speaking, only to waves that propagate from the strong magnetic field. For waves propagating in the opposite direction, the geometric-optics series apparently diverges.

Although the longitudinal inhomogeneity of the magnetic field in tokamaks is weak (it is connected with the poloidal

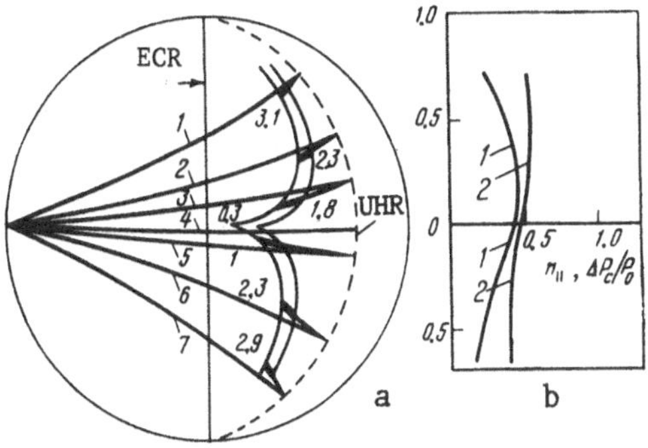


Fig. 13. Ray trajectories of extraordinary waves: a) projection of ray trajectories on the minor section of the torus. Trajectory segments where more than 99% of the Bernstein-wave power is absorbed are drawn as thick lines. The characteristic value of $n_\parallel$ in the absorption region is marked on the endpoint of each trajectory ($q_0 = 0.6$, $u_0 = 1$, $\delta = 48°$): 1) $\alpha = 27°$; 2) $\alpha = 13°$; 3) $\alpha_0 = 6°$; 4) $\alpha = -3°$; 5) $\alpha = -6°$; 6) $\alpha = -13°$; 7) $\alpha = 27°$. b) Dependence of the relative absorption of the extraordinary-wave power $1-e^{-\Gamma}$ in the electron-cyclotron resonance region (ECR, curve 1) and of the value of $n_\parallel$ on the position of the intersection point of the cyclotron-resonance line and the ray trajectory (curve 2).

field), the effect considered may, generally speaking, turn out to be substantial. One might think, however, that at $n_\perp \gtrsim n_\parallel$ it will not affect adversely the heating efficiency, since a wave that is bounded (along the z axis) can propagate as before toward a resonance in a direction transverse to the magnetic field.

## 10. Ray Trajectories and Wave Absorption in Tokamaks

As already noted, many numerical calculations have been made by now of ray trajectories and of wave damping under real conditions of experiments on electron-cyclotron heating of a plasma [5, 6, 30-39], and a sufficiently complete pattern was traced for the wave phenomena in a tokamak plasma in a given frequency band. The main features of this pattern were reported already in the first papers on

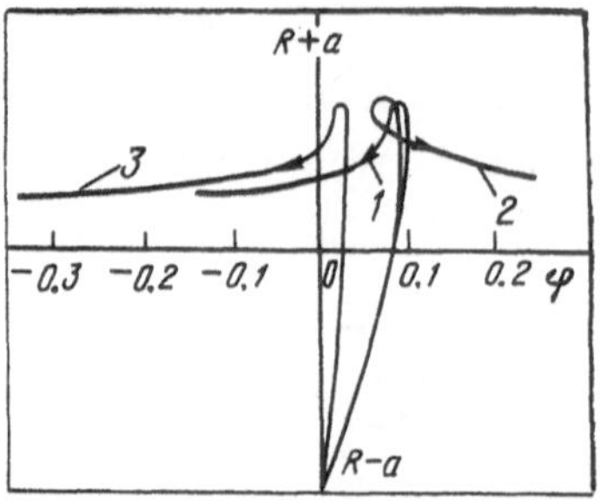


Fig. 14. Projection of ray trajectories on the equatorial cross-section of the torus at $q_0 = 0.6$: 1) $\alpha = -6°$, $\delta = 48°$; 2) $\alpha = 6°$, $\delta = 48°$; 3) $\alpha = -3°$, $\delta = 83°$. The thick line identifies the trajectory segments where more than 99% of the Bernstein-wave power is absorbed.

this topic [30, 31], and various details were mainly made more precise in later papers.

We shall not analyze here these numerous data. We report instead, by way of illustration, the results of one of the papers [37] and use them as an example to discuss the main features of wave propagation and absorption in tokamaks. Equations (57) were integrated in that paper in the "toroidal" coordinate frame $r, \theta, \varphi$, where $r$ is the radius in the minor cross-section of the tokamak, while $\theta$ and $\varphi$ are angles measured along the minor and major circuits of the torus, respectively. Far from the hybrid resonance, the electromagnetic waves were described in the cold-plasma approximation, and the Bernstein waves in Eq. (47). Their mutual transformation in the vicinity of the hybrid resonance was described in the framework of the weak spatial dispersion approximation. The wave damping was calculated by integrating the imaginary part of $\mathbf{k}$ along the trajectory. $\mathrm{Im}\,k$ was obtained from (13) for electromagnetic waves and from (47) for Bernstein waves. The plasma parameters were assumed to be distributed in the minor cross-section of the torus in accordance with the equations

$$q = \frac{\omega_p^2}{\omega^2} = q_0\left(1 - \frac{r^2}{a^2}\right)^{P_1}, \qquad T_e = T_0 \exp\left(-P_2\frac{r^2}{a^2}\right),$$

$$B_\theta = \frac{8\pi^2}{cr}\int_0^r j(r')\,r'\,dr',$$

$$j = j_0 \exp\left(-\frac{3}{2}P_2\frac{r^2}{a^2}\right) \sim T_e^{3/2}, \qquad B_\varphi = B_{\varphi 0}\left(1 + \frac{r\cos\theta}{R}\right)^{-1}.$$

If the parameters $P_1$ and $P_2$ were properly chosen, these expressions described well enough the experimental data obtained with the apparatus for which the calculations were performed.

Figure 13 shows the ray trajectories of the extraordinary waves launched from the periphery of the plasma from the inner side of the torus, at various initial angles to the magnetic field [$\alpha = \tan^{-1}(k_0\theta/k_{0r})$ is the launching angle in the minor section relative to the minor radius $\delta = \tan^{-1}(k_0 r/k_{0\varphi})$]. The calculation was carried out for the parameters $R$ = 62 cm, $a$ = 15 cm, $B_{\varphi_0} = 10^4$ G, $T_0$ = 300 eV, and $J_p$ = 30 kA. The trajectory pattern, however, is quite sensitive to the parameters of the apparatus and is determined mainly by the position of the electron-cyclotron-resonance (ECR) line and by the value of $q_0$. Some asymmetry of the trajectory pattern relative to the equatorial plane is due to the influence of the poloidal magnetic field, with allowance for which the medium becomes inhomogeneous along the magnetic-field force lines. This leads to variation of $n_{||}$ along the magnetic-field lines along trajectories that are different in the upper and lower halves of the torus. This effect is immaterial for electromagnetic waves, while for Bernstein waves it is quite important, since a noticeable increase of $n_{||}$ along the trajectory shifts substantially the region of their absorption away from the cyclotron resonance.

In contrast to the trajectories, the damping of the waves depends very strongly on the apparatus parameters and on the plasma temperature. Thus, under T-15 conditions the considered waves ($\delta = 48^\circ$) would be absorbed without reaching the ECR region.

Note that the pattern of the propagation of the extraordinary waves in the transformation waves is quite whimsical, while the Bernstein waves propagate at $n_{||} \neq 0$ at a small angle to the magnetic field (Fig. 14).

The character of the wave propagation changes radically on going to a dense plasma $q_0 > 2$ (Fig. 15). In this

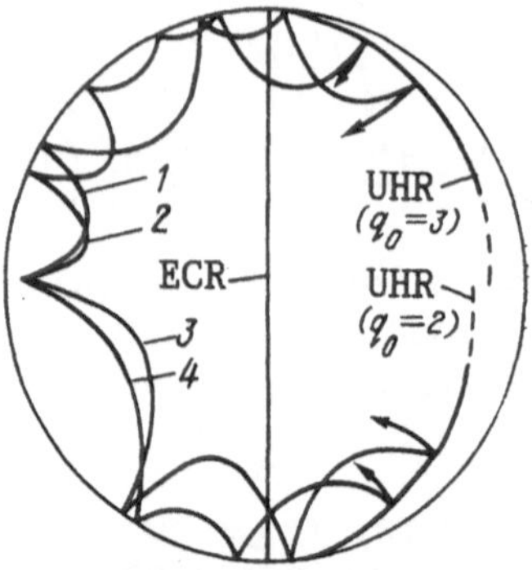


Fig. 15. The same as in Fig. 13a, at a higher plasma density, with allowance for wave reflection from the chamber walls: 1) $q_0 = 3$, $\alpha = 21°$, $\delta = 48°$; 2) $q_0 = 3$, $\alpha = 30°$, $\delta = 48°$; 3) $q_0 = 2$, $\alpha = -21°$, $\delta = 48°$; 4) $q_0 = 2$, $\alpha = 35°$, $\delta = -35°$.

case an important role is played by wave reflection from the chamber walls. It was assumed in the calculations that on reflection from the wall the quantity $n_{\|}$ is preserved and that the quantity $n_{\varphi}$ changes in random fashion, since a shallow corrugated wall leads apparently to diffuse reflection of the waves in the region of the major circuit of the torus. The wave polarization was also assumed to vary randomly on reflection.

The picture of propagation of an ordinary wave from the weak-field side is noticeably simpler (Fig. 16). It is important that under conditions of large tokamaks (e.g., T-10 and larger) the relativistic ordinary-wave absorption corresponding to $n_{\|} = 0$ is sufficient for total absorption of its energy within one pass through the resonant layer [2]. When resonance is reached, the ordinary wave is primarily absorbed by low-energy electrons [40]. Thus, the use of an ordinary wave is very convenient from the standpoint of plasma heating. However, launching an extraordinary wave from the inner side of the torus permits energy to be fed to the central part of the plasma pinch at relatively high (about double) plasma densities [35].

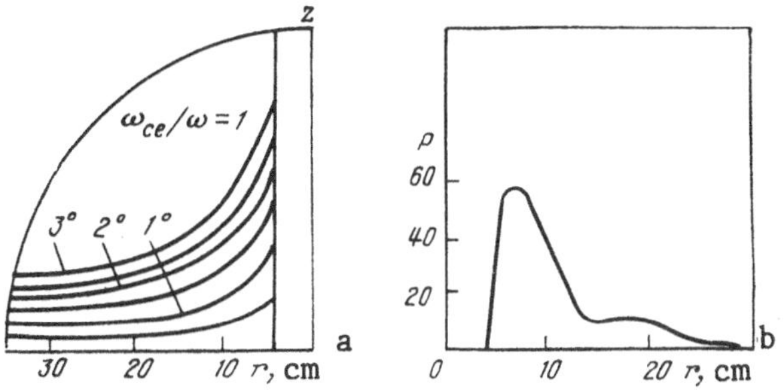


Fig. 16. Projections of ray trajectories on the cross-section of a plasma pinch when ordinary waves are launched from the outer side of the torus: $q_0 = 1$, $\delta = 0$, $T_{e0} = 1$ keV (a), and distribution of the absorbed microwave-power density (in relative units) (b).

## 11. Role of Nonlinear Effects in Electron-Cyclotron Heating

The entire theory expounded so far was linear, i.e., it took no account of the action of the absorbed microwave power on the properties of the medium. Yet it is clear that electron interaction with intense electromagnetic radiation can alter substantially the electron distribution function, and by this token also the wave-damping rates. To describe these changes, one can attempt to use the standard quasi-linear-theory approach [41-43], even though the main premise of this theory, the assumption that the electromagnetic field is random (noiselike), does not hold, strictly speaking, in heating experiments, where the wave frequency is set by the generator. The applicability of the quasilinear approximation can be based on the fact that the antennas frequently employed emit a whole spectrum of waves with different values of $k_{\parallel}$, so that the spatial distribution of the fields becomes quite complicated. Additional relaxations of the particle-rotation phase relative to the electric field are caused by the spatial inhomogeneity of the magnetic field. One can advance a hypothesis that both indicated circumstances can be explained in an adequate physical manner by assuming the medium to be spatially homogeneous and the field to be a random function of the coordinates; the latter can be conveniently represented in the form

$$\mathbf{E}(\mathbf{r}, t) = \frac{1}{2} \int_{-\infty}^{\infty} dk_{\parallel} \mathbf{E}_{k_{\parallel}} \exp\left[ik_{\perp} \cdot x + i k_{\parallel} z - i\omega t\right] + \mathrm{c.c.},$$

where the phase of the $\mathbf{E}_{k\parallel}$ amplitudes are assumed random. Dividing next, in the spirit of the quasilinear theory, the distribution function into a slowly varying part F(r, v, t) and an oscillating part f that satisfies the linearized kinetic equation, we can represent the rate of change $(\partial F/\partial t)_w$ of F on account of the interaction with the waves in the form

$$\left(\frac{\partial F}{\partial t}\right)_w = \frac{\partial}{\partial \mathbf{v}} \overleftrightarrow{D} \frac{\partial}{\partial \mathbf{v}} F, \qquad (62)$$

where the diffusion tensor $\overleftrightarrow{D}$ depends on the components $v_\perp$ and $v_\parallel$ of the particle velocity. An explicit expression for $\overleftrightarrow{D}$ in the case of interaction of electrons with electromagnetic waves was obtained in [44-46], and with allowance for relativistic corrections in [47]. Analysis shows that in electron-cyclotron resonance the principal role in (62) is played by diffusion along $v_\perp$.* The corresponding diffusion coefficient is of the form

$$D_\perp = \frac{\pi e^2}{2m^2} \int dk_\parallel \left| \overleftrightarrow{\Pi}_s \mathbf{E} \right|^2 \delta(\omega - s\omega_B - k_\parallel v_\parallel),$$

where

$$\Pi_{sx} = \frac{sJ_s(\rho)}{\rho}, \qquad \Pi_{sy} = iJ_s'(\rho),$$

$$\Pi_{sz} = \frac{v_\parallel}{v_\perp} J_s(\rho), \qquad \rho = \frac{k_\perp v_\perp}{\omega_B}, \qquad s = 1, 2.$$

If it is assumed that at s = 1 the main contribution to the diffusion is made by the resonant component of the electric field and that the wave spectrum constitutes a narrow pack-

---

*This conclusion is obvious for the case of interaction with an extraordinary wave. In perpendicular propagation of an ordinary wave, when $\mathbf{E}$ is directed along z, the electric field exerts on the particles a force $F_z = eE \exp\left\{ i \frac{k_\perp v_T}{\omega_B} \sin \omega_B t - i\omega t \right\}$; at $\omega = \omega_B$ the force contains a resonant (time-independent) component $F_{res} \cong eEk_\perp v_\perp/\omega$ [11]. It is easily seen that it is completely offset by the z-th projection of the Lorentz force $\frac{e}{c}[\mathbf{v}\tilde{\mathbf{B}}]$ ($\tilde{\mathbf{B}}$ is the magnetic field of the wave). This leaves only the perpendicular component of this force, which causes $v_\perp$ to change.

et of width $\Delta k$ about the central values $k_{\|0}$, then $D_\perp$ reduces at $\rho \ll 1$ to the expression

$$D_\perp = \frac{\pi e^2 |\overline{E}_1|^2}{m^2 v_\| \Delta k}. \tag{63}$$

Here, $|\overline{E}_1|^2$ is the mean squared field intensity, defined by the relation

$$|\overline{E}_1|^2 = \int |E_{k_\|}|^2 dk_\| \approx |\overline{E}_{k_\|}|^2 \Delta k;$$

$\Delta k$ is the spectral width of the waves emitted by the antenna. Equation (63) can be obtained, accurate to a coefficient of order unity, from the following lucid considerations that explain the physical meaning of the assumptions made. Assume that the wave field is of the form $E_1(z)\exp[ik_\perp x + ik_{\|0}z - i\omega t]$, where $E_1(z)$ is a slowly varying random function. The time $\tau$ of the coherent interaction of the electron with the wave is then given by the relation $\tau \sim \ell_c/v_\|$, where $\ell_c$ is the amplitude correlation length. For the diffusion coefficient of the resonant particles we then obtain

$$D_\perp \sim \left(\frac{e}{m} E_1 \tau\right)^2 / \tau.$$

Since $\ell_c \sim 1/\Delta k$, this leads to (63). The complete kinetic equation for the electron-distribution function must take into account also the electron–electron and electron–ion collisions:

$$\frac{\partial F}{\partial t} = \frac{1}{v_\perp} \frac{\partial}{\partial v_\perp} (v_\perp D_\perp) \frac{\partial F}{\partial v_\perp} + \left(\frac{\partial F}{\partial t}\right)_{\text{coll}}. \tag{64}$$

The collision term is usually chosen in the form proposed by Landau [48], or in some other equivalent form. Equation (64) permits an easy estimate of the role of the quasilinear diffusion. Multiplying (62) by $mv^2/2$ and integrating over the velocities, we obtain the power absorbed per unit volume of the plasma, $P_0 = 2n_{res}mD_\perp$ (where $n_{res}$ is the density of the resonant particles), and the total absorbed power $P = P_0V_{heat}$, where $V_{heat}$ is the volume of the layer of magnetic surfaces that cross the wave-absorption region. If it is assumed that the HF heating is the principal one, we have in the stationary state $P \sim NT/\tau_E$, where $\tau_E$ is the energy lifetime, and $D_\perp \sim (N/N_{res})(v_T^2/\tau_E)$ ($N_{res}$ is the number of

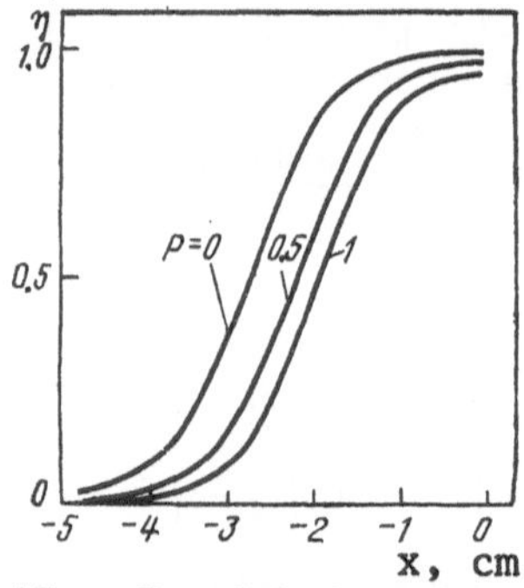


Fig. 17. Spatial damping of the ordinary wave at various levels of the introduced power P (MW). The point x = 0 corresponds to resonance. The values assumed in the calculations were $n_{e0}$ = $5\cdot10^{13}$ $cm^{-3}$, $T_{e0}$ = 2 keV, B = 30 kG, $R_0$ = 130 cm, and $a$ = 40 cm.

particles participating in the absorption and N is the total number of plasma electrons). For the ratio of the collision term in (64) to the quasilinear term we now obtain

$$\left(\frac{\partial F}{\partial t}\right)_{\text{coll}} \Big/ \left(\frac{\partial F}{\partial t}\right)_{\omega} \approx \frac{N_{\text{res}}}{N}\, \nu(v_{\text{res}})\, \tau_E,$$

where $\nu(v_{res}) \sim \nu_{ee}(v_{res}) \sim \nu_{ei}(v_{res})$ is the frequency of the Coulomb collisions at the characteristic velocity $v_{res}$ of the resonant electrons. Since $\nu(v_T)\,\tau_E \gg 1$, the quasilinear distortion of the distribution function will be insignificant if the volume fraction in which the absorbed power is directly distributed is not too small, and the absorbing particles are not too fast. The situation can change strongly if $v_{res}$ is several times larger than the thermal velocity. This case was specially studied in [49] for an extraordinary wave. It was found that even for strong distortions of the distribution function the wave-damping rate differs little from its linear value. The authors have concluded that the equations of the linear theory can almost always be used in the damping calculations of the ray trajectories.

The case of the ordinary wave in quasiperpendicular propagation was considered in [47]. Figure 17 shows an example taken from that paper and illustrating the dependence of the damping on the power-input level for a wave incident from the side of the strong magnetic field (it was assumed, in fact, that $\omega < \omega_B$). The observed changes of the absorbed-energy spatial distribution are not very significant and cannot substantially influence the picture obtained from the linear analysis. When the wave is incident from the weak-field side, the dependence on the power is altogether negligibly small. This difference is due to the asymmetry of the relativistic Doppler effect [see (17)]. At $\omega > \omega_B$ and $n_{\parallel} \ll 1$, resonance is possible only for slow particles, so that the energy is absorbed by the bulk of the electrons and the distribution function changes little. At $\omega < \omega_B$ there is no such restriction, and at high temperature the wave is absorbed already far from resonance by a small group of fast particles. Naturally, under these conditions, the quasilinear distortion of the distribution function is more noticeable. The general conclusion that can be drawn from the published numerical examples and from earlier estimates [31] is that no absorption saturation sets in at the powers needed for HF heating of a plasma in large tokamaks, and the linear theory can be used for the calculations.

Quasilinear effects play a decisive role in another problem that is attracting great interest at present – that of excitation of stationary current in a tokamak by electromagnetic waves, particularly in the electron-cyclotron resonance range [49-51]. This problem, however, is outside the scope of the present review.

The authors thank A. N. Savel'ev for the numerical calculations needed to plot some of the figures.

## REFERENCES

1. V. E. Golant, "High-frequency heating in toroidal plasmas," in: Proceedings of the Joint Varenna–Grenoble International Symposium on Heating in Toroidal Plasmas (1978), Vol. 2, p. 149.
2. V. V. Alikaev, "HF and microwave methods of plasma heating," Itogi Nauki Tekh., Ser. Fiz. Plazmy, 1, Part 2, 80 (1981).
3. K. N. Stepanov et al., "Mode conversion and wave damping in the low-frequency range," in: Proceedings 2nd Joint Varenna–Grenoble International Symposium, Como (1980), Vol. 1, p. 519.

4. V. E. Golant and A. D. Piliya, "Linear transformation and absorption of waves," Usp. Fiz. Nauk, 104, No. 3, 413 (1971).
5. A. G. Litvak, G. V. Permitin, E. V. Suvorov, and A. A. Fraiman, Pis'ma Zh. Tekh. Fiz., 1, No. 18, 858 (1975).
6. V. V. Alikaev, Yu. N. Dnestrovskii, V. V. Parail, and B. V. Pereverzev, Preprint IAE-2610, Kurchatov Atomic Energy Institute, Moscow (1976).
7. V. D. Shafranov, in: Reviews of Plasma Physics, Vol. 3, M. A. Leontovich (ed.), Consultants Bureau, New York (1967).
8. A. I. Akhiezer, I. A. Akhiezer, R. V. Polovin, et al., Plasma Electrodynamics [in Russian], Nauka, Moscow (1974).
9. A. V. Timofeev and G. N. Chulkov, Fiz. Plazmy, 5, No. 6, 1271 (1979).
10. I. P. Shkarofsky, Phys. Fluids, 9, 561, 570 (1966).
11. E. V. Suvorov and A. A. Fraiman, Izv. Vyssh. Uchebn. Zaved. Radiofiz., 20, 67 (1977).
12. M. Tanaka, M. Fujiwara, and M. Ikegami, Nagoya University Research Report IPPJ-427 (1979).
13. H. Weitzner and O. B. Betchelor, Phys. Fluids, 23, No. 7, 1359 (1980).
14. A. V. Timofeev, Usp. Fiz. Nauk, 110, 329 (1973).
15. V. L. Ginzburg, Propagation of Electromagnetic Waves in Plasma, Gordon and Breach, New York (1961).
16. J. Preinhaelter and V. Kopecky, J. Plasma Phys., 10, Part 1 (1973).
17. J. Preinhaelter, Czech. J. Phys., B25, 39 (1975).
18. T. Mackawa, S. Tanaka, Y. Terumichi, and Y. Hamada, Nagoya University Research Report IPPJ-313 (1977).
19. T. Antonsen and W. M. Manheimer, Phys. Fluids, 21 (12), 2295 (1978).
20. A. V. Zvonkov and A. V. Timofeev, Fiz. Plazmy, 6, No. 6, 1219 (1980).
21. V. I. Fedorov, Pis'ma Zh. Tekh. Fiz., 6, No. 21, 1307 (1980).
22. Yu. F. Baranov and V. I. Fedorov, in: Proceedings 10th European Conference on Controlled Fusion and Plasma Physics (1981), Vol. 1, paper H-13.
23. A. D. Piliya and V. I. Fedorov, Zh. Eksp. Teor. Fiz., 57, 1198 (1969).
24. I. B. Bernstein, Phys. Fluids, 18, No. 3, 320 (1975).
25. S. Weinberg, Phys. Rev., 126, 1399 (1962).

26. F. De Luca, C. Maroli, and V. Petrillo, Nuovo Cimento, 53B, No. 2, 181 (1979).
27. A. K. Nekrasov and A. V. Timofeev, Nucl. Fusion, 10, 377 (1970).
28. A. V. Timofeev, Usp. Fiz. Nauk, 110, No. 3, 329 (1973).
29. N. S. Erokhin and S. S. Moiseev, Usp. Fiz. Nauk, 109, No. 2, 225 (1973).
30. V. V. Alikaev, Yu. N. Dnestrovskii, V. V. Parail, and G. V. Pereverzev, Fiz. Plazmy, 3, 230 (1977).
31. A. G. Litvak, G. V. Permitin, E. V. Suvorov, and A. A. Frayman, Nucl. Fusion, 17, 659 (1977).
32. I. Fidone, G. Granata, G. Ramponi, and R. L. Meyer, Phys. Fluids, 21, 645 (1978).
33. E. Ott, B. Hui, and K. R. Chu, Phys. Fluids, 23, 23, 1031 (1980).
34. O. Eldridge, W. Namkung, and O. C. England, Oak Ridge Natl. Lab. Rep. ORNL/TM-6052 (1977).
35. F. De Luca, C. Maroli, and V. Petrilo, Nuovo Cimento, 53B, No. 2, 195 (1979).
36. Yu. F. Baranov and V. I. Fedorov, Pis'ma Zh. Tekh. Fiz., 7, No. 10, 608 (1981).
37. Yu. F. Baranov and V. I. Fedorov, Fiz. Plazmy, 9, No. 4, 677 (1983).
38. S. M. Walfe, D. R. Cohn, R. J. Temkin, and K. Kreisoher, Nucl. Fusion, 19, 389 (1979).
39. T. Mackawa, S. Tanaka, Y. Terumichi, and Y. Hamada, J. Phys. Soc. Japan, 48, No. 1, 247 (1980).
40. K. Baumgartel, Nucl. Fusion, 19, No. 11, 1543 (1979).
41. A. A. Vedenov, E. P. Velikhov, and R. Z. Sagdeev, Nucl. Fusion, 1, 82 (1961).
42. W. E. Drummond and D. Pines, Nucl. Fusion Suppl., Part 3, 1049 (1962).
43. A. A. Vedenov, in: Reviews of Plasma Physics, Vol. 3, M. A. Leontovich (ed.), Consultants Bureau, New York (1967).
44. V. L. Yakimenko, Zh. Eksp. Teor. Fiz., 44, 1534 (1963).
45. J. Retslands, V. L. Sizonenko, and K. N. Stepanov, Zh. Eksp. Teor. Fiz., 50, 994 (1966).
46. C. F. Kennel and F. Engelman, Phys. Fluids, 9, 2371 (1966).
47. I. Fidone, G. Granata, and R. L. Meyer, Fontenay-aux-Roses Preprint EUR-CEA-FC-1053 (1980).
48. L. D. Landau, Zh. Eksp. Teor. Fiz., 7, 203 (1937).
49. C. F. F. Karney and N. J. Fish, Nucl. Fusion, 21, 1549 (1981).

50. J. G. Gordey, T. Edlington, and D. Star, Culham Lab. Preprint CML-P636 (1981).
51. V. V. Parail and G. V. Pereverzev, Fiz. Plasmy, 8, 45 (1982).